Chemical Properties, Nutritional Quality, and Bioactive Components of Horticulture Food

Chemical Properties, Nutritional Quality, and Bioactive Components of Horticulture Food

Editors

Alessandra Durazzo
Massimo Lucarini

MDPI • Basel • Beijing • Wuhan • Barcelona • Belgrade • Manchester • Tokyo • Cluj • Tianjin

Editors
Alessandra Durazzo
CREA-Research Centre for
Food and Nutrition
Rome, Italy

Massimo Lucarini
CREA-Research Centre for
Food and Nutrition
Rome, Italy

Editorial Office
MDPI
St. Alban-Anlage 66
4052 Basel, Switzerland

This is a reprint of articles from the Special Issue published online in the open access journal *Horticulturae* (ISSN 2311-7524) (available at: https://www.mdpi.com/journal/horticulturae/special_issues/Chemical_Nutritional_Bioactive).

For citation purposes, cite each article independently as indicated on the article page online and as indicated below:

LastName, A.A.; LastName, B.B.; LastName, C.C. Article Title. *Journal Name* **Year**, *Volume Number*, Page Range.

ISBN 978-3-0365-8374-7 (Hbk)
ISBN 978-3-0365-8375-4 (PDF)

Contents

About the Editors

Alessandra Durazzo

Alessandra Durazzo was awarded a master's degree in Chemistry and Pharmaceutical Technology cum laude in 2003 and a PhD in Horticulture in 2010. She is a Senior Researcher at the CREA-Research Centre for Food and Nutrition. The core of her research is the study of the chemical, nutritional, and bioactive components of food, with regard to the wide spectrum of substance classes and their nutraceutical features. For several years, she was involved in national and international research projects on the evaluation of several factors (agronomic practices, processing, etc.) that affect food quality, the levels of bioactive molecules and the total antioxidant properties, as well as their possible impact on the biological role played by bioactive components in human physiology. Particular attention is given to the study of alternative sources of nutraceutical compounds such as agro-food waste and the application of nanotechnologies in pharmaceutical and nutraceutical compounds. Her research activities also address the development, management, and updating of databases of bioactive compounds, nutraceuticals, and dietary supplements; particular attention is paid to the harmonization of analytical procedures and the classification and codification of dietary supplements.

Massimo Lucarini

Massimo Lucarini was awarded a master's degree in Industrial Chemistry "cum laude" at the University of Rome "La Sapienza", Italy (1992), and a PhD in Chemistry (University of Rome "La Sapienza"). He is a Senior Researcher at the CREA-Research Centre for Food and Nutrition. His research activity is mainly aimed at the evaluation of nutrient content, molecules with biological and anti-nutrient activity in foods and diets, and studies on the stability and technological treatments of food products using specific process markers. Particular interest is given to the evaluation of the nutritional quality of foods, the bioavailability of nutrients and bioactive components and their interaction with the food matrix (using in vitro models and cellular models), and applications in the nutraceutical field; recent attention has been paid to the exploitation of waste from the agri-food industry, with a view to sustainable agri-food production. An integral part of Dr. Lucarini's research is linked to institutional activity, including food composition tables, dietary guidelines for healthy eating, and the evaluation of fraud risk in the agri-food system. In relation to the food production system, the effects of technological treatments on molecules of nutritional interest are also evaluated; he is also interested in using natural substances with strong antioxidant properties to improve the shelf-life of food products. His research activity is also aimed at the development of new analytical methods, the exchange of scientific information, and the acquisition of new skills both at the national and international levels, through training courses and participation in congresses and seminars. His work is disseminated through the production of scientific articles, interviews released in national journals and broadcasting systems, the creation of web pages, and participation in congresses and educational and informative activities.

Preface to "Chemical Properties, Nutritional Quality, and Bioactive Components of Horticulture Food"

In this Special Issue, chemical properties, nutritional quality, and bioactive components of horticultural food are explored. A strong linkage is present between food quality, territory, and environment, and several factors can affect food quality. The research of Wang, X. and collaborators [1] addresses the construction of volatile profiles of eight popular peach cultivars produced in Shanghai using GC-MS and GC-IMS, providing a reference for the future design of breeding to improve fruit flavor. A detailed comparative study on some of the physicochemical properties, volatile compositions, fatty acids, and mineral profiles of different almond (Prunus dulcis L.) varieties was conducted by Levent et al. [2], who highlighted how different almond varieties showed significant differences depending on the genotypes as well as on the origins. Seasonal variations in the starch properties of sweet potato cultivars were investigated by dos Santos et al. [3]. Rakocevic et al. [4] studied the variations in yield, berry distribution, and chemical attributes of Coffea arabica beans among the canopy strata of four genotypes cultivated under differing water regimes . In a study by Lee et al. [5], they investigated the growth, physicochemical, nutritional, and postharvest qualities of leaf lettuce (Lactuca sativa L.) affected by the cultivar and amount of applied nutrient solution, whereas the study by Mauro et al. [6] is focused on the yields and compositional profiles of eggplant fruits affected by phosphorus supply, genotype, and grafting.

Moving towards the characterization of derived products and how several factors can affect their quality, a study by Wang, Z. et al. [7] was focused on the quality characteristics of Oolong tea products in different regions and the contribution of thirteen phytochemical components to its taste; the construction of the Oolong tea flavor wheel could aid in the qualitative and quantitative sensory evaluation of Oolong tea from different origins and contribute to the quality identification and directional improvement of Oolong tea products. Another example is given by Berwal et al. [8], who studied the bioactive compounds and fatty acid profiles of bitter apple seed oil obtained from hot, arid environments. Phan et al. [9] studied the effects of yeast product additions and fermentation temperatures on the lipid composition, taste, and mouthfeel characteristics of Pinot Noir Wine.

It is worth mentioning the review by Jacobo-Velázquez et al. [10] of UVA and UVB radiation as innovative and effective tools with which to biofortify horticultural crops with nutraceuticals. We would like to acknowledge the efforts of the authors of the publications in this Special Issue.

1. Wang, X.; Li, X.; Su, M.; Zhang, M.; Hu, Y.; Du, J.; Zhou, H.; Zhang, X.; Ye, Z.; Yang, X. The Construction of Volatile Profiles of Eight Popular Peach Cultivars Produced in Shanghai Using GC-MS and GC-IMS. *Horticulturae* **2023**, *9*, 382. https://doi.org/10.3390/horticulturae9030382.

2. Levent, O. A Detailed Comparative Study on Some Physicochemical Properties, Volatile Composition, Fatty Acid, and Mineral Profile of Different Almond (*Prunus dulcis* L.) Varieties. *Horticulturae* **2022**, *8*, 488. https://doi.org/10.3390/horticulturae8060488.

3. dos Santos, T.P.R.; Leonel, M.; de Oliveira, L.A.; Fernandes, A.M.; Leonel, S.; da Silva Nunes, J.G. Seasonal Variations in the Starch Properties of Sweet Potato Cultivars. *Horticulturae* **2023**, *9*, 303. https://doi.org/10.3390/horticulturae9030303.

4. Rakocevic, M.; dos Santos Scholz, M.B.; Pazianotto, R.A.A.; Matsunaga, F.T.; Ramalho, J.C. Variation in Yield, Berry Distribution and Chemical Attributes of *Coffea arabica* Beans among the Canopy Strata of Four Genotypes Cultivated under Contrasted Water Regimes. *Horticulturae* **2023**, *9*, 215. https://doi.org/10.3390/horticulturae9020215.

5. Lee, J.-S.; Chandra, D.; Son, J. Growth, Physicochemical, Nutritional, and Postharvest Qualities of Leaf Lettuce (*Lactuca sativa* L.) as Affected by Cultivar and Amount of Applied Nutrient Solution. *Horticulturae* **2022**, *8*, 436. https://doi.org/10.3390/horticulturae8050436.

6. Mauro, R.P.; Stazi, S.R.; Distefano, M.; Giuffrida, F.; Marabottini, R.; Sabatino, L.; Allevato, E.; Cannata, C.; Basile, F.; Leonardi, C. Yield and Compositional Profile of Eggplant Fruits as Affected by Phosphorus Supply, Genotype and Grafting. Horticulturae 2022, 8, 304. https://doi.org/10.3390/horticulturae8040304.

7. Wang, Z.; Gan, S.; Sun, W.; Chen, Z. Quality Characteristics of Oolong Tea Products in Different Regions and the Contribution of Thirteen Phytochemical Components to Its Taste. *Horticulturae* **2022**, *8*, 278. https://doi.org/10.3390/horticulturae8040278.

8. Berwal, M.K.; Ram, C.; Gurjar, P.S.; Gora, J.S.; Kumar, R.; Verma, A.K.; Singh, D.; Basile, B.; Rouphael, Y.; Kumar, P. The Bioactive Compounds and Fatty Acid Profile of Bitter Apple Seed Oil Obtained in Hot, Arid Environments. Horticulturae 2022, 8, 259. https://doi.org/10.3390/horticulturae8030259.

9. Phan, Q.; DuBois, A.; Osborne, J.; Tomasino, E. Effects of Yeast Product Addition and Fermentation Temperature on Lipid Composition, Taste and Mouthfeel Characteristics of Pinot Noir Wine. Horticulturae 2022, 8, 52. https://doi.org/10.3390/horticulturae8010052.

10. Jacobo-Velázquez, D.A.; Moreira-Rodríguez, M.; Benavides, J. UVA and UVB Radiation as Innovative Tools to Biofortify Horticultural Crops with Nutraceuticals. *Horticulturae* **2022**, *8*, 387. https://doi.org/10.3390/horticulturae8050387.

Alessandra Durazzo and Massimo Lucarini
Editors

horticulturae

MDPI

Editorial

Chemical Properties, Nutritional Quality, and Bioactive Components of Horticulture Food

Alessandra Durazzo * and Massimo Lucarini *

CREA—Research Centre for Food and Nutrition, Via Ardeatina 546, 00178 Rome, Italy
* Correspondence: alessandra.durazzo@crea.gov.it (A.D.); massimo.lucarini@crea.gov.it (M.L.)

Introduction

The Special Issue "Chemical Properties, Nutritional Quality, and Bioactive Components of Horticulture Food" is here presented. Horticultural crops mainly include fruits, vegetables, and ornamental trees [1]. Horticultural crops and products are an important source of compounds of nutritional and nutraceutical interest for human nutrition. Physiological, nutritional, nutraceutical quality and potential of horticultural products should be explored.

To give a current overview of the interest of this topic, a search throughout the Scopus online database has been carried out using the string TITLE-ABS-KEY ("horticulture" OR "horticulturae" AND "food quality *"). The functions of the Scopus web online platform named "Analyze" and "Create Citation Report" were utilized for carrying out basic analyses, whereas the "full records and cited references" have been exported for further processing to the VOSviewer software (version 1.6.16, 2020; www.vosviewer.com, accessed on 15 November 2021) [2–4].

The search returned 144 documents covering the period 1997 to 2021, and a total of 83 terms were identified and visualized as a term map in Figure 1. The main terms are: horticulture, food quality, fruit, crop yield, harvesting, cultivar, nonhuman, food storage, fruit quality, yield, quality, vegetable, agriculture, fruits, orchard (Figure 1). The most recent document is published by Biel et al. [5] addressed on-farm reduced irrigation and fertilizer doses, and arbuscular mycorrhizal fungal inoculation improve water productivity in tomato production [5]. Among the "Review" category, some examples are as follows: (i) current understanding and use of quality characteristics of horticulture products [6]; (ii) salinity as eustressor for enhancing quality of vegetables [7]; (iii) the cuticle as a key factor in the quality of horticultural crops [8]; (iv) preharvest factors influencing bruise damage of fresh fruits—a review [9]; (v) use of light quality manipulation in improvement vegetable quality at harvest and postharvest [10].

Narrowing the search exploring chemical properties, nutritional quality, and bioactive components features using the string TITLE-ABS-KEY ("horticulture" OR "horticulturae" AND "food quality*" AND "bioactive compound*" OR "bioactive component*" OR "bioactive molecule*" OR "nutrient*" OR "nutritional" or "chemical*"), the search resulted in 50 documents covering the time period 1997–2021 mainly *Agricultural and Biological Sciences* area.

The oldest work is by Haglund et al. [11] in 1997 on sensory quality of tomatoes cultivated with ecological fertilizing systems, whereas the most recent is by Mwinuka et al. [12] and it addresses optimizing water and nitrogen application for neglected horticultural species in tropical sub-humid climate areas, with particular regards of a case of African eggplant.

Among the most cited documents: —a review of antimicrobial and antioxidative activities of chitosans in food [13]; —a study on how vine growth, yield, berry quality attributes and leaf nutrient content of grapevines can be influenced by seaweed extract (*Ascophyllum nodosum*) and nanosize fertilizer pulverizations [14]; —a study on how rootstocks can enhance tomato growth and quality characteristics at low potassium supply [15].

Citation: Durazzo, A.; Lucarini, M. Chemical Properties, Nutritional Quality, and Bioactive Components of Horticulture Food. *Horticulturae* **2022**, *8*, 3. https://doi.org/10.3390/horticulturae8010003

Received: 15 November 2021
Accepted: 26 November 2021
Published: 21 December 2021

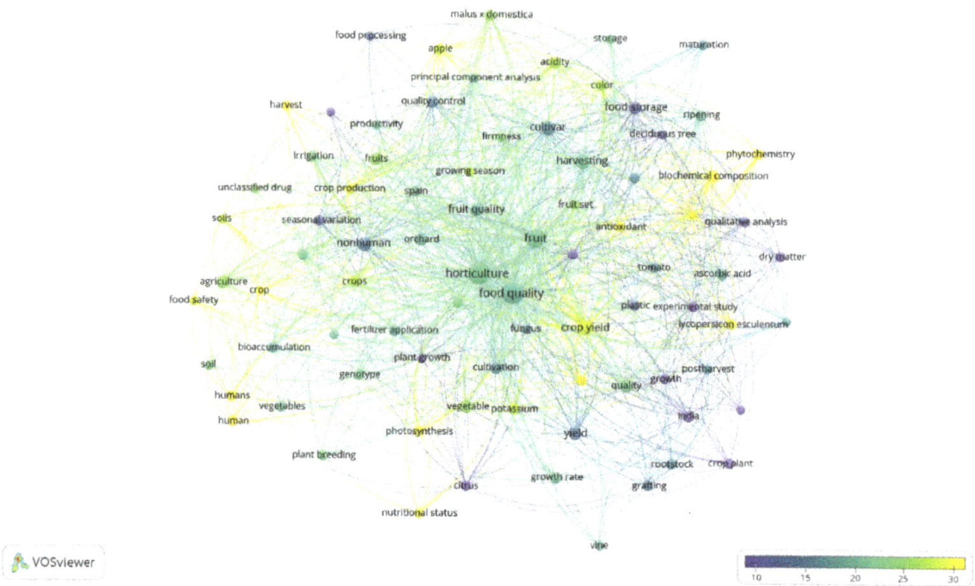

Figure 1. Term map for the relationship of horticulture/horticulturae and food quality research. Bubble size represents the number of publications. Bubble color represents the citations per publication (CPP). Two bubbles are closer to each other if the terms co-appeared more frequently (Bibliometric data were extracted from the Scopus online database and elaborated by the VOSviewer software).

In this context, the investigation on pre- and postharvest factors, storage and processing conditions under the food product quality perspective is a key issue to investigate and it represents one of the main focuses of the Special Issue. Studies on different factors influencing the food quality of horticultural crops under different agricultural production systems are welcome; particularly emerging strategies to maximize the content of bioactive components are considered.

Noteworthy research should be addressed towards the conventional and innovative, modern methodologies and technologies in light of scientific progress and green and eco-friendly approaches. In this regard, the book published by Bauddh et al. [16] focused on ecological and practical applications for sustainable agriculture is here mentioned.

New models of horticultural commodities should be analyzed, and new monitoring parameters should be proposed. Studies on classification of horticultural products should be promoted.

The use of deep learning and chemometrics for management of large datasets represents a new goal of horticultural research [17,18].

Moreover, studies on the use of nanotechnologies for improving productivity and quality of food represent new frontiers [19]

The quality of horticultural products should be explored from local and rural communities to urban ones, and also territories reconversion should be taken into account. The work of Ansari et al. [20] is addressed on converting primary forests to cultivated lands, with a focus on long-term effects on the vertical distribution of soil carbon and biological activity in the foothills of Eastern Himalaya. Studies on the use and applications of horticultural products and waste/byproducts are welcome. Utilization of high-value horticultural waste should be promoted [21]. Attention should be given to sustainable crop production systems, in the perspective of the tight bridge between food quality, territory and biodiversity [22].

Horticulture meets consumer demand through new solutions and challenges. Studies of promotion of underutilized horticultural crops and diversification for food security should be taken into account.

The effect of climate change on horticultural production and product quality should be considered.

An emerging and modern image of horticultural systems and product status should be based on a multidisciplinary approach that involves agronomy, chemical science, food science, and nutrition. Trends and challenges on horticultural status will be delineated and promoted.

Author Contributions: A.D. and M.L. have made a substantial, direct, and intellectual contribution to the work and approved it for publication. All authors have read and agreed to the published version of the manuscript.

Funding: This research received no external funding.

Conflicts of Interest: The authors declare no conflict of interest.

References

1. Herklots, G.A.C.; Janick, J.S. Patrick Millington and Perrott, Roy. horticulture. In *Encyclopedia Britannica*; Britannica Group: Chicago, IL, USA, 2021; Available online: https://www.britannica.com/science/horticulture (accessed on 7 November 2021).
2. Van Eck, N.J.; Waltman, L. Software survey: VOSviewer, a computer program for bibliometric mapping. *Scientometrics* **2010**, *84*, 523–538. [CrossRef] [PubMed]
3. Van Eck, N.J.; Waltman, L. Text mining and visualization using VOSviewer. *arXiv* **2011**, arXiv:1109.2058.
4. Waltman, L.; van Eck, N.J.; Noyons, E.C.M. A unified approach to mapping and clustering of bibliometric networks. *J. Inf.* **2010**, *4*, 629–635. [CrossRef]
5. Biel, C.; Camprubí, A.; Lovato, P.E.; Calvet, C. On-farm reduced irrigation and fertilizer doses, and arbuscular mycorrhizal fungal inoculation improve water productivity in tomato production. *Sci. Hortic.* **2021**, *288*, 110337. [CrossRef]
6. Schreiner, M.; Korn, M.; Stenger, M.; Holzgreve, L.; Altmann, M. Current understanding and use of quality characteristics of horticulture products. *Sci. Hortic.* **2013**, *163*, 63–69. [CrossRef]
7. Rouphael, Y.; Petropoulos, S.A.; Cardarelli, M.; Colla, G. Salinity as eustressor for enhancing quality of vegetables. *Sci. Hortic.* **2018**, *234*, 361–369. [CrossRef]
8. Tafolla-Arellano, J.C.; Báez-Sañudo, R.; Tiznado-Hernández, M.E. The cuticle as a key factor in the quality of horticultural crops. *Sci. Horticult.* **2018**, *232*, 145–152. [CrossRef]
9. Hussein, Z.; Fawole, O.A.; Opara, U.L. Preharvest factors influencing bruise damage of fresh fruits—A review. *Sci. Hortic.* **2018**, *229*, 45–58. [CrossRef]
10. Ilić, Z.S.; Fallik, E. Light quality manipulation improves vegetable quality at harvest and postharvest: A review. *Environ. Exper. Botan.* **2017**, *139*, 79–90. [CrossRef]
11. Haglund, Å.; Johansson, L.; Gäredal, L.; Dlouhy, J. Sensory quality of tomatoes cultivated with ecological fertilizing systems. *Swed. J. Agricult. Rese* **1997**, *27*, 135–145.
12. Mwinuka, P.R.; Mbilinyi, B.P.; Mbungu, W.B.; Mahoo, H.F.; Schmitter, P. Optimizing water and nitrogen application for neglected horticultural species in tropical sub-humid climate areas: A case of African eggplant (*Solanum aethiopicum* L.). *Sci. Hortic.* **2021**, *276*, 109756. [CrossRef]
13. Friedman, M.; Juneja, V.K. Review of antimicrobial and antioxidative activities of chitosans in food. *J. Food Protect.* **2010**, *73*, 1737–1761. [CrossRef] [PubMed]
14. Sabir, A.; Yazar, K.; Sabir, F.; Yazici, M.A.; Goksu, N. Vine growth, yield, berry quality attributes and leaf nutrient content of grapevines as influenced by seaweed extract (*Ascophyllum nodosum*) and nanosize fertilizer pulverizations. *Sci. Horticult.* **2014**, *175*, 1–8. [CrossRef]
15. Schwarz, D.; Öztekin, G.B.; Tüzel, Y.; Brückner, B.; Krumbein, A. Rootstocks can enhance tomato growth and quality characteristics at low potassium supply. *Sci. Hortic.* **2013**, *149*, 70–79. [CrossRef]
16. Bauddh, K.; Kumar, S.; Singh, R.P.; Korstad, J. *Ecological and Practical Applications for Sustainable Agriculture*; Springer Nature: Singapore, 2020; pp. 1–470.
17. Yang, B.; Xu, Y. Applications of deep-learning approaches in horticultural research: A review. *Horticult. Res.* **2021**, *8*, 123. [CrossRef] [PubMed]
18. Lu, Y.; Saeys, S.; Kim, M.; Peng, Y.; Lu, R. Hyperspectral imaging technology for quality and safety evaluation of horticultural products: A review and celebration of the past 20-year progress. *Postharvest Biol. Technol.* **2020**, *170*, 111318. [CrossRef]
19. Rana, R.A.; Siddiqui, M.N.; Skalicky, M.; Brestic, M.; Hossain, A.; Kayesh, E.; Popov, M.; Hejnak, V.; Gupta, D.R.; Mahmud, N.U.; et al. Prospects of Nanotechnology in Improving the Productivity and Quality of Horticultural Crops. *Horticulturae* **2021**, *7*, 332. [CrossRef]

20. Ansari, M.A.; Choudhury, B.U.; Mandal, S.; Jat, S.L.; Meitei, C.B. Converting primary forests to cultivated lands: Long-term effects on the vertical distribution of soil carbon and biological activity in the foothills of Eastern Himalaya. *J. Environm. Man.* **2022**, *301*, 113886. [CrossRef] [PubMed]
21. Lucarini, M.; Durazzo, A.; Bernini, R.; Campo, M.; Vita, C.; Souto, E.B.; Lombardi-Boccia, G.; Ramadan, M.F.; Santini, A.; Romani, A. Fruit Wastes as a Valuable Source of Value-Added Compounds: A Collaborative Perspective. *Molecules* **2021**, *26*, 6338. [CrossRef] [PubMed]
22. Durazzo, A. The Close Linkage between Nutrition and Environment through Biodiversity and Sustainability: Local Foods, Traditional Recipes, and Sustainable Diets. *Sustainability* **2019**, *11*, 2876. [CrossRef]

horticulturae

MDPI

Article

The Construction of Volatile Profiles of Eight Popular Peach Cultivars Produced in Shanghai Using GC-MS and GC-IMS

Xia Wang [1,2,†], Xiongwei Li [2,3,†], Mingshen Su [2], Minghao Zhang [2], Yang Hu [2], Jihong Du [2,4], Huijuan Zhou [2], Xianan Zhang [2], Zhengwen Ye [2,*] and Xuelian Yang [1,*]

1 College of Agriculture, Guizhou University, Guiyang 550025, China
2 Forest and Fruit Tree Research Institute, Shanghai Academy of Agricultural Sciences, Shanghai 201403, China
3 Shanghai Runzhuang Agricultural Science and Technology Co., Ltd., Shanghai 201415, China
4 Shanghai Co-Elite Agricultural Sci-Tech (Group) Co., Ltd., Shanghai 201106, China
* Correspondence: yezhengwen1300@163.com (Z.Y.); yxl4684221@126.com (X.Y.)
† These authors contributed equally to this work.

Abstract: Peach (*Prunus persica* L.) is an economically important fruit crop worldwide due to its pleasant flavor. Volatile organic compounds (VOCs) are vital factors for assessing fruit quality. Here, we constructed the VOC profiles for the top eight popular commercial peach cultivars produced in Shanghai by combining gas chromatography-mass spectrometry (GC-MS), odor activity value and gas chromatograph-ion mobility spectrometry (GC-IMS). Seventy VOCs were detected using GC-MS, of which twenty-three were commonly found in eight peach cultivars, including hexanal, nonanal, benzaldehyde, 2-hexenal, butyl acetate, hexyl acetate, (Z)-3-hexen-1-yl acetate, linalool, β-myrcene, D-limonene, 1-hexanol, 3-hexenol, 2-hexenol, 2-ethyl-1-hexanol, γ-octalactone, δ-decalactone, γ-hexalactone, γ-decalactone, γ-dodecalactone, β-ionone, 2-octanone, 2-ethyl furan and 2,4-ditert-butyl phenol. A total of 17 VOCs were screened on the basis of OAV ≥ 1 and the top 5 of this contribution were γ-decalactone, β-ionone, hexanal, 2-hexenal and linalool. Lactones had the highest OAV in HJML and terpenoids had the highest OAV in JC. JXIU had the lowest OAV of lactones and terpenoids. Based on the range of their OAV values, the flavor evaluation standard of Shanghai high-quality peach cultivars can be established, which is also a reference for breeding excellent offspring. Twenty-six VOCs were detected using GC-IMS, and the largest proportion were aldehydes. Principal component analysis (PCA) showed that Hikawa Hakuho (HH) and Jinchun (JC) were distant from the other samples, indicating that their volatiles were more distinct. These results provide a foundation for improving our understanding of aroma compositions in these high-quality peach cultivars, which might also provide a reference for future design breeding to improve fruit flavor.

Keywords: peach; commercial cultivars; VOCs; GC-MS; GC-IMS; OAV

Citation: Wang, X.; Li, X.; Su, M.; Zhang, M.; Hu, Y.; Du, J.; Zhou, H.; Zhang, X.; Ye, Z.; Yang, X. The Construction of Volatile Profiles of Eight Popular Peach Cultivars Produced in Shanghai Using GC-MS and GC-IMS. *Horticulturae* **2023**, *9*, 382. https://doi.org/10.3390/horticulturae9030382

Academic Editor: Anita Sønsteby

Received: 16 January 2023
Revised: 23 February 2023
Accepted: 28 February 2023
Published: 15 March 2023

1. Introduction

The peach (*Prunus persica* (L.) Batsch) has become an important economic crop due to its soft and juicy flesh, attractive aroma and enriched nutrition, and it has been grown extensively at home and abroad [1]. As the originating country of the peach, China holds a large number of wild relatives and landraces [2]. 'Chinese Cling', the most influential founder in peach-breeding history, has derived many excellent cultivars such as 'Elberta', 'J.H. Hale', 'Okubo' and 'Hakuho' [3]. The elite commercial peach cultivars planted in Shanghai, such as 'Hujingmilu' (HJML), 'Xinfengmilu' (XFML) and 'Jinxiu' (JX), are also excellent offspring of 'Chinese Cling'. These cultivars are popular in Shanghai for their high quality and were cultivated in 3100 hectares in 2021, accounting for approximately 82% of the total peach-cultivated area in Shanghai (data source: Shanghai Agricultural Technology Extension Service Center).

Aroma is an essential factor to determine consumers' preference, which has received great attention from geneticists and breeders in recent years [4,5]. The aroma profiles in fruit result from the interactions of many volatiles. Currently, more than 100 volatiles have been identified in peaches, mainly including aldehydes, alcohols, esters, lactones, terpenes and ketones [6,7]. However, the compositions and content of volatiles are influenced by genetic backgrounds, fruit maturity and cultivation conditions. For instance, aldehydes and alcohols are the major contributors of VOCs in immature fruits and their contents decrease gradually during maturation, while the contents of lactones increase with peach fruit maturation [8]. Studies on several varieties have shown many differences in the composition and content of volatiles. Wang et al. have reported that the Chinese wild peach 'Wutao' has the highest content of total volatiles, while 'Ruipan 14' and 'Babygold 7' show high contents of lactones, and cultivars of American and European origin mainly contain high levels of linalool [7]. Several studies have also shown that cultivation conditions can also influence the composition and content of volatiles, such as fertilization and bagging [9–11]. Although more than 100 VOCs can be detected in peaches, only around 25 of them significantly contribute to flavor quality, and these are recognized as characteristic compounds [12]. The odor activity value (OAV) is commonly used to evaluate the contribution of a compound to the overall aroma, and it represents the ratio of the actual concentration in the sample to the threshold value in water, and was first proposed by Rothe [13]. Compounds with OAV ≥ 1 are not only considered major contributors to flavor, but also good indicators for the breeding of good-quality peaches.

Gas chromatography-mass spectrometry (GC-MS) is a commonly used technique that takes advantage of the high separation ability of GC and the superiority of mass spectrometry in the identification of substances, which allows it to be applied with high sensitivity and precision in the qualitative and quantitative analysis of VOCs in samples [14]. However, due to the complexity of most sample matrices, this method requires tedious and time-consuming sample pretreatment before analysis while working under vacuum conditions, making it unsuitable for rapid characterization studies of VOCs in complex samples [15,16]. Ion mobility spectrometry (IMS) is an atmospheric analytical chemistry method for the detection and identification of different types of substances based on the mobility of gas-phase ions in a weak electric field [15]. GC-IMS integrates the high separation ability of GC with the high sensitivity and fast response of IMS. Compared to GC-MS, it has the advantages of having a lower detection limit, no pre-treatment, convenience and can be operated at atmospheric pressure, which can maximize the authenticity of flavor in samples and reflect the original flavor information [16,17]. The method of combining various techniques to study VOCs in foods has become a hot topic, which can establish a more comprehensive and scientific aroma fingerprint. As far as we know, comparison analysis of volatiles in different commercial peach cultivars based on GC-MS and GC-IMS is rarely reported.

In the present study, the volatiles of eight commercial peach cultivars from Shanghai were analyzed by GC-MS, OAV and GC-IMS to comprehensively investigate their volatile compounds and aroma profiles. The results could improve our understanding of aroma compositions in these high-quality peach cultivars, which might also provide a reference for future design breeding to improve fruit flavor.

2. Materials and Methods

2.1. Plant Materials

Eight elite peach cultivars derived from 'Chinese Cling' were selected in this study, and all of them had a melting texture and displayed strong aromas. The characteristic information of each cultivar is listed in Figure 1 and Table 1. All trees were planted in the Peach Germplasm Repository of the Shanghai Academy of Agricultural Sciences, China (30°89′ N, 121°39′ E), and the orchard management procedures such as irrigation, pruning, disease control and fertilization were the same for all cultivars. Thirty fruits of each cultivar were picked at the commercial harvest stage in the summer of 2022 according to grounded skin color, fruit firmness and recorded maturity time. The fruits were picked from three trees and immediately transported to the laboratory. Fifteen fruits of each cultivar with uniform size were selected in each biological replicate. The mesocarp was cut into pieces, frozen immediately in liquid nitrogen, grounded into powder and stored at −80 °C for use.

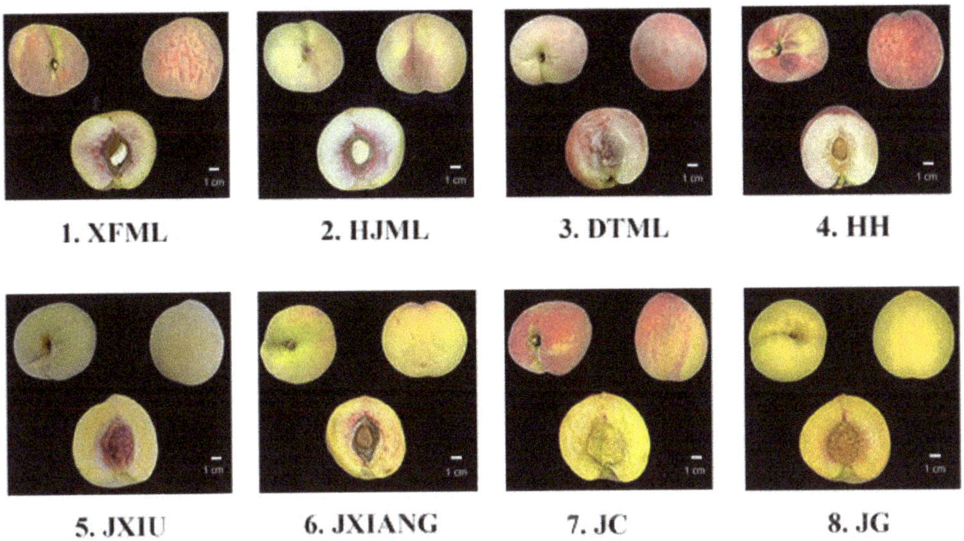

Figure 1. Representative images of eight peach cultivars used in this study. The codes under each image are the abbreviation names for each cultivar.

Table 1. The characteristic information of peach materials.

No.	Cultivar Name	Abbreviation Code	Origin	Flesh Color	Fruit Texture	Brix (%)	Acid	Harvest Date
1	Xinfengmilu	XFML	Shanghai, China	White	Soft-melting	14.9	Sub-acid	Mid-July
2	Hujingmilu	HJML	Jiangsu, China	White	Soft-melting	13.2	Sub-acid	Mid-July
3	Datuanmilu	DTML	Shanghai, China	White	Soft -melting	14.3	Sub-acid	Early-July
4	Hikawa Hakuho	HH	Japan	White	Soft-melting	10.2	Sub-acid	Mid-June
5	Jinxiu	JXIU	Shanghai, China	Yellow	Hard-melting	18.5	Sub-acid	Mid-August
6	Jinxiang	JXIANG	Shanghai, China	Yellow	Hard-melting	10.1	Sub-acid	Early-July
7	Jinchun	JC	Shanghai, China	Yellow	Hard-melting	11.0	Sub-acid	Early-June
8	Jinguan	JG	Shanghai, China	Yellow	Hard-melting	14.6	Sub-acid	Mid-July

2.2. Chemicals

All qualification and quantitative standards were purchased from Sigma Aldrich (Shanghai, China), and their purity was above 95%.

2.3. GC-MS Analysis of Volatile Compounds

The method of sample pretreatment referred to a previous study [4], with the difference that each headspace vial in this study contained three grams of sample. A fiber (65 μm, PDMS/DVB, Supelco, Bellefonte, PA, USA) was inserted into the headspace of a vial to extract the volatile compounds at 40 °C for 30 min. At the end of extraction, the fiber was desorbed into the injection port of the GC for 5 min. The time of solvent delay was 3 min. The samples were analyzed using the Agilent gas chromatography mass spectrometry (GC-MS) instrument (7890–5975) with a DB-WAX (60 m × 250 μm × 0.25 μm, Agilent

122–7062) under splitless injection mode. Helium (99.999%) was used as a carrier gas at a constant flow (1 mL/min).

The GC oven temperature was initially held at 40 °C for 2 min, increased to 100 °C at 3 °C/min and then increased to 230 °C at 5 °C/min and held for 5 min. The mass detector was used in an electron impact mode at 70 eV, an ion source temperature of 230 °C and a mass spectrometry scan range of m/z 30–350. According to the NIST 08 (National Institute of Standards and Technology) mass spectrometry database, the volatile compound was identified by matching degree, retention time, retention index (RI) and standard chemicals. Compounds with matching degrees greater than 80% were selected as effective aroma components. In order to quantify the components, semi-quantification was performed using an internal standard method to calculate the relative concentration of volatile compounds in the samples. Measurements were repeated three times for each sample.

2.4. GC-IMS Analysis of Volatile Compounds

The VOCs of different peach cultivars were detected using HS-GC-IMS (FlavorSpec ®, G.A.S, Dortmund, Germany). Two grams of grounded sample from each cultivar were transferred into 20 mL headspace bottles. The headspace injection condition was set at the following parameters: incubation time was 15 min, incubation temperature was 40 °C, injection needle temperature was 85 °C and injection volume was 500 µL. The GC conditions were as follows: the chromatographic column was MXT-5 (15 m × 0.53 mm, 1 µm, Agilent Technology, Palo Alto, CA, USA)). Nitrogen of 99.99% purity was used as a carrier gas at a programmed flow as follows: 2 mL/min for 2 min, 10 mL/min for 8 min and 100 mL/min for 10 min. The IMS conditions were as follows: the drift tube temperature was 45 °C, and the drift gas velocity was 150 mL/min. All tests were repeated three times. Volatile compounds were preliminarily identified by comparing the RI and ion drift times (the time in milliseconds it takes for an ion to reach the collector through the drift tube) of standards in the GC-IMS library and the NIST database.

2.5. Identification of Characteristic Aroma Compounds in Samples by the Odor Activity Value Method

The OAVs of compounds were calculated using the following formula: $OAV = C_i/OT_i$, where C_i indicates the concentration of a compound and OT_i indicates the odor threshold. The odor thresholds of all compounds were obtained from the published literature [18]. For a compound with multiple thresholds, the standard we chose was its threshold in water and the most recent data.

2.6. Statistical Analysis

The experimental data derived from GC-MS were analyzed using Origin 2021 software (Microcal Software, Inc., Northampton, MA, USA). Analysis of variance (ANOVA) and Duncan's multiple range test ($p < 0.05$) were used to analyze the significant differences among samples. Cluster analysis was performed using TB tools software (versions v1.1043). For the data obtained using GC-IMS, four software programs from the instrument were used for analyzing: (1) VOCal was used for viewing analytical spectra and qualitative and quantitative data; the NIST database and IMS database built into the application software can be used for qualitative analysis of compounds. (2) The Reporter plug-in for directly comparing the spectral differences between samples. (3) The Gallery Plot plug-in was used for fingerprint profile comparison. (4) The Dynamic PCA plug-in was used for dynamic principal component analysis.

3. Results and Discussion

3.1. Volatile Profile of Different Peach Cultivars Using GC-MS

3.1.1. The Construction of VOCs Profiling for Eight Peach Cultivars

By searching the NIST database and the chemical standard, a total of seventy VOCs were tentatively identified and quantified for eight peach cultivars using GC-MS. They were categorized into ten groups, including nine aldehydes, ten esters, eleven terpenoids, nine alcohols, six lactones, five ketones, five alkanes, eleven aromatic hydrocarbons, two furans and two other compounds (Figure 2, Table 2). Twenty-three VOCs were commonly identified in all cultivars, including hexanal, nonanal, benzaldehyde, 2-hexenal, butyl acetate, hexyl acetate, linalool, β-myrcene, D-limonene, 1-hexanol, 3-hexenol, 2-hexenol, 2-ethyl-1-hexanol, γ-octalactone, δ-decalactone, γ-hexalactone, γ-decalactone, γ-dodecalactone, β-ionone, 2-octanone, 2-ethyl furan and 2,4-ditert-butyl phenol. More than 40 VOCs were identified in HJML (44), DTML (42), HH (42) and JC (43). Fewer than 35 volatile compounds were identified in XFML (35), JXIU (29), JXIANG (35) and JG (32). Seventeen unique compounds were identified only in one cultivar. Differences were observed in the total content of volatile compounds among the eight peach cultivars (Table 2 and Figure 2).

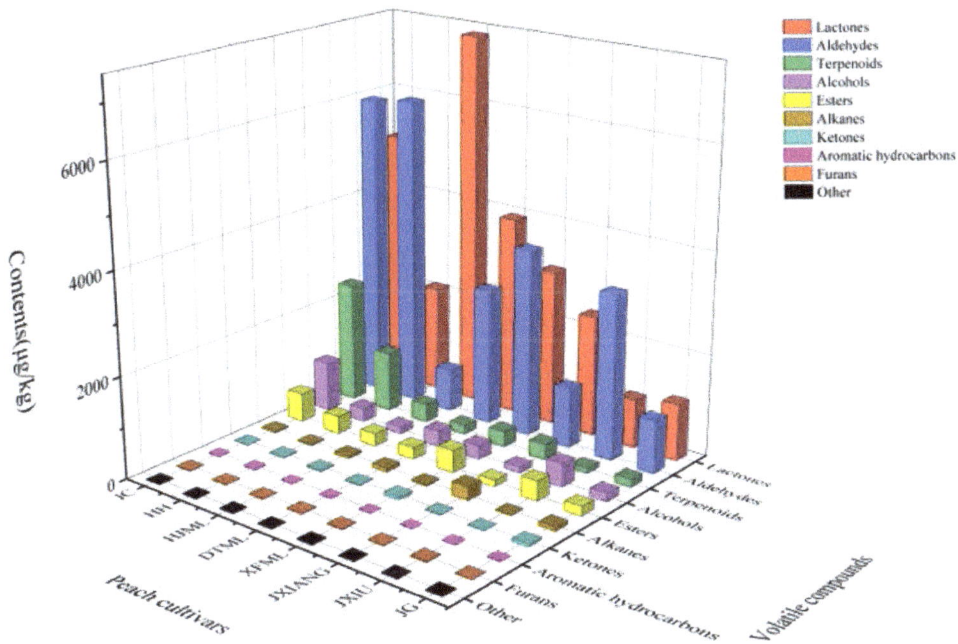

Figure 2. The total contents of each category of VOCs in eight peach cultivars. 3.1.2 Cluster Analysis of VOCs of eight peach cultivars.

Aldehydes were regarded as one of the most significant contributors to the peaches' green and grassy flavor, and have been reported to exist widely in immature plums, apples and pears [19,20]. The aldehydes detected in this experiment include hexanal, nonanal, benzaldehyde, 2-hexenal, 3-hexenal, (E)-2-nonenal, (E)-2-hexenal, (E)-2-heptenal and heptanal (Table 1). Among these, hexanal, nonanal, 2-hexenal and benzaldehyde were identified as major components and were detected in all the samples, which is consistent with previous studies [20,21]. Notably, the highest content of 2-hexenal among all samples was 5132.24 μg/kg in HH, while the lowest was 589.96 μg/kg in HJML; therefore, this compound may be the most important factor responsible for the differences in aldehydes among the samples. It is also clear from Figure 2 that the total content of aldehydes is highest in HH and lowest in HJML.

Alcohols are a group of compounds other than aldehydes that give peach fruit a green aroma. In this study, nine alcohols were found, namely 2-ethyl-1-hexanol, 3-methyl-3-nonanol, carbitol, 3-hexenol, hexanol, 2-hexenol, 3-methyl-3-buten-1-ol, α-terpineol and geraniol. Among them, 3-hexenol, hexanol, 2-hexenol and 2-ethyl-1-hexanol were detected in all samples at high levels (Table 2). JC had the highest level of alcohols and JXIANG had the lowest (Figure 2).

Table 2. The absolute and relative contents of VOCs in eight peach cultivars tested using GC-MS.

Compound	CAS	Identification	Concentration (μg/kg)							
			XFML	HJML	DTML	HH	JXIU	JXIANG	JC	JG
Aldehydes										
Benzaldehyde	100-52-7	MS, RI, Std	46.70 ± 34.23a	29.62 ± 1.96a	14.33 ± 2.70a	17.66 ± 0.32a	20.14 ± 7.80a	37.13 ± 35.21a	10.64 ± 2.87a	11.74 ± 0.77a
2-Hexenal	505-57-7	MS, RI, Std	3127.39 ± 512.19b	589.96 ± 109.27d	2228.96 ± 151.92bc	5132.24 ± 276.90a	2713.64 ± 1240.09b	882.17 ± 142.51cd	4782.91 ± 1407.01a	816.73 ± 193.16cd
Hexanal	66-25-1	MS, RI, Std	484.13 ± 173.46b	148.91 ± 21.65c	420.23 ± 42.53bc	891.07 ± 52.4a	490.39 ± 277.91b	240.58 ± 40.05bc	1134 ± 328.54a	123.66 ± 13.28c
Nonanal	124-19-6	MS, RI	19.62 ± 4.07b	37.81 ± 8.21a	19.55 ± 0.59b	34.23 ± 1.26a	32.82 ± 1.81a	17.42 ± 0.91b	23.84 ± 2.93b	37.69 ± 1.1a
3-Hexenal	4440-65-7	MS, RI	ND	ND	1.34 ± 0.12a	ND	ND	ND	ND	ND
(E)-2-Nonenal	18829-56-6	MS, RI	3.60 ± 0.50ab	ND	ND	ND	ND	ND	4.36 ± 0.19a	3.5 ± 0.33b
(E)-2-Hexenal	6728-26-3	MS, RI	ND	22.15 ± 3.52b	ND	ND	ND	ND	ND	52.82 ± 13.12a
(E)-2-Heptenal	18829-55-5	MS, RI	ND	ND	ND	2.85 ± 0.10a	ND	ND	2 ± 0.3	ND
Heptanal	111-71-7	MS, RI	ND	ND	ND	3.64 ± 0.12a	ND	ND	ND	ND
Esters										
Butyl acetate	123-86-4	MS, RI, Std	151.68 ± 54.97b	1.49 ± 0.09c	132.75 ± 15.77b	274.47 ± 18.42a	153.13 ± 86.01b	53.1 ± 46.12bc	353.26 ± 107.81a	1.0 ± 0.1c
Isopentyl acetate	123-92-2	MS, RI, Std	ND	ND	ND	5.49 ± 0.52ab	ND	6.29 ± 1.09a	3.61 ± 1.04b	ND
Hexyl acetate	142-92-7	MS, RI, Std	3.25 ± 0.69d	14.9 ± 0.37d	30.49 ± 2.28c	5.05 ± 0.16d	72.28 ± 11.88b	5.43 ± 0.39d	91.17 ± 17.61a	1.56 ± 0.14d
Heptyl acetate	112-06-1	MS, RI, Std	ND	ND	1.47 ± 1.10a	ND	ND	ND	ND	ND
(Z)-3-Hexen-1-yl acetate	3681-71-8	MS, RI, Std	274.56 ± 28a	223.62 ± 11.88b	37.06 ± 2.32de	58.64 ± 11.23d	156.08 ± 31.53c	7.9 ± 0.48e	9.78 ± 0.68e	180.94 ± 30.04c
2-Ethyl hexyl acetate	103-09-3	MS, RI	ND	1.22 ± 0.03cd	0.89 ± 0.12e	1.52 ± 0.03b	1.07 ± 0.02d	1.32 ± 0.16cd	1.97 ± 0.02a	ND
(E)-2-Hexen-1-yl acetate	2497-18-9	MS, RI	6.73 ± 0.72d	ND	19.47 ± 1.30c	ND	ND	45.06 ± 3.44b	99.57 ± 10.23a	2.86 ± 0.2d
(Z)-2-Penten-1-yl acetate	42125-10-0	MS, RI	ND	5.00 ± 0.07b	5.82 ± 0.73a	ND	ND	ND	ND	ND
Nonyl acetate	143-13-5	MS, RI	ND	1.5 ± 0.12	ND	ND	ND	ND	ND	ND
(Z)-3-Nonen-1-yl acetate	13049-88-2	MS, RI	ND	ND	ND	ND	2.47 ± 0.42 d	ND	ND	ND
Terpenoids										
Linalool	78-70-6	MS, RI, Std	57.66 ± 3.46d	128.13 ± 7.35c	4.03 ± 0.38d	339.26 ± 3.37b	3.76 ± 0.50 d	181.32 ± 3.05c	820.73 ± 88.58a	3.81 ± 0.18d
β-Myrcene	123-35-3	MS, RI, Std	5.84 ± 1.95d	9.86 ± 6.78d	3.64 ± 1.5d	42.55 ± 3.51b	3 ± 0.92d	31.8 ± 3.4c	112.76 ± 5.6a	3.42 ± 0.62d
D-Limonene	5989-27-5	MS, RI, Std	225.5 ± 18.37c	209.86 ± 2.36c	155.22 ± 15.7c	772.21 ± 17.36b	93.57 ± 16.62c	100.17 ± 122.38c	1427.32 ± 390.57a	150.66 ± 8.18c
(Z)-2-Dodecene	7206-26-0	MS, RI	ND	ND	2.81 ± 2.79a	1.58 ± 0.08a	ND	ND	4.37 ± 0.25a	ND
1-Dodecene	112-41-4	MS, RI	4.95 ± 0.68	ND	ND	ND	ND	ND	ND	ND
1-Tetradecene	1120-36-1	MS, RI	9.07 ± 9.07a	ND	8.32 ± 8.32a	9.42 ± 9.42a	ND	ND	ND	ND
(E)-3-Dodecene	7206-14-6	MS, RI	ND	ND	1.61 ± 0.16a	2.32 ± 0.9a	ND	ND	ND	ND
(Z)-β-Ocimene	3338-55-4	MS, RI	ND	1.37 ± 0.09	ND	ND	ND	ND	ND	ND
2-Carene	554-61-0	MS, RI	ND	2.19 ± 0.08	ND	ND	ND	ND	ND	ND
(E)-β-Ocimene	3779-61-1	MS, RI	ND	ND	ND	ND	ND	1.53 ± 0.02b	4.8 ± 0.68a	ND
Terpinolene	586-62-9	MS, RI	ND	ND	ND	3.58 ± 0.85	ND	ND	ND	ND
Alcohols										
1-Hexanol	111-27-3	MS, RI, Std	46.55 ± 24.93bcd	14.19 ± 1.38d	36.18 ± 4.16bcd	63.73 ± 7.92bc	73.98 ± 15.75b	20.85 ± 4.51cd	263.24 ± 57.24a	9.72 ± 0.61d
3-Hexenol	928-96-1	MS, RI, Std	60.97 ± 13.64c	45.55 ± 1.6cd	91.23 ± 8.47b	33.9 ± 17.32de	120.7 ± 20.7a	14.4 ± 2.75e	72.08 ± 20.03bc	46.99 ± 0.51cd
2-Hexenol	928-95-0	MS, RI, Std	28.52 ± 3.98c	23.22 ± 1.44c	60.43 ± 16.52c	110.88 ± 13.45bc	215.88 ± 45.15b	33.27 ± 22.51c	556.7 ± 157.73a	22.63 ± 4.48c
3-Methyl-3-buten-1-ol	763-32-6	MS, RI, Std	45.26 ± 11.06a	ND	67.79 ± 26.14a	ND	ND	ND	ND	ND
2-Ethyl-1-hexanol	104-76-7	MS, RI	71.51 ± 0.5a	43.33 ± 0.73c	43.39 ± 1.07c	46.44 ± 1.14b	41.71 ± 1.00c	43.01 ± 1.48c	38.93 ± 0.95d	73.3 ± 0.59a
3-Methyl-3-nonanol	21078-72-8	MS, RI	9.67 ± 0.36a	8.28 ± 0.08a	ND	8.89 ± 0.17a	8.02 ± 0.11cd	7.86 ± 0.19d	ND	ND
Carbitol	111-90-0	MS, RI	16.26 ± 1.70a	ND	3.68 ± 0.25c	ND	1.86 ± 0.08c	ND	ND	12.54 ± 0.92b
α-Terpineol	98-55-5	MS, RI	ND	ND	ND	ND	ND	ND	15.4 ± 0.98	ND
Geraniol	106-24-1	MS, RI	ND	ND	ND	ND	ND	ND	3.45 ± 0.35	ND

Table 2. *Cont.*

Compound	CAS	Identification	Concentration (μg/kg)							
			XFML	HJML	DTML	HH	JXIU	JXIANG	JC	JG
Lactones										
γ-Octalactone	104-50-7	MS, RI, Std	31.82 ± 2.90cd	152.72 ± 5.82a	57.2 ± 23.84b	30.87 ± 1.43cd	14.66 ± 1.60d	43.03 ± 3.20bc	55.31 ± 2.45b	27.07 ± 2.77cd
γ-Decalactone	706-14-9	MS, RI, Std	566.29 ± 25.71c	1878.4 ± 101.12a	847.91 ± 57.66b	335.06 ± 12.00d	111.9 ± 7.80e	361.77 ± 17.98d	811.07 ± 22.98b	149.08 ± 17.74e
δ-Decalactone	705-86-2	MS, RI, Std	2125.73 ± 422.74c	4334.12 ± 517.04a	2203.62 ± 211.66c	1097.68 ± 119.79d	254.40 ± 37.91e	981.25 ± 65.47d	3060.85 ± 107.94b	276.04 ± 54.89e
γ-Hexalactone	695-0-7	MS, RI, Std	272.53 ± 28.29e	891.48 ± 37.54b	760.62 ± 50.23c	633.98 ± 29.58cd	575.32 ± 62.65d	958.80 ± 69.18b	1099.58 ± 122.39a	647.4 ± 48.79cd
γ-Heptalactone	105-21	MS, RI, Std	ND	17.03 ± 2.98b	7.91 ± 2.06c	6.07 ± 1.70c	ND	28.54 ± 2.62a	15.07 ± 1.97b	ND
γ-Dodecalactone	2305-05-7	MS, RI, Std	67.77 ± 1.13b	83.76 ± 5.08a	43.17 ± 2.88c	1.48 ± 0.77g	4.27 ± 0.25g	10.27 ± 0.23f	35.05 ± 0.85d	14.8 ± 0.57e
Ketones										
β-Ionone	79-77-6	MS, RI, Std	0.43 ± 0.07e	1.95 ± 0.14a	0.90 ± 0.05c	2.03 ± 0.15a	0.67 ± 0.06d	0.65 ± 0.07d	1.16 ± 0.11b	0.64 ± 0.06d
2-Octanone	111-13-7	MS, RI	44.97 ± 1.75b	2.49 ± 0.02c	3.06 ± 0.42c	3.63 ± 0.22c	2.80 ± 0.68c	6.67 ± 6.09c	2.64 ± 0.32c	50.81 ± 1.23a
Dihydro-beta-ionone	17283-81-7	MS, RI	11.79 ± 1.1c	16.01 ± 0.07b	11.64 ± 0.06c	24.22 ± 1.33a	ND	ND	ND	ND
2-Nonanone	821-55-6	MS, RI	ND	0.87 ± 0.07	ND	ND	ND	ND	ND	ND
Geranylacetone	3796-70-1	MS, RI	ND	ND	ND	ND	ND	ND	3.93 ± 0.64	ND
Alkanes										
Tridecane	629-50-5	MS, RI, Std	ND	11.39 ± 7.91b	43.06 ± 34.94b	17.53 ± 9.64b	ND	227.41 ± 102.94a	ND	36.73 ± 12.76b
Dibromochloro-methane	124-48-1	MS, RI	ND	2.38 ± 0.41a	2.90 ± 1.12a	2.15 ± 0.60a	2.27 ± 0.98a	2.13 ± 0.34a	1.96 ± 0.37a	ND
Octamethyl cyclotetrasiloxane	556-67-2	MS, RI	20.64 ± 16.52a	26.63 ± 4.95a	24.87 ± 3.86a	ND	ND	33.00 ± 8.39a	27.25 ± 10.18a	35.70 ± 3.37a
Bromoform	75-25-2	MS, RI	ND	4.17 ± 0.06a	4.37 ± 0.37a	ND	2.88 ± 0.77c	ND	3.68 ± 0.49b	ND
Dodecane	112-40-3	MS, RI	ND	3.82 ± 0.16b	ND	ND	ND	5.36 ± 3.04a	2.29 ± 0.26c	ND
Aromatic hydrocarbons										
Styrene	100-42-5	MS, RI, Std	1.04 ± 0.46a	1.82 ± 0.19a	1.45 ± 0.28a	ND	4.52 ± 4.45a	2.65 ± 0.35a	ND	ND
Ethylbenzene	100-41-4	MS, RI	ND	2.75 ± 0.03bc	1.85 ± 0.28c	5.54 ± 0.81a	ND	ND	2.96 ± 0.42b	ND
Toluene	108-88-3	MS, RI	3.13 ± 0.58b	ND	ND	5.36 ± 0.64a	ND	ND	5.99 ± 1.13a	1.82 ± 0.11b
Meta-xylene	108-38-3	MS, RI	ND	1.99 ± 0.11b	3.51 ± 0.81a	ND	ND	ND	ND	2.35 ± 0.29b
Ortho-cymene	527-84-4	MS, RI	ND	0.50 ± 0.10	ND	ND	ND	ND	ND	ND
Hemimellitene	526-73-8	MS, RI	ND	1.13 ± 0.09	ND	ND	ND	ND	ND	ND
α-Ionene	475-03-6	MS, RI	ND	ND	ND	2.53 ± 0.34b	ND	ND	2.76 ± 0.03a	ND
Edulan	41678-29-9	MS, RI	ND	ND	ND	ND	ND	ND	3.78 ± 0.23	ND
p-Xylene	106-42-3	MS, RI	ND	ND	ND	7.47 ± 5.84	ND	ND	ND	ND
o-Xylene	95-47-6	MS, RI	ND	ND	ND	6.13 ± 1.64	ND	ND	ND	ND
1-Methyl naphthalene	90-12-0	MS, RI	ND	ND	ND	0.94 ± 0.08	ND	ND	ND	ND
Furans										
2-ethyl furan	3208-16-0	MS, RI, Std	15.22 ± 2.83ab	5.85 ± 2.32c	18.17 ± 2.66ab	9.16 ± 3.85ab	20.89 ± 17.51a	5.39 ± 1.28c	5.89 ± 3.46c	5.16 ± 3.86c
2-Pentyl furan	3777-69-3	MS, RI	ND	ND	ND	ND	ND	2.12 ± 0.09	ND	ND
Other										
2,4-Ditert-butyl phenol	96-76-4	MS, RI	14.09 ± 0.57a	3.32 ± 0.15e	3.62 ± 0.25de	4.34 ± 0.39cd	3.82 ± 0.22cde	4.71 ± 0.49cde	4.12 ± 0.51cde	13.03 ± 0.76b
Neroloxide	1786-08-9	MS, RI	ND	ND	ND	ND	ND	2.93 ± 0.41a	1.50 ± 0.06b	ND

ND represents not detected; the meanings of a–e in the same row with different superscripts represent significant differences ($p < 0.05$). The "Std" in the identification column indicates that these compounds were quantified by a chemical standard.

Esters are the main flavor components of most ripe fruits and provide the fruity notes, and particularly (Z)-3-hexen-1-yl acetate and hexyl acetate are considered to contribute significantly to the aroma characteristics of peaches [6,22,23]. A total of ten esters were detected in the eight peach cultivars. Among these, butyl acetate, (Z)-3-hexen-1-yl acetate and hexyl acetate was detected in all samples and the levels were significantly different between samples. (E)-2-hexen-1-yl acetate was also detected at high levels in most cultivars. The remaining unmentioned esters were present at less than 10 μg/kg in each sample, so these components were not the focus of this study.

A total of 11 terpenoids were identified, of which linalool, β-myrcene and D-limonene were detected in all peach cultivars with significant differences. Horvat et al. [24] identified linalool as one of the key aroma compounds of peach fruit that increased significantly with fruit ripening and has floral aroma properties [25,26]. The order of linalool content in all samples was as follows: JC > HH > JXIANG > HJML > XFML> DTML > JG > JXIU. Surprisingly, the linalool content in JC was more than twice that of HH. Previous studies have shown that nectarines contain higher levels of linalool than peaches [7]. Although the material in this study did not include nectarines, JC was derived from peach Jinxiu and nectarine Huyou 018, which may be the reason for its highest linalool content among the eight cultivars. β-myrcene is reported as a precursor of linalool and is one of the common terpenoids in peaches and nectarines [27,28]. Here, the content of β-myrcene ranked consistently with linalool in the eight cultivars.

More than 10 lactones have been found in peaches [6,27,29]. Here, we identified six lactones: γ-hexalactone, γ-decalactone, γ-octalactone, γ-dodecalactone, γ-heptalactone and δ-decalactone. Consistent with previous studies [7], γ-decalactone, δ-decalactone and γ-hexalactone were the leading lactones with the highest content in these elite cultivars, with the difference that δ-decalactone was the most abundant in the present study. As reported, mid- and later-ripening cultivars have higher lactone content than early-ripening cultivars [30]. However, in this study, some early- and mid-maturing cultivars such as HJML and JC had the highest total lactone content, while the late-maturing cultivar JXIU had the lowest content (Figure 2), which might be related to its genetic background and cultivation conditions.

Five ketones were detected, and β-ionone and 2-octanone were detected in all samples. The content of total ketone was highest in 'XFML' and lowest in 'JX' (Figure 2). Interestingly, dihydro-β-ionone was detected in all white flesh peaches, but not in all yellow flesh peaches; this result is similar to that of several researchers [31]. This is due to the fact that C-13-norisoprenoids such as dihydro-β-ionone in peach fruit are mainly synthesized via the isoprenoid pathway, and carotenoids are precursors for their synthesis. Carotenoids in white-flesh peaches are cleaved to disubstituted carotenoids via the action of carotenoid cleavage dioxygenase (CCD4), which in turn forms the volatile compounds such as dihydro-β-ionone, while the CCD4 enzyme gene in yellow-flesh peaches is mutated and cannot degrade carotenoids, so less dihydro-β-ionone is synthesized than in white-fleshed peaches [32,33].

To further understand the differences of volatiles in different peach cultivars, a cluster analysis heat map was constructed (Figure 3). The vertical axis of the figure represents different volatiles and the horizontal axis represents different peach cultivars. The red-to-blue and large-to-small circles indicate the abundance of volatiles from highest to lowest. According to the content of each of the VOCs, cluster analysis clustered JXIU, DTML, JG and XFML into one group. This indicates that the composition and content of their volatiles are similar. Horizontally, the numerous volatiles were roughly divided into three clusters. Cluster 1 covers most of the aldehydes and alcohols. In general, this cluster of compounds is relatively higher in XFML and JG. The substances in cluster 2 mainly include esters and lactones, and the HH and HJML shows a high abundance in this cluster. The most abundant compounds in cluster 3 are terpenoids and aromatic hydrocarbons, and JC shows the highest abundance in this cluster. Cluster analysis plots visually show the abundance of various substances in the sample.

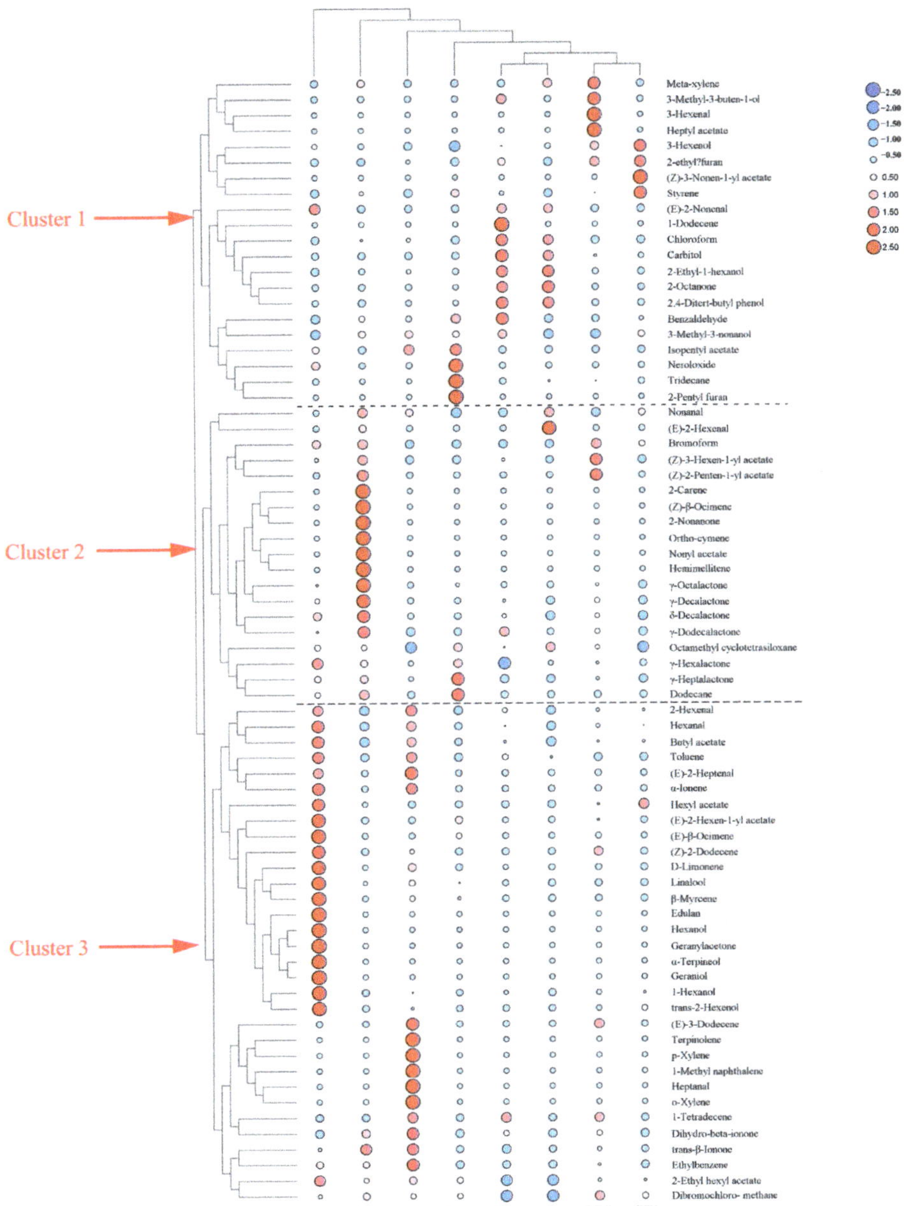

Figure 3. A clustered heatmap of VOCs among eight peach cultivars.

3.1.2. Analysis of Odor Activity Values of VOCs in Eight Peach Cultivars

OAV is the main method of aroma characterization and it is also an indicator of whether human can perceive VOCs in samples by olfactory. Here, by calculating the OAV for each compound, we grasped the compounds that contribute to the peaches. The sensory thresholds of each compound and their OAV in the samples are listed in Table 3, and considering that OAV ≥ 1 is required to have a contribution to flavor, we only analyzed this part of the data.

Table 3. Comparison of odor activity values of different VOCs.

Compound	Aroma Description [1]	Threshold	XFML	HJML	DTML	HH	JXIU	JXIANG	JC	JG	Range of OAV
Aldehydes											
2-Hexenal	fruity, vegetable	30	104.25	19.67	74.3	171.07	90.45	29.41	159.43	27.22	19.67–171.07
Hexanal	green, grassy	5	96.83	29.78	84.05	178.21	98.08	48.12	226.8	24.73	24.73–226.8
Nonanal	cucumber	1.1	17.84	34.37	17.77	31.12	29.84	15.84	21.67	34.26	15.84–34.37
Octanal	orange peel, green	0.587	ND	15.03	11.31	15.47	13.37	8.18	9.86	22.37	8.18–22.37
Esters											
Butyl acetate	sweet, ripe banana	58	2.62	0.03	2.29	4.73	2.64	0.92	6.09	0.02	0.02–6.09
Isopentyl acetate	sweet, fruity	0.15	ND	ND	ND	36.6	ND	41.93	24.07	ND	24.07–41.93
(Z)-3-Hexen-1-yl acetate	fruity, apple and pear	31	8.86	7.21	1.2	1.89	5.03	0.25	0.32	5.84	0.25–8.86
Terpenoids											
Linalool	floral, woody	0.22	262.09	582.41	18.32	1542.09	17.09	824.18	3730.59	17.32	17.09–3730.59
β-Myrcene	woody, vegetative	1.2	4.87	8.22	3.03	35.46	2.5	26.5	93.97	2.85	2.5–93.97
D-Limonene	sweet, orange, citrus	34	6.63	6.17	4.57	22.71	2.75	2.95	41.98	4.43	2.75–41.98
Alcohols											
1-Hexanol	green, fruity	5.6	8.31	2.53	6.46	11.38	13.21	3.72	47.01	1.74	1.74–47.01
3-Hexenol	fresh, green	1.9	32.09	23.97	48.02	17.84	63.53	7.58	37.94	24.73	7.58–63.53
Lactones											
γ-Octalactone	fruity, creamy	12	2.65	12.73	4.77	2.57	1.22	3.59	4.61	2.26	1.22–12.73
γ-Decalactone	fruity, peach	1.1	514.81	1707.64	770.83	304.6	101.73	328.88	737.34	135.53	101.73–1707.64
δ-Decalactone	creamy, fatty, buttery	66	32.21	65.67	33.39	16.63	3.85	14.87	46.38	4.18	3.85–65.67
γ-Dodecalactone	fruity, peach	2	33.89	41.88	21.59	0.74	2.14	5.14	17.53	7.4	0.74–41.88
Ketones											
β-Ionone	floral, violet	0.007	61.43	278.57	128.57	290	95.71	92.86	165.71	91.43	61.43–290

[1] The aroma description of the compounds come from the website: http://www.thegoodscentscompany.com/ (accessed on 12 January 2023).

By statistical analysis, only 17 volatiles with OAV ≥ 1 were retained, including 4 aldehydes, 3 esters, 3 terpenoids, 1 ketones, 2 alcohols and 4 lactones. The importance of these compounds was determined by ranking them according to the minimum OAV in each of the eight cultivars: γ-decalactone > β-ionone > hexanal > 2-hexenal > linalool > nonanal > 3-hexenol > δ-decalactone > D-limonene > β-myrcene > 1-hexanol > γ-octalactone > γ-dodecalactone > (Z)-3-hexen-1-yl acetate > butyl acetate > isopentyl acetate > octanal. Although γ-decalactone was not the most abundant in quantity, it showed the highest OAV due to the low threshold value. As the main contributor of peach-like aroma, it is also an essential factor that distinguishes peaches from other fruits [34,35]. The OAV of γ-decalactone was highest in HJML. β-ionone and linalool are the two most dominant floral compounds in peach fruits with extremely low-threshold values. They can exhibit rose and violet flavor despite their low content. In our result, both the amount and OAV of linalool were high in the eight peaches. The OAV ranged from 17.09 to 3730.59, and the highest value was present in JC. Due to the 0.007 μg/kg threshold value, the range of OAV of β-ionone was from 61.43 to 290, and the highest value exhibited in HH. Regarding the grass/green-notes aldehydes and C6 alcohols, hexanal and 3-hexenol were the most active flavor compounds in most investigated cultivars. Their OAV ranged from 24.73 to 226.8 and 7.58 to 63.53. The highest value of these two compounds was observed in JC and JXIU, respectively.

3.2. Volatile Profile of Different Peach Cultivars Using GC-IMS

3.2.1. Differential Analysis of the Topographic Plots of VOCs in Eight Cultivars Using GC-IMS

The VOC profiles of eight peaches were also constructed using GC-IMS. The 3D topographic plot of the peach volatiles showed that the composition and signal intensities of VOCs varied considerably between different peach cultivars (Figure 4a). To observe the data more conveniently and intuitively, we normalized the ion migration time and the reaction ion peak, and obtained a two-dimensional top-view plot of the ion migration spectrum (Figure 4b). The red line at the horizontal coordinate 1.0 in the graph is the RIP peak. Each point to the right of the RIP peak represents a volatile and the color of the point represents the concentration of the compound, with white to red indicating low-to-high concentrations [36]. Most of the signal is concentrated between drift times of 1.0–1.9 ms and retention times of 100–700 s. It is clear from the graph that the signal intensity of VOCs appearing between retention times of 500–700 s varies considerably between samples. To compare the differences more obviously between samples, a difference comparison model can be used, where the spectrum of JC is selected as a reference and the spectra of the other samples are deducted from the reference. If the VOCs of both are the same, the background after deduction is white, while red means that the concentration of the substance is higher than the reference, and blue means that the concentration of the substance is lower than the reference [37]. JXIU, HJML and XFML showed significantly higher concentrations of VOCs in the retention time range of 500–600 s than the other samples. It is clearly seen that VOCs in the retention time range of 350–450 s and ion drift times of 1.6 ms are specific in the HH. The substances in the green-dotted box are significantly more concentrated in the HJML than in the other samples. The location and quantities of VOC peaks in the samples are approximately the same in the spectra, but there are differences in peak strength, indicating that the content of VOCs is determined by the peach cultivars.

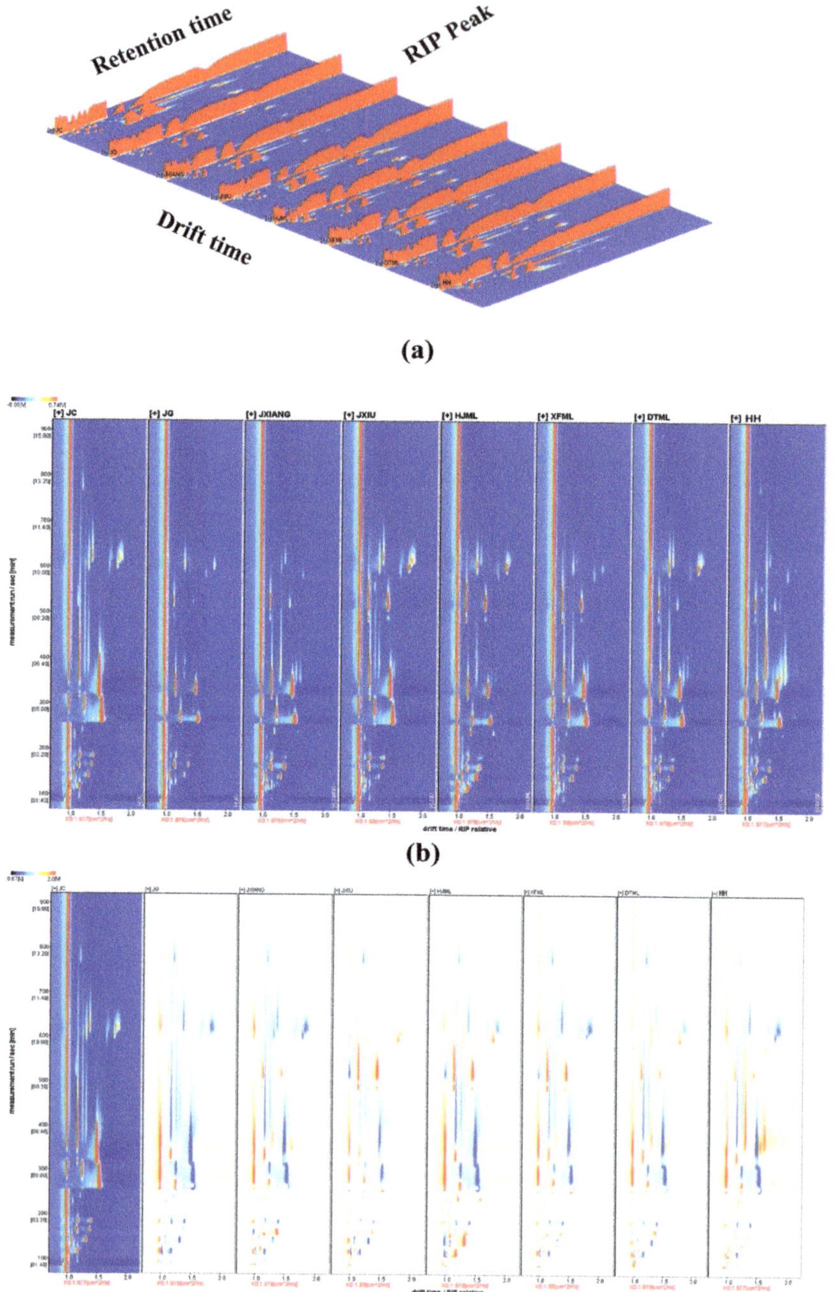

Figure 4. VOCs of eight peach cultivars determined by GC-IMS. (**a**) The 3D topographic plot of the peach VOCs. (**b**) The 2D topographic plots. (**c**) The difference comparison topographic plots.

3.2.2. The Volatile Fingerprints of Eight Cultivars Using GC-IMS

The results of the qualitative analysis of the VOCs of the sample using GC-IMS are shown in Table 4. According to the NIST gas phase retention index database and the IMS migration time database built into the flavor analyzer (FlavourSpec®) software, 53 signal peaks were detected and 26 typical volatiles were identified, including 11 aldehydes, 5 esters, 4 alcohols, 2 ketones, 2 terpenes and 2 furans. There are 12 substances that were not identified. The most abundant substances detected in peach fruit using GC-IMS in this study were aldehydes, while other researchers [31,38] have detected esters, which are closely related to factors such as peach ripeness and variety selection. The compounds of 2-hexenal, 1-hexanal, acetic acid ethyl ester, benzaldehyde, 1-hexanol, methyl acetate, 2-butanone, pentanal, 2-methylbutanal, 3-methyl butanal, (Z)-3-hexenyl acetate, 2-methylpropyl acetate and (E)-2-pentenal exhibited more than two peaks. This is due to the fact that during ionic drift, when the concentration of a substance increases, two or more molecules will share a proton or electron, forming a dimer or monomers [14,39]. The 53 signals obtained were used to draw fingerprint profiles based on peak intensities to compare the differences in volatiles between samples, which can give clearer results (Figure 5). Each row of the figure indicates the signal peaks of all volatiles in the same sample, and each column indicates the signal intensity of the same compounds in different samples. The areas where the signal is more pronounced are boxed out for easier analysis.

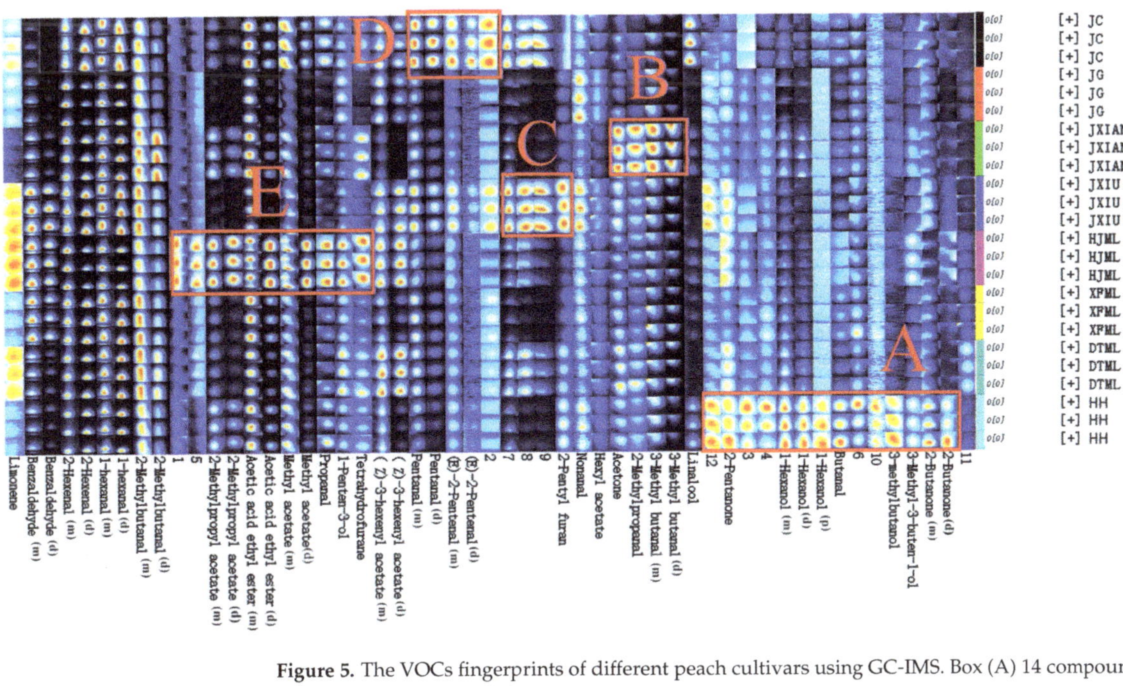

Figure 5. The VOCs fingerprints of different peach cultivars using GC-IMS. Box (A) 14 compounds highly accumulated in cv. 'HH'; Box (B) four compounds highly accumulated in 'JXIANG'; Box (C) four compounds highly accumulated in 'JXIIU'; Box (D) five compounds highly accumulated in cv. 'JC'; Box (E) eleven compounds highly accumulated in 'HJML'.

Table 4. Identification of VOCs in eight peach cultivars using GC-IMS.

Compound	CAS	RI [1]	Rt [2]	Dt [3]	Peak Volume							
					JC	JG	JXIANG	JXIU	HJML	XFML	DTML	HH
Aldehydes												
(E)-2-Pentenal(d)	C1576870	758.7	227.87	1.36	51.97 ± 4.05a	22.12 ± 0.50d	26.16 ± 3.97cd	38.57 ± 2.19b	24.88 ± 2.37cd	21.81 ± 3.64d	27.52 ± 1.39cd	31.11 ± 4.68c
(E)-2-Pentenal(m)	C1576870	758.9	227.98	1.10	336.98 ± 31.57a	139.48 ± 2.95f	157.38 ± 3.59ef	280.05 ± 6.04b	204.01 ± 5.93c	177.96 ± 1.76de	196.39 ± 2.51cd	191.34 ± 14.43cd
1-Hexanal(d)	C66251	801.9	268.96	1.55	30,604.71 ± 561.98a	8987.07 ± 662.56g	13,835.73 ± 88.57e	23,730.42 ± 141.44b	8637.36 ± 337.60g	18,301.17 ± 439.93c	16,176.68 ± 225.65d	11,927.3 ± 1846.66f
1-Hexanal(m)	C66251	796.4	263.50	1.27	7287.71 ± 346.03a	4782.78 ± 103.52e	5988.45 ± 81.33c	7405.57 ± 55.69a	4583.58 ± 65.67e	6670.8 ± 235.55b	6219.04 ± 48.74c	5230.7 ± 369.12d
2-Hexenal	C505577	854.7	326.70	1.51	43,619.63 ± 710.51a	9204.41 ± 2304.39g	19,841.39 ± 151.32e	32,700.99 ± 133.9b	11,936.95 ± 667.43f	25,817.46 ± 417.42c	22,247.55 ± 306.11d	20,911.24 ± 2001.96de
2-Hexenal(d)	C505577	865.5	339.97	1.18	1600.33 ± 144.88de	1606.4 ± 64.49de	1997.49 ± 29.93a	1837.04 ± 34.42b	1487.00 ± 16.23e	1765.89 ± 14.43bc	2010.28 ± 8.72a	1688.35 ± 74.85cd
2-Methylbutanal(d)	C96173	674.6	166.00	1.40	12,233.73 ± 215.63a	8206.57 ± 633.12de	9785.16 ± 187.90c	12,317.08 ± 220.07a	8804.16 ± 339.86d	11,064.99 ± 303.33b	9597.46 ± 161.29c	7740.01 ± 454.24e
2-Methylbutanal(m)	C96173	677.9	167.61	1.17	877.81 ± 221.07eg	796.89 ± 84.01g	2773.17 ± 85.98a	1423.53 ± 115.26c	1174.23 ± 19.75de	1292.4 ± 13.07cd	1987.4 ± 37.99b	1021.39 ± 135.35ef
2-Methylpropanal	C78842	577.4	125.97	1.28	42.23 ± 9.50de	33.31 ± 2.27e	152.23 ± 2.62a	74.87 ± 9.72c	68.22 ± 2.22c	49.41 ± 0.93d	91.89 ± 14.92b	72.12 ± 8.52c
Butanal	C123728	612.1	138.80	1.29	312.37 ± 12.32c	155.28 ± 12.96e	261.22 ± 10.32d	174.22 ± 6.66e	419.05 ± 7.74b	287.09 ± 25.66cd	195.58 ± 2.36e	546.85 ± 49.49a
Nonanal	C124196	1104.8	774.37	1.47	150.22 ± 18.46cd	230.37 ± 0.78a	122.16 ± 6.07d	223.34 ± 2.33a	160.43 ± 5.25c	155.95 ± 9.30c	149.23 ± 15.95cd	193.6 ± 29.89b
Pentanal(d)	C110623	705.2	184.54	1.42	968.25 ± 121.94a	130.00 ± 8.01e	244.36 ± 10.61cd	645.23 ± 32.61b	288.21 ± 5.09c	306.74 ± 12.17c	298.92 ± 2.21c	166.87 ± 44.9de
Pentanal(m)	C110623	704.5	184.00	1.20	1871.51 ± 76.8a	830.14 ± 36.13e	1113.94 ± 16.97cd	1557.64 ± 31.15b	1060.65 ± 16.75d	1189.49 ± 20.90c	1169.46 ± 10.15cd	843.09 ± 143.07e
3-Methyl butanal(d)	C590863	661.9	160.09	1.41	205.63 ± 79.89e	281.10 ± 71.31de	1761.39 ± 46.38a	541.67 ± 18.76b	471.98 ± 29.5bc	552.2 ± 117.72b	584.41 ± 62.46b	384.04 ± 58.43cd
3-Methyl butanal(m)	C590863	657.8	158.21	1.19	317.96 ± 87.56d	445.8 ± 75.54bc	1029.99 ± 21.67a	502.81 ± 12.27b	353.13 ± 17.07cd	466.57 ± 77.21b	470.14 ± 20.84b	529.67 ± 45.99b
Benzaldehyde(d)	C100527	968.7	501.50	1.47	578.38 ± 111.68d	619.73 ± 67.97d	1443.43 ± 126.43d	8603.64 ± 1495.6a	6885.61 ± 537.68b	5497.32 ± 225.66c	1555.85 ± 96.07d	1225.19 ± 314.58d
Benzaldehyde(m)	C100527	969.9	503.84	1.15	1757.8 ± 382.67f	2795.57 ± 223.01e	4466.92 ± 176.88cd	10,020.69 ± 940.46a	10,399.49 ± 241.07a	8684.41 ± 97.20b	4594.33 ± 211.88c	3617.75 ± 542.84de
Propanal	C123386	538.5	113.51	1.15	468.16 ± 30.05b	104.94 ± 23.59d	317.07 ± 8.66c	144.57 ± 1.8d	820.09 ± 9.05a	258.12 ± 2.33c	263.93 ± 73.16c	419.19 ± 84.39b
Esters												
(Z)-3-hexenyl acetate(d)	C3681718	1021.5	601.44	1.81	1714.33 ± 223.69c	1175.14 ± 118.31d	160.16 ± 24.78f	2639.57 ± 61.82b	3100.4 ± 36.46a	603.68 ± 86.84e	3106.75 ± 168.24a	361.33 ± 60.58f
(Z)-3-hexenyl acetate(m)	C3681718	1022.8	603.79	1.32	3152.18 ± 153.7d	2704.95 ± 176.68e	1068.08 ± 61.83g	3537.69 ± 43.29c	4131.79 ± 27.19b	2045.67 ± 146.72f	4537.70 ± 10.96a	2524.31 ± 44.88e
2-Methylpropyl acetate(d)	C110190	771.2	239.36	1.61	62.12 ± 9.66c	29.38 ± 3.92d	148.21 ± 18.50b	29.35 ± 8.12d	457.21 ± 6.19a	49.65 ± 7.83c	54.31 ± 2.84c	165.64 ± 4.44b
2-Methylpropyl acetate(m)	C110190	771.8	239.88	1.23	375.34 ± 21.76d	100.63 ± 7.69g	655.74 ± 31.80c	155.76 ± 12.85f	1207.12 ± 11.83a	336.15 ± 21.58e	374.42 ± 10.72d	742.15 ± 7.55b
Acetic acid ethyl ester(d)	C141786	627	144.89	1.34	3001.99 ± 89.14b	733.20 ± 31.31f	2266.37 ± 128.24d	476.77 ± 7.82g	11,109.44 ± 39.89a	2685.42 ± 63.79c	1226.58 ± 16.69e	2630.65 ± 67.8c
Acetic acid ethyl ester(m)	C141786	627	144.89	1.10	2598.11 ± 12.56b	1650.35 ± 66.16e	2382.52 ± 37.21c	1337.23 ± 26.8f	2697.34 ± 3.36a	2600.02 ± 61.28b	1984.03 ± 16.38d	2425.16 ± 24.8c
Hexyl acetate	C142927	1015.7	590.78	1.41	241.27 ± 5.73b	294.96 ± 7.52a	143.01 ± 6.74d	287.98 ± 19.06a	252.94 ± 5.62b	200.91 ± 23.16c	237.13 ± 6.59b	162.53 ± 9.66d
Methyl acetate(d)	C79209	545	115.50	1.19	2781.88 ± 273.34b	655.77 ± 121.18e	923.26 ± 23.47d	634.88 ± 46.62e	7076.44 ± 26.91a	771.28 ± 9.32de	957.98 ± 92.82d	1284.63 ± 126.41c
Methyl acetate(m)	C79209	543.2	114.96	1.04	3104.27 ± 93.55a	1188.14 ± 77.32e	1861.66 ± 27.21d	1826.83 ± 17.01d	2819.23 ± 22.61b	1724.07 ± 19.18d	2305.57 ± 124.32c	2298.63 ± 98.43c
Alcohols												
1-Hexanol(d)	C111273	887.6	368.84	1.64	1323.93 ± 140.84c	405.12 ± 88.78f	1594.08 ± 53.78b	850.42 ± 39.26de	757.35 ± 34.80e	1043.38 ± 68.98d	1637.77 ± 75.13b	4895.53 ± 264.62a
1-Hexanol(m)	C111273	893.2	376.65	1.33	3967.25 ± 553.26de	2153.19 ± 734.17f	5842.45 ± 443.38c	3288.28 ± 260.43e	3820.28 ± 51.39de	4707.34 ± 223.45d	7033.03 ± 370.91b	12,508.58 ± 521.32a
1-Hexanol(p)	C111273	889.8	371.96	1.98	434.94 ± 29.41b	244.16 ± 19.66d	331.93 ± 28.52c	418.18 ± 12.56b	253.31 ± 24.32d	350.39 ± 44.11c	322.64 ± 15.26c	538.73 ± 26.7a
1-Penten-3-ol	C616251	697.4	178.90	0.94	411.06 ± 57.52cd	489.01 ± 26.97b	379.93 ± 14.22d	364.81 ± 9.91d	677.31 ± 18.63a	446.93 ± 27.38bc	631.23 ± 7.78a	453.1 ± 10.38bc
3-Methyl-3-buten-1-ol	C763326	720.6	196.05	1.17	92.47 ± 16.12c	72.95 ± 6.08d	139.63 ± 2.58b	103.79 ± 4.23c	144.53 ± 5.64b	92.15 ± 7.96c	93.56 ± 3.21c	162.17 ± 2.94a
3-Methylbutanol	C123513	748	218.42	1.24	23.03 ± 1.93bc	25.77 ± 3.09bc	30.39 ± 0.98b	21.71 ± 3.07bc	18.91 ± 1.99c	20.93 ± 1.16c	20.66 ± 2.67c	90.71 ± 10.66a
Ketones												
2-Butanone(d)	C78933	599.2	133.77	1.25	31.98 ± 2.22c	19.15 ± 1.68c	24.05 ± 2.07c	27.19 ± 1.13c	87.91 ± 6.46b	28.79 ± 1.21c	21.3 ± 0.93c	110.06 ± 20.68a
2-Butanone(m)	C78933	598.5	133.50	1.06	319.61 ± 14.41d	360.83 ± 12.34c	321.68 ± 4.50d	482.58 ± 13.03b	160.1 ± 0.44e	336.18 ± 4.64cd	316.86 ± 2.98d	557.61 ± 41.51a
Acetone	C67641	575.8	125.42	1.12	364.17 ± 71.04e	389.55 ± 40.35e	1092.99 ± 8.34a	751.46 ± 46.56b	482.36 ± 5.97d	610.86 ± 28.07c	729.81 ± 54.62b	590.19 ± 24.6c
Terpenoids												
Limonene	C138863	1025.7	609.11	1.22	120.52 ± 18.56bc	105.33 ± 13.46c	63.63 ± 2.70d	117.98 ± 15.92bc	173.58 ± 9.04a	109.2 ± 11.04c	138.42 ± 9.07b	96.33 ± 13.65c
Linalool	C78706	1105.3	775.33	1.22	1186.93 ± 63.28a	187.44 ± 30.65e	360.36 ± 9.68c	181.43 ± 7.50e	348.96 ± 18.96c	262.68 ± 43.69d	172.45 ± 23.91e	606.47 ± 16.55b
Furans												
2-Pentyl furan	C3777693	996.7	557.68	1.25	230.71 ± 37.63b	136.53 ± 12.88c	119.58 ± 14.26cd	422.35 ± 7.37a	97.53 ± 5.06d	133.59 ± 14.86c	218.49 ± 15.8b	237.38 ± 12.38b
Tetrahydrofurane	C109999	574.9	125.13	1.23	142.53 ± 22.09bc	43.25 ± 3.04e	163.44 ± 1.57b	78.88 ± 6.29d	318.79 ± 2.48a	70.99 ± 2.85d	162.15 ± 18.57b	138.77 ± 10.35c
Unknown												
1	*	685.1	171.09	1.09	117.25 ± 5.5b	82.43 ± 4.12e	93.33 ± 11.25cde	104.67 ± 2.36bcd	375.05 ± 5.42a	105.09 ± 2.08bcd	107.42 ± 5.46bc	92.37 ± 12.96de
2	*	798.5	265.59	1.40	413.68 ± 65.07a	259.3 ± 62.85b	195.78 ± 11.76b	259.33 ± 9.99b	253.49 ± 14.52b	207.40 ± 3.70b	196.21 ± 17.12b	238.52 ± 17.63b
3	*	816.2	283.45	1.47	102.35 ± 2.7d	150.87 ± 7.31bc	127.98 ± 3.29cd	106.11 ± 5.90d	166.64 ± 18.91b	136.19 ± 13.21bcd	131.14 ± 11.05bcd	287.51 ± 46.04a
4	*	813.2	280.43	1.81	77.86 ± 5.71cd	69.75 ± 13.53d	92.80 ± 7.10bcd	115.27 ± 3.63bc	63.20 ± 4.48d	129.57 ± 6.95ab	115.98 ± 9.41bc	158.38 ± 52.02a
5	*	917	412.17	0.94	297.44 ± 31.73c	315.20 ± 25.56c	283.00 ± 7.370c	294.40 ± 21.60c	661.04 ± 12.12a	381.65 ± 32.88b	419.70 ± 25.18b	230.25 ± 6.78d
6	*	1012.2	584.68	1.67	470.66 ± 60.53b	464.72 ± 22.45b	478.73 ± 36.63b	554.72 ± 20.11b	521.88 ± 55.17b	655.55 ± 137.57ab	529.27 ± 62.05b	789.95 ± 210.28a
7	*	1034.5	625.61	1.39	3870.5 ± 210.86b	1001.12 ± 115.81f	1368.3 ± 87.42e	4268.32 ± 34.83a	1887.88 ± 51.23d	885.33 ± 51.50f	2823.64 ± 8.95c	1237.11 ± 73.81e
8	*	1033.5	623.68	1.85	2420.1 ± 342.75b	292.2 ± 44.66e	163.82 ± 17.50e	2841.04 ± 55.39a	1225.76 ± 54.02d	210.16 ± 19.39e	1774.59 ± 78.03c	269.1 ± 0.7e
9	*	1031.5	619.81	1.89	1184.83 ± 163.65b	102.19 ± 2.83e	135.6 ± 10.61e	1426.78 ± 56.07a	267.01 ± 15.53d	89.92 ± 8.87e	497.00 ± 22.86c	135.14 ± 19.9e
10	*	1155	901.20	1.80	125.8 ± 13.95c	137.23 ± 16.59bc	139.42 ± 17.29bc	147.90 ± 3.48bc	160.7 ± 32.14abc	168.21 ± 36.69ab	155.05 ± 7.89abc	193.68 ± 10.28a
11	*	780	247.77	1.29	31.42 ± 5.58cd	36.91 ± 2.77cd	30.15 ± 1.72d	40.15 ± 3.23c	49.24 ± 0.59b	55.47 ± 5.87b	101.64 ± 8.16a	38.82 ± 2.25cd
12	*	751.4	221.36	1.17	24.2 ± 5.97d	33.01 ± 7.70c	25.65 ± 2.22cd	42.27 ± 0.77b	20.35 ± 1.23d	21.52 ± 2.94d	25.50 ± 3.30cd	56.07 ± 4.55a

[1] indicates retention index; [2] denotes retention time; [3] denotes migration time; d, m and p following compounds represents dimer, monomer and polymer, respectively. * indicates unknown compounds without CAS number.

As shown in box A, some alcohols, ketones and unknown compounds were the most abundant in the HH, and these were the main reasons for the differences in the overall fingerprint profile of the HH compared to the other samples. Aldehydes, namely (E)-2-pentenal, 3-methyl butanal, 2-methylpropanal and pentanal, dominated in JC and JXIANG (box B and D), which differed from the GC-MS, where aldehydes were mainly dominated by 2-hexenal and hexanal. Among the overall peak signals of JXIU, box C had the strongest signals of substances that were not characterized; it is therefore necessary to characterize these unknown compounds in future research. Box E demonstrates that HJML had high concentrations of methyl acetate, acetic acid ethyl ester, 2-methylpropyl acetate, and the flavor of these compounds is discriminatingly fruity. None of these were found in GC-MS. As we can see, 12 unknown compounds have strong peak signals in the fingerprints, indicating that these compounds were present at high levels in the samples, and characterizing these substances is essential for the establishment of a complete GC-IMS-based peach fingerprint.

Combining Tables 2 and 4, the types of VOCs detected by the two techniques differed significantly, with GC-MS detecting more than GC-IMS. Lactones, typical VOCs in peaches, were not found in GC-IMS due to their high molecular weight, which makes them difficult to be detected. Although both techniques are used in conjunction with GC, their focus is dissimilar. GC-MS works under a vacuum and at a high temperature and focuses on the accuracy of qualitative and quantitative information, while GC-IMS performs detection at atmospheric pressure and normal temperature and focuses more on reflecting the authenticity of the sample flavor [40,41].

3.2.3. Similarity Analysis of Samples between Different Peach Cultivars Based on PCA

PCA was performed on the volatile compound content (signal peak volume) of different cultivars of peaches using the Dynamic PCA plug-in, and the data were visualized to obtain a spatial–dimensional plot containing three principal components, where the X, Y and Z axes indicate the first, second and third principal components, respectively (Figure 6). The cumulative contribution rate of PC1, PC2 and PC3 was 65.7%. The results show that the eight peach cultivars were clearly distinguished based on the identified VOCs using GC-IMS. JC and JXIU are located on the right side of the PC1-axis (Figure 6a), indicating that the content of characteristic aroma compounds in these two cultivars were more similar, which might be caused by the fact that JC was an F1 offspring of JXIU. DTML, XFML, JG and JXIANG are close together and can be grouped together, while HH and HJML are located at the ends of the PC3 and PC2 axes, which means that they are relatively different from the other samples based on the VOC performance (Figure 6a). The difference between samples in a group is small, while the differences between groups are more obvious. Overall, the aroma profiles of JC, HJML and HH are more specific compared to the other samples. This result is highly consistent with that of the cluster analysis in Figure 3. The PCA loading plot showed that (E)-2-pentenal, 2-hexenal and pentanal highly contributed to the JC's aroma, and 1-hexanol, 3-methylbutanol highly contributed to the HH aroma. These results are consistent with the GC-IMS fingerprints. These findings suggest that the aroma profiles of these eight popular peaches are quite different for selecting strong/specific aroma cultivars.

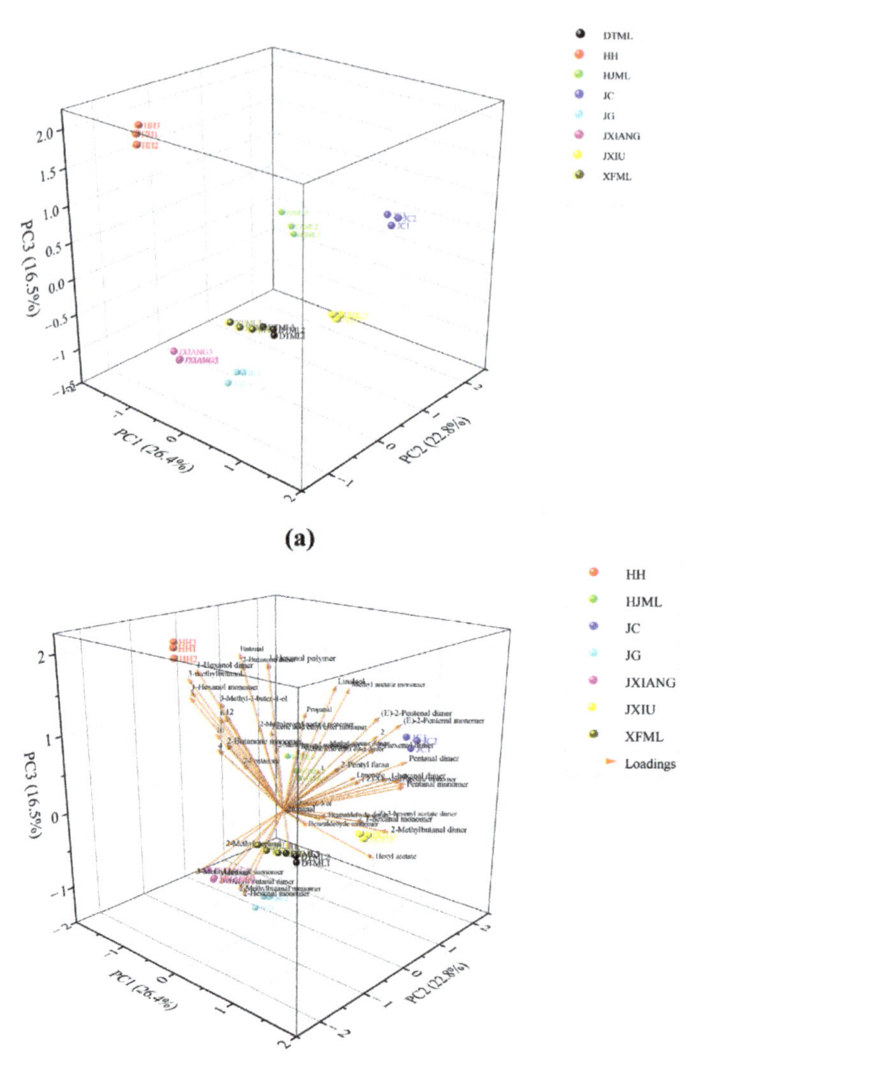

Figure 6. Principal component analysis (PCA) score plot (**a**) and PCA loading plot of different peach cultivars (**b**).

4. Conclusions

The aroma profiles of eight popular peach cultivars produced in Shanghai were comprehensively analyzed using GC–MS, OAV determination and GCIMS. Seventy VOCs were detected using GC-MS, including nine aldehydes, ten esters, eleven terpenoids, nine alcohols, six lactones, five ketones, five alkanes, eleven aromatic hydrocarbons, two furans and two other compounds, of which twenty-three VOCs were commonly identified in eight peach cultivars. The most VOCs were detected in HJML and the least in JG. Cluster analysis classified XFML, JG, DTML and JXIU into one cluster, which means that their aroma profiles are more similar.

In addition, based on an OAV ≥ 1, a total of 17 key aroma compounds were screened, and their contributions were ranked as follows: γ- decalactone> β- ionone > hexanal > 2-hexenal > linalool > nonanal > 3-hexenol > δ-decalactone > D-limonene > β-myrcene

> 1-hexanol > γ-octalactone > γ-dodecalactone > (Z)-3-hexen-1-yl acetate > butyl acetate > isopentyl acetate > octanal. The OAV of VOCs differed greatly among the cultivars. Lactones had the highest OAV in HJML, which indicates that HJML is the cultivar that contains the most fruity notes. Terpenoids had the highest OAV in JC, meaning that JC had the strongest floral notes. It is obvious that JXIU was the cultivar with the lightest aroma.

Furthermore, a total of twenty-six VOCs were detected using GC-IMS, among which aldehydes accounted for the largest proportion. GC-IMS has the advantage of data visualization, which can visualize the differences of volatiles among samples. It can be directly observed in the fingerprints that aldehydes have the strongest signals in samples. The PCA results showed that XFML, JG, DTML and JXIU were close to each other, whereas HJML, HH and JC were distant from other samples, which is similar to the results of the cluster analysis. The differences in focus and working conditions between GC-MS and GC-IMS led to significant differences in the composition of the VOCs detected by them. The combination of GC-MS and GC-IMS can capture relatively complete flavor information of samples, which is a hot research topic in the field of flavor at present and will be for a long time in the future.

Author Contributions: Conceptualization, X.L., X.Y. and Z.Y; methodology, X.W. and X.L.; software, X.W.; validation, X.L.; formal analysis, X.W. and X.L.; investigation, X.L.; resources, M.S., M.Z., J.D., H.Z. and X.Z.; data curation, X.W. and Y.H.; writing—original draft preparation, X.W. and X.L.; writing—review and editing, X.L., Z.Y. and X.Y.; visualization, X.W.; supervision, Z.Y. and X.Y.; project administration, X.L. and Z.Y.; funding acquisition, X.L. and Z.Y. All authors have read and agreed to the published version of the manuscript.

Funding: This work was supported by funds from the National Key Research and Development Program of China (2019YFD1000801), the Shanghai Science and Technology committee Rising-Star Program (19QB1404600), Hu Nong Ke Chuang Zi-2021-1-1, the Shanghai Municipal Science and Technology Commission 2019 Science and Technology innovation Action Plan Agricultural Project (19391903400), the Outstanding Team Program of Shanghai Academy of Agricultural Science (Grant No. ZY2022–004), the Pangao Plan of Shanghai Academy of Agricultural Science, the Qiankehe Basics [2019]1408 and the Qiankehe Platform Talent [2018]5781.

Data Availability Statement: Not applicable.

Conflicts of Interest: The authors declare no conflict of interest.

References

1. Li, X.; Meng, X.; Jia, H.; Yu, M.; Ma, R.; Wang, L.; Cao, K.; Shen, Z.; Niu, L.; Tian, J.; et al. Peach genetic resources: Diversity, population structure and linkage disequilibrium. *BMC Genet.* **2013**, *14*, 84. [CrossRef] [PubMed]
2. Li, Y.; Wang, L. Genetic resources, breeding programs in China, and gene mining of peach: A review. *Hortic. Plant J.* **2020**, *6*, 205–215. [CrossRef]
3. Aranzana, M.; Abbassi, E.; Howad, W.; Arús, P. Genetic Variation, population structure and linkage disequilibrium in peach commercial varieties. *BMC Genet.* **2010**, *11*, 69. [CrossRef] [PubMed]
4. Li, X.; Jiang, J.; Zhang, L.; Yu, Y.; Ye, Z.; Wang, X.; Zhou, J.; Chai, M.; Zhang, H.; Arús, P.; et al. Identification of volatile and softening-related genes using digital gene expression profiles in melting peach. *Tree Genet. Genomes.* **2015**, *11*, 71. [CrossRef]
5. Zhang, X.; Su, M.; Du, J.; Zhou, H.; Li, X.; Zhang, M.; Hu, Y.; Ye, Z. Analysis of the free amino acid content and profile of 129 peach (*Prunus Persica* (L.) Batsch) germplasms using LC-MS/MS without derivatization. *J. Food Compos. Anal.* **2022**, *114*, 104811. [CrossRef]
6. Aubert, C.; Milhet, C. Distribution of the volatile compounds in the different parts of a white-fleshed peach (*Prunus Persica* L. Batsch). *Food Chem.* **2007**, *102*, 375–384. [CrossRef]
7. Wang, Y.; Yang, C.; Li, S.; Yang, L.; Wang, Y.; Zhao, J.; Jiang, Q. Volatile characteristics of 50 peaches and nectarines evaluated by HP–SPME with GC–MS. *Food Chem.* **2009**, *116*, 356–364. [CrossRef]
8. Zhang, B.; Shen, J.; Wei, W.; Xi, W.; Xu, C.; Ferguson, I.; Chen, K. Expression of genes associated with aroma formation derived from the fatty acid pathway during peach fruit ripening. *J. Agric. Food Chem.* **2010**, *58*, 6157–6165. [CrossRef]
9. Jia, H.; Araki, A.; Okamoto, G. Influence of fruit bagging on aroma volatiles and skin coloration of 'Hakuho' peach (*Prunus Persica* Batsch). *Postharvest Biol. Technol.* **2005**, *35*, 61–68. [CrossRef]
10. Toselli, M.; Baldi, E.; Marangoni, B.; Noferini, M.; Fiori, G.; Bregoli, A.; Ndagijimana, M. Effects of mineral and organic fertilization and ripening stage on the emission of volatile organic compounds and antioxidant activity of "Stark redgold" nectarine. *Acta Hortic.* **2010**, *868*, 381–388. [CrossRef]

11. Wang, Y.; Yang, C.; Liu, C.; Xu, M.; Li, S.; Yang, L.; Wang, Y. Effects of bagging on volatiles and polyphenols in "Wanmi" peaches during endocarp hardening and final fruit rapid growth stages. *J. Food Sci.* **2010**, *75*, S455–S460. [CrossRef]
12. Grosch, W. Evaluation of the key odorants of foods by dilution experiments, aroma models and omission. *Chem. Senses.* **2001**, *26*, 533–545. [CrossRef]
13. Rothe, M.; Thomas, B. Aromastoffe des Brotes. *Eur. Food Res. Technol.* **1963**, *119*, 302–310. [CrossRef]
14. Ruiz, M.; Rodríguez, M.; Flores, G.; Blanch, G. New method based on solid phase microextraction and multidimensional gas chromatography-mass spectrometry to determine pesticides in strawberry jam. *LWT* **2019**, *99*, 283–290. [CrossRef]
15. Cavanna, D.; Zanardi, S.; Dall, C.; Suman, M. Ion mobility spectrometry coupled to gas chromatography: A rapid tool to assess eggs freshness. *Food Chem.* **2019**, *271*, 691–696. [CrossRef]
16. Qi, H.; Ding, S.; Pan, Z.; Li, X.; Fu, F. Characteristic volatile fingerprints and odor activity values in different citrus-tea by HS-GC-IMS and HS-SPME-GC-MS. *Molecules* **2020**, *25*, 6027. [CrossRef]
17. Wang, S.; Chen, H.; Sun, B. Recent progress in food flavor analysis using gas chromatography–ion mobility spectrometry (GC–IMS). *Food Chem.* **2020**, *315*, 126158. [CrossRef]
18. Van Gemert, L.J. *Odour Thresholds-Compilations of Odour Threshold Values in Air, Water and Other Media*; Oliemans Punter & Partners: Utrecht, The Netherlands, 2003.
19. Hadi, M.; Zhang, F.; Wu, F.; Zhou, C.; Tao, J. Advances in fruit aroma volatile research. *Molecules* **2013**, *18*, 8200–8229. [CrossRef]
20. Mihaylova, D.; Popova, A.; Vrancheva, R.; Dincheva, I. HS-SPME-GC–MS volatile profile characterization of peach (*Prunus Persica* L. Batsch) Varieties grown in the eastern Balkan Peninsula. *Plants* **2022**, *11*, 166. [CrossRef]
21. Xi, W.; Zheng, Q.; Lu, J.; Quan, J. Comparative Analysis of three types of peaches: Identification of the key individual characteristic flavor compounds by integrating consumers' acceptability with flavor quality. *Hortic. Plant J.* **2017**, *3*, 1–12. [CrossRef]
22. Zhu, J.; Xiao, Z. Characterization of the key aroma compounds in peach by gas chromatography–olfactometry, quantitative measurements and sensory analysis. *Eur. Food Res. Technol.* **2019**, *245*, 129–141. [CrossRef]
23. Tan, F.; Wang, P.; Zhan, P.; Tian, H. Characterization of key aroma compounds in flat peach juice based on gas chromatography-mass spectrometry-olfactometry (GC-MS-O), odor activity value (OAV), aroma recombination, and omission experiments. *Food Chem.* **2022**, *366*, 130604. [CrossRef] [PubMed]
24. Horvat, R.; Chapman, G.; Robertson, J.; Meredith, F.; Scorza, R.; Callahan, A.; Morgens, P. Comparison of the volatile compounds from several commercial peach cultivars. *J. Agric. Food Chem.* **1990**, *38*, 234–237. [CrossRef]
25. Robertson, J.; Meredith, F.; Horvat, R.; Senter, S. Effect of cold storage and maturity on the physical and chemical characteristics and volatile constituents of peaches (*Cv. Cresthaven*). *J. Agric. Food Chem.* **1990**, *38*, 620–624. [CrossRef]
26. Chapman, G.; Horvat, R.; Forbus, W. Physical and chemical changes during the maturation of peaches (*Cv. Majestic*). *J. Agric. Food Chem.* **1991**, *39*, 867–870. [CrossRef]
27. Eduardo, I.; Chietera, G.; Bassi, D.; Rossini, L.; Vecchietti, A. Identification of key odor volatile compounds in the essential oil of nine peach accessions. *J. Sci. Food Agric.* **2010**, *90*, 1146–1154. [CrossRef]
28. Brodkorb, D.; Gottschall, M.; Marmulla, R.; Lüddeke, F.; Harder, J. Linalool dehydratase-isomerase, a bifunctional enzyme in the anaerobic degradation of monoterpenes. *J. Biol. Chem.* **2010**, *285*, 30436–30442. [CrossRef]
29. Engel, K.; Flath, R.; Buttery, R.; Mon, T.; Ramming, D.; Teranishi, R. Investigation of volatile constituents in nectarines. 1. analytical and sensory characterization of aroma components in some nectarine cultivars. *J. Agric. Food Chem.* **1988**, *36*, 549–553. [CrossRef]
30. Mohammed, J.; Belisle, C.; Wang, S.; Itle, R.; Adhikari, K.; Chavez, D. Volatile Profile Characterization of Commercial Peach (*Prunus Persica*) Cultivars Grown in Georgia, USA. *Horticulturae* **2021**, *7*, 516. [CrossRef]
31. Sun, P.; Xu, B.; Wang, Y.; Lin, X.; Chen, C.; Zhu, J.; Jia, H.; Wang, X.; Shen, J.; Feng, T. Characterization of volatile constituents and odorous compounds in peach (*Prunus Persica* L.) fruits of different varieties by gas chromatography–ion mobility spectrometry, gas chromatography–mass spectrometry, and relative odor activity value. *Front. Nutr.* **2022**, *9*, 965796. [CrossRef]
32. Brandi, F.; Bar, E.; Mourgues, F.; Horváth, G.; Turcsi, E.; Giuliano, G.; Liverani, A.; Tartarini, S.; Lewinsohn, E.; Rosati, C. Study of "Redhaven" peach and its white-fleshed mutant suggests a key role of *CCD4* carotenoid dioxygenase in carotenoid and norisoprenoid volatile metabolism. *BMC Plant Biol.* **2011**, *11*, 24. [CrossRef] [PubMed]
33. Adami, M.; Defranceschi, P.; Brandi, F.; Liverani, A.; Giovannini, D.; Rosati, C.; Dondini, L.; Tartarini, S. Identifying a carotenoid cleavage dioxygenase (*ccd4*) gene controlling yellow/white fruit flesh color of peach. *Plant Mol. Biol. Report.* **2013**, *31*, 1166–1175. [CrossRef]
34. Spencer, M.; Pangborn, R.; Jennings, W. Gas chromatographic and sensory analysis of volatiles from cling peaches. *J. Agric. Food Chem.* **1978**, *26*, 725–732. [CrossRef]
35. Aubert, C.; Günata, Z.; Ambid, C.; Baumes, R. Changes in physicochemical characteristics and volatile constituents of yellow- and white-fleshed nectarines during maturation and artificial ripening. *J. Agric. Food Chem.* **2003**, *51*, 3083–3091. [CrossRef] [PubMed]
36. Li, M.; Yang, R.; Zhang, H.; Wang, S.; Chen, D.; Lin, S. Development of a flavor fingerprint by HS-GC–IMS with PCA for volatile compounds of Tricholoma Matsutake Singer. *Food Chem.* **2019**, *290*, 32–39. [CrossRef]
37. Feng, T.; Sun, J.; Song, S.; Wang, H.; Yao, L.; Sun, M.; Wang, K.; Chen, D. Geographical differentiation of Molixiang table grapes grown in China based on volatile compounds analysis by HS-GC-IMS coupled with PCA and sensory evaluation of the grapes. *Food Chem. X* **2022**, *15*, 100423. [CrossRef]

38. Leng, P.; Hu, H.; Cui, A.; Tang, H.; Liu, Y. HS-GC-IMS with PCA to analyze volatile flavor compounds of honey peach packaged with different preservation methods during storage. *LWT* **2021**, *149*, 111963. [CrossRef]
39. Rodríguez, R.; Vyhmeister, E.; Meisen, S.; Rosales, A.; Kuklya, A.; Telgheder, U. Identification of terpenes and essential oils by means of static headspace gas chromatography-ion mobility spectrometry. *Anal. Bioanal. Chem.* **2017**, *409*, 6595–6603. [CrossRef]
40. Louw, S. Recent trends in the chromatographic analysis of volatile flavor and fragrance compounds: Annual review 2020. *Anal. Sci. Adv.* **2021**, *2*, 157–170. [CrossRef]
41. Li, S.; Yang, H.; Tian, H.; Zou, J.; Li, J. Correlation analysis of the age of brandy and volatiles in brandy by gas chromatography-mass spectrometry and gas chromatography-ion mobility spectrometry. *Microchem. J.* **2020**, *157*, 104948. [CrossRef]

horticulturae

MDPI

Article

A Detailed Comparative Study on Some Physicochemical Properties, Volatile Composition, Fatty Acid, and Mineral Profile of Different Almond (*Prunus dulcis* L.) Varieties

Okan Levent

Department of Food Engineering, Faculty of Engineering, Inonu University, Malatya 44090, Turkey; okan.levent@inonu.edu.tr

Abstract: In the present investigation, the main purpose of the research was to reveal the differences among the almond genotypes in terms of their physicochemical properties, volatile composition, fatty acid, and mineral profile. For that reason, ten different almond genotypes originated from different countries were subjected to relevant analysis. The results showed that the total oil, protein, and ash levels of the almond samples ranged between 30.84–41.43%, 17.43–22.72%, and 2.90–3.40%, respectively. Additionally, total phenolic content of the samples was in the range of 38.7–101.03 mg GAE/100 g sample. It was revealed that the almond samples were rich in monounsaturated fatty acids, and oleic acid was the major one with levels of 61.22–77.63%. For all samples, potassium, magnesium, and phosphorus were the major minerals, and the highest concentration was for potassium with levels of 6192.08–11,046.05 mg/kg. Volatile profile analysis showed that the toluene, 4-octanone, pinacol, and 2-methylpentanal were the dominant volatile compounds for all almond genotypes. The results revealed that the different almond varieties showed significant differences depending on the origin.

Keywords: almond; aroma compounds; fatty acids; minerals; phenolic

Citation: Levent, O. A Detailed Comparative Study on Some Physicochemical Properties, Volatile Composition, Fatty Acid, and Mineral Profile of Different Almond (*Prunus dulcis* L.) Varieties. *Horticulturae* **2022**, *8*, 488. https://doi.org/10.3390/horticulturae8060488

Academic Editors: Alessandra Durazzo and Massimo Lucarini

Received: 25 April 2022
Accepted: 29 May 2022
Published: 31 May 2022

1. Introduction

Almond (*Prunus dulcis*) knows as also *Amygdalus communis*, which is the most widely cultivated nut crop of Mediterranean area and shows many benefits for human health, is a tree species that together with peach is included in the subgenus Amygdalus. In the world, almond is one of the most consumed nuts, especially in high-income economies, and it was reported that the total production of almonds in the whole world was more than two million tons. In the production of almonds, USA was the first country with more than one-million-ton production for 2020, according to the records of FAOSTAT. It is one of the most important nut species that has increased production and consumption ratios day by day. Karaat [1] stated that the almond-growing areas started to increase in the last decades because of an increased demand for almond due to its natural and healthy food ingredients. Şimsek [2] informed that the almond is a good material in the diet due to providing micronutrients, macronutrients, and various bioactive constituents. Jenkins et al. [3] also stated that the almond kernel is used to treat the heart and autoimmune system, rheumatoid arthritis, and cancer in recent years. Barreca et al. [4] reported that the nut has many nutritious ingredients such as fatty acids, lipids, amino acids, proteins, carbohydrates, vitamins, and minerals, as well as secondary metabolites. Almonds contain lipids (around 50%), proteins (around 25%), and carbohydrates (around 20%), and have a low moisture content and diverse minor bioactive compounds. The beneficial effects of almond consumption are associated with its composition of macro- and micronutrients [5].

Almond has an acceptable and desired aroma and flavor, and it is accepted as a nutritious and delicious fruit due to its protein, fat, mineral matter, fibre, and vitamin E content [6]. Almond proximate composition is generally influenced by the by ecological conditions, location, and technical and cultural practices [7,8]. There are different investigations performed for the characterization of the different almond genotypes. Beyhan et al. [6] investigated the fatty acid compositions of some important almond (*Prunus amygdalus* L.) varieties selected from Turkey, and they reported that there was no huge significance among the samples. Kodad et al. [9] studied the variability of oil content and major fatty acid of the different almond genotypes and reported that the differences were significant. Kalita [2018] reviewed the cardiovascular effects of almond, and they reported that the almonds have a functional role for the regulation cardiovascular problems. In addition, Mathpal and Rathore [2021] reported the health benefits of almonds, and they informed that almonds are highly rich in vitamin E, copper manganese, fiber magnesium, phosphorus, monounsaturated fatty acids, and riboflavin protein, among other nutrients.

In the current study, the aim was to characterize the different almond varieties originated from different countries (Ferragnes, Ferraduel, Marta, and Lauranne from France; Texas and Nonpareil from USA; Guara from Spain; Yaltinskii from Russia; and Nurlu and Acıbadem from Turkey) in terms of their main proximate composition, fatty acid and mineral profile, and volatile compound composition. Additionally, a cluster analysis was performed to compare the almond genotypes and to determine the similar varieties by their characterized properties.

2. Materials and Methods

2.1. Materials

The different almond genotypes were provided from Agricultural Credit Cooperative Adıyaman Kahta Almond-Pistachio Enterprise (harvest year 2021). During the harvest year, the average temperature for the growing area was 29.99 °C, the lowest temperature was in the range of −2.2 and 16 °C in December–January, and the highest temperature ranged between 22–41.7 °C for July and August. In addition to that, monthly total precipitation ratios were 34.25 mm, while the highest precipitation level was 199.8 mm in January and the minimum precipitation was zero in July. Average monthly total sunbathing time was 283 h. The provided almond varieties were Ferragnes (FS), Ferraduel (FL), and Marta (MA) (Spanish); Lauranne (LU) (France); Texas (TX) and Nonpareil (NL) (USA); Guara (GU) (Spain); Yaltinskii (YS) (Russia); and Nurlu (NU) and Acıbadem (BA) (Turkey). The kernel mass values of the almond genotypes were in the range of 1.88–4.89 g. The highest kernel was for the genotypes of Ferraduel, and the average kernel size was 3.02. After harvesting, they were kept at room temperature for about 1 month, and the provided samples were stored at 4 °C until the related analysis.

2.2. Methods

2.2.1. Analysis of Proximate Composition

For the proximate composition, total oil, total protein, and total ash contents of the almond samples were determined according to Nizamlıoğlu and Nas [10]. For the determination of total oil content of the samples, the powdered almond samples were exposed to extraction by n-hexane for approximately 6 h using the Soxhlet extraction system. At the end of the duration, the oil content was calculated depending on the mass balance. For the determination of the total protein content of the samples, Kjeldahl protein analysis methodology was followed, and the determined nitrogen content was multiplied by 6.25 to calculate the final total protein content of the samples. Total ash level of the almond samples was determined as following: for this aim, the powdered almond kernels were exposed to carbonization at firstly 200 °C for 24 h and then 6 h at 600 °C. At the end of the carbonization, the ash levels of the samples were calculated according to the mass balance [1]. All measurements were performed as replicates with three repetitions.

2.2.2. Total Phenolic Content Analysis

For the extraction of the almond samples, 0.5 g of powdered almond was mixed with 30 mL of ethanol (96%) and stirred for 30 min using a magnetic stirrer (Clifton, Nickel

Elctro MSH1, Thermo Fisher Scientific, UK). At the end of the mixing, the suspension was filtered using a Whatman no. 1 filter paper, and the filtrate was exposed to concentration until 10 mL left as a volume using rotary evaporator (Bibby RE 100, By Bibby Sterilin Ltd., Stone, UK) at 40 °C. The prepared final solution was used as the almond extract for the total phenolic content analysis. To determine the total phenolic content of the samples, the Folin–Ciocalteu reagent method was followed as suggested by Yıldırım et al. [11] and Durmaz and Alpaslan [12]. For this purpose, 100–500 µL of the samples was diluted to 1 mL using ethanol, and this volume was adjusted to 46 mL with incorporation of distilled water. Then, 1 mL of the Folin–Ciocalteu reagent was added, and this mixture was left to stand for 3 min. After that, 3.0 mL of sodium carbonate (2% *w*/*v*) was added to the mixture, the mixture was mixed well by vortex and the tubes were incubated at room conditions for 120 min. At the end of the duration, the absorbance of the samples was recorded at 700 nm using a spectrophotometer (UV-1700, Shimadzu, Kyoto, Japan). To calculate the total phenolic content of the samples, a calibration curve was prepared by using gallic acid standard at different concentration, and the results were given as mg gallic acid equivalent (GAE)/100 g sample. All measurements were performed as replicates with three repetitions.

2.2.3. Analysis of Mineral Composition of the Samples

Macro and micro element compositions of the almond samples were determined according to the suggested methods by Mertens [13,14]. Totally, 250 mg of powdered almond sample was exposed to digestion process by using 4 mL of HNO3 (65%) in a speed wave microwave digestion equipment (Berghof, MSW-4, Eningen, Germany), and the digested samples were subjected to elemental analyses. Contents of P, Mg, Ca, K, Zn, Mn, Cu, Fe, and Na were determined according to standard stock solution (1000 mg/L) of each element using an inductively coupled plasma mass spectrometry (ICP-MS) (NexION 350X, Perkin Elmer, Waltham, MA, USA). All measurements were performed as replicates with three repetitions.

2.2.4. Fatty Acid Profile of Almond Oil Samples

The fatty acid composition of the almond sample oil was determined according to the method described by Karaat [1]. At first, the almond oil samples were subjected to methylation process. For this process, 100 mg of oil sample was weighed and 100 µL of 2 M KOH (prepared with methanol) was incorporated into oil for saponification and mixed well by a vortex for a while. Then, 3 mL of n-hexane was added to the mixture and the mixture was vigorously shaken with a vortex for 1 min, and then the tubes were centrifuged at 2516× *g* for 5 min at 25 °C. The supernatant was placed into vials and the samples were exposed to GC-FID analysis (Shimadzu, QP2010 ULTRA) equipped with a flame ionization detector and Rtx-5 MS capillary column. The oven temperature of the GC was programmed as follows: 180 °C for 2 min, increased to 200 °C at 2 °C/min, held at 200 °C for a further 10 min, and then increased to 250 °C at 2 °C/min and kept there for 10 min. The injector and detector temperatures were set as 200 and 250 °C, respectively. The carrier gas was helium at a flow rate of 1.5 mL/min. The results were given as percentages of each fatty acid with regard to total oil content. All measurements were performed as replicates with three repetitions.

2.2.5. Volatile Profile of Almond Samples

Volatile profile of almond samples was characterized according to the methodology reported by Doguer et al. [15]. For the identification of the volatile compounds of each almond sample, solid phase microextraction (SPME) process was performed by using Divinylbenzene/Carboxen/Polydimethylsiloxane (50/30 µm coating thickness; 2 cm length; Supelco, Bellefonte, PA, USA) fiber. For this purpose, 3 g of samples was transferred into 15 mL of SPME vials (Supelco, Bellefonte, PA, USA). Then, 10 µL of two internal standards (2-methyl-3-heptanone and 2-methyl-pentanoic acid) were added into the vials. Vials were placed on a heater at the temperature of 40 °C for 30 min to provide the accumulation of the volatiles up to headspace. After that, fiber was injected in a vial to absorb volatile compounds for 30 min. Desorption process was conducted at a temperature of 250 °C in MS sampler. A total of 3 g of the powdered almond sample was immediately placed into SPME vials (Supelco, Bellefonte, PA, USA). Then, the fiber was placed on a Shimadzu GC-2010 gas chromatography-QP-2010 mass spectrometry system (Shimadzu Corporation, Kyoto, Japan) to desorb the extracted volatiles. The separation process of extracted volatiles from the almond samples was achieved by application of DBWax column (60 m × 0.25 mm × 0.25 mm; J&W Scientific, Folsom, CA, USA). The volatile compounds were then identified according to retention index (RI) by using an n-alkane series (C10-C26) under the same conditions as mentioned above. WILEY 8 and NIST 05 mass spectral libraries were used to identify peaks.

2.2.6. Statistical Analysis

Statistical analysis was conducted using SPSS 22.0 software (SPSS Inc., Chicago, IL, USA). All values were recorded and expressed as mean ± standard deviation (SD). The significant differences between mean values of the almond samples were determined by analysis of variance (one way-ANOVA) using Tukey's HSD (Honestly Significant Difference) test at a significance level of $p < 0.05$. Pearson correlation coefficients were calculated with the OriginPro version 2020b (OriginLab, Northampton, MA, USA). Principal component analysis (PCA) and cluster analysis (Ward method and hierarchical) were performed using the JMP (12.2.0 SAS Institute, Inc., Cary, NC, USA).

3. Results

3.1. *Some Physicochemical Properties of Almond Samples*

Table 1 shows some basic physicochemical properties and total phenolic content of the almond samples. As is seen, protein content of the samples was in the range of 17.43–22.72% while the lowest protein level was for the sample of YS and the highest level was determined for the sample of NL. It was seen that the averaged protein level was 19.27%. The variation of the protein content among the samples was determined as significant ($p < 0.05$). For the oil content of the different almond samples, it was also observed that the oil levels of the samples changed significantly depending on the almond type ($p < 0.05$). It was determined that the oil levels of the almond samples ranged between 30.84–41.43%. Among all the samples, the highest oil level was determined for the sample of FS (41.43%) while the lowest oil content was monitored for the sample of LU (30.84%). Additionally, it was calculated that the average oil level of the samples was 35.37%. Ash content of the samples was also tabulated in Table 1. As is seen, the highest ash level was determined for the sample of LU (3.4%) and the lowest value was for the sample of GU (2.9%). It was observed that the difference in the ash content of the samples was found as significant ($p < 0.05$) but most of the ash values for the samples were determined as very similar to each other. It was reported that the proximate composition of almond was affected by the addition to soil, climate, and growing conditions (irrigation, fertilization, etc.) as well as differences in geographical origin, all of which have been cited as the reason for the reason of compositional change [8,10,16]. Summo et al. [2018] studied the effects of harvest time and cultivar on the chemical and nutritional characteristics of almonds, and they reported that the lipid content increased during ripening, while both protein and carbohydrate content decreased. The fatty acid composition showed a not univocal behavior during ripening and was highly influenced by the cultivar when total phenolic compounds and antioxidant activity varied among cultivars. It was also reported that the highest components found are fat and protein according to the chemical composition values of different almond varieties grown in California. It was shown that the change in composition of vitamin E and fat content and fatty acids was dependent on the harvest year, horticulture, and mainly on almond genotype [8]. Nizamlıoğlu and Nas [10] also reported that the total oil, total ash, and protein content of the almond (Akbadem) were 52.32, 3.15 and 20.57%, respectively.

Table 1. Some physicochemical and bioactive properties of almond samples.

Samples [δ]	Protein (%)	Oil (%)	Ash (%)	TPC (mg GAE/100 g)
FS	18.85 ± 0.04 [bcd]	41.43 ± 1.19 [a]	3.15 ± 0.07 [ab]	79.21 ± 1.23 [b]
FL	18.55 ± 0.47 [cd]	33.78 ± 1.09 [cde]	3.3 ± 0.14 [a]	101.03 ± 1.96 [a]
MA	17.64 ± 0.66 [d]	32.99 ± 0.33 [cde]	3.25 ± 0.07 [ab]	71.41 ± 1.17 [c]
LU	19.61 ± 0.07 [ab]	30.84 ± 1.34 [e]	3.4 ± 0.14 [a]	65.07 ± 1.5 [cd]
YS	17.43 ± 0.34 [d]	39.49 ± 2.57 [ab]	3.15 ± 0.07 [ab]	38.7 ± 1.63 [g]
NL	22.72 ± 0.32 [a]	33.67 ± 0.5 [cde]	3.05 ± 0.07 [ab]	56.41 ± 0.88 [ef]
TX	20.11 ± 0.16 [b]	36.92 ± 1.01 [abcd]	3.1 ± 0.00 [ab]	41.51 ± 1.27 [g]
GU	17.7 ± 0.42 [d]	37.76 ± 1.22 [abc]	2.9 ± 0.14 [b]	81.1 ± 1.65 [b]
NU	20.23 ± 0.36 [b]	32.1 ± 1.58 [de]	3.05 ± 0.07 [ab]	50.18 ± 0.76 [f]
BA	19.85 ± 0.48 [ab]	34.29 ± 1.09 [bcde]	3.05 ± 0.07 [ab]	62.35 ± 2.92 [de]

[δ] Ferragnes (FS), Ferraduel (FL), Marta (MA), Lauranne (LU), Texas (TX), Nonpareil (NL), Guara (GU), Yaltinskii (YA), Nurlu (NU), and Acıbadem (BA). Different small letters in each column show the statistical significance ($p < 0.05$).

Garcia-Lopez et al. [17] reported that the lipid contents of the different almond genotypes were in the range of 53.10–61.70% for the almond samples (4 from USA, 3 from Italy, 7 from Spain, and 1 from Australia). In another study, the total lipid level of the almond samples ranged between 52.50–57.00% for four different almond genotypes from Italy [18] while the lipid concentrations of five almond genotypes (Mission, Nonpareil, Carmel, Neplus, and Peerless) from USA were in the range of 53.6–56.1%. For the Turkish almond genotypes grown in Isparta, total lipid, protein, and ash content were reported to be in the range of 44.25–55.68%, 21.23–35.27%, and 2.75–3.81%, respectively [11]. In another study, two different almond genotypes named as Akbadem and Nonpareil were compared in terms of their proximate composition and fatty acid profile. It was resulted that the total oil, ash, and protein content of both almond genotypes were 52.32 and 52.43%, 20.57 and 21.54%, and 3.15–3.26%, respectively [19]. Karaat [1] investigated the proximate composition and fatty acid profile of organic vs. conventional almonds (Ferragnes and Ferraduel) and it was revealed that the total oil, total protein, and total ash levels of organic and conventional almond were 44.5 and 46.7%, 20.9 and 20.8%, and 3.2 and 3.2%, respectively, for the Ferraduel variety. As is obviously seen from the results, the ash and protein contents were close to the reported results, but the oil content of the almond samples revealed in the current study was quite lower than that of the reported values in the literature.

Total phenolic content, which is one of the main key factors of the bioactive performance of the almond samples, was characterized and the values of total phenolic levels were also given in Table 1. As is seen from the table, it can be said that the almonds had a bioactivity and the total phenolic level ranged between 38.7–101.03 mg GAE/100 g sample. The lowest total phenolic level was observed for the sample of FL (38.7 mg GAE/100 g) and the highest total phenolic content was determined for the sample of YS (101.03 mg GAE/100 g sample). It was monitored that the differences among the sample in terms of the total phenolic levels were determined as significant ($p < 0.05$). Milbury [20] reported the total phenolic content of the different almond varieties from California were in the range of 127–241 mg GAE/100 g sample. In another study performed by Esfahlan et al. [21], total phenolic content of the almond samples from Iran ranged between 75.9–122.2 mg GAE/g for the shell and 18.1–46.6 mg GAE/g for the almond flour while the total phenolic level of the almond was 3.74 mg GAE/g in the study reported by Pinelo [22].

3.2. Fatty Acid Profile of Almond Oil Samples

After extraction of the oils from the almond samples, fatty acid profiles of each oil sample were determined, and the results were tabulated in Table 2. As is clearly seen from Table 2, major fatty acids detected in the almond oil samples were palmitic acid (C16:0), palmitoleic acid (C16:1), stearic acid (C18:0), oleic acid (C18:1), and linoleic acid (C18:2) Among the fatty acids, oleic acid was determined as the main and dominant unsaturated fatty acid for the almond oils, while the palmitoleic acid was the minor unsaturated fatty acid having lower than 1% concentration in the fatty acid profile. For the palmitic acid levels of the samples, the highest level was determined for the sample of NL (8.15%), while the lowest was monitored for the sample of NU (5.59%). It was seen that there was a significant difference among the samples ($p < 0.05$). Palmitoleic acid was the minor fatty acid for all samples, with levels ranging between 0.41–0.76%. The results were similar to the palmitic acid level for the samples, and the differences were determined as significant ($p < 0.05$). Stearic acid levels also were in the range of 1.36–3.73% for the samples, and the highest stearic acid level was determined for the sample of GU and the lowest one was in the sample of TX. The level of oleic acid, which is the major and dominant fatty acid, ranged between 61.22–77.63%. The sample of NU oil, having the lowest palmitic acid and palmitoleic acid concentration, showed the highest level of oleic acid concentration (77.63%), and NL oil having the highest palmitic acid and palmitoleic acid showed the lowest oleic acid concentration (61.22%). Linoleic acid content of the almond oil samples was also given in Table 2, and the level of the linoleic acid, which is the only one polyunsaturated fatty acid in the structure of the almond oil, ranged between 13.38–27.69%. The highest linoleic acid level was for the sample of NL oil having the lowest oleic acid content and the lowest linoleic acid level was determined in the sample of FS oil. For the almond oil samples, saturated fatty acid levels (SFA), monounsaturated fatty acid levels (MUSFA), and polyunsaturated fatty acid levels (PUSFA) were calculated, and the results were also tabulated in Table 2. As is seen from the table, the lowest SFA concentration was determined for the samples of TX (7.85%) almond type, and the highest saturation level in the oil was seen in the sample of GU (10.60%). The saturation level of NL oil (10.33%) was also determined to be quite similar to the saturation level of the oil of GU almond type. Monounsaturated fatty acid (MUSFA) levels of the samples were in the range of 61.98–78.04%, while the polyunsaturated fatty acid (PUSFA) ranged between 13.38–27.69%. The highest and lowest MUSFA levels were determined for the almond oil of NU and NL samples, respectively. Thus, it can be easily said that the samples having the lowest oleic acid showed the lowest MUSFA level and vice versa. For the oil samples, PUSFA levels were equal to the levels of linoleic acid concentrations because it was the only fatty acid having double bounds in the oil samples. Thus, the PUSFA levels of the oil samples ranged between 13.38–27.69% as similar to the linoleic acid variation. Yildirim et al. [23] reported that the predominant unsaturated fatty acids of Amygdalus communis were oleic and linoleic acids, and predominant saturated fatty acid was recorded as palmitic acid. The highest oleic acid level was measured for Glorieta in both 2008 (83.35%) and 2009 (72.74%) while the highest linoleic acid content was recorded in Picantili (26.08%) in 2008 and Yaltinskii (30.01%) in 2009.

Table 2. Some physicochemical and bioactive properties of almond samples (%).

Samples $^{\delta}$	PA (C16:0)	PLA (C16:1)	SA (C18:0)	OA (C18:1)	LA (C18:2)	SFA	MUSFA	PUSFA
FS	6.57 ± 0.04^{ab}	0.75 ± 0.05^{a}	2.13 ± 0.03^{bcd}	77.18 ± 0.75^{ab}	13.38 ± 0.73^{b}	8.70	77.93	13.38
FL	6.64 ± 0.06^{ab}	0.74 ± 0.06^{a}	1.97 ± 0.02^{cd}	76.16 ± 1.00^{abc}	14.51 ± 0.98^{b}	8.60	76.89	14.50
MA	6.62 ± 0.03^{ab}	0.61 ± 0.01^{ab}	2.28 ± 0.04^{bc}	71.23 ± 0.37^{bcd}	19.28 ± 0.31^{bc}	8.89	71.83	19.28
LU	6.93 ± 0.05^{ab}	0.65 ± 0.01^{ab}	2.36 ± 0.04^{b}	69.96 ± 0.99^{d}	20.12 ± 1.00^{abc}	9.28	70.61	20.11
YS	7.48 ± 1.36^{b}	0.53 ± 0.04^{bc}	1.87 ± 0.19^{de}	70.8 ± 2.11^{cd}	19.33 ± 3.71^{bc}	9.34	71.33	19.33
NL	8.15 ± 0.85^{a}	0.76 ± 0.07^{a}	2.18 ± 0.11^{bcd}	61.22 ± 2.6^{e}	27.69 ± 3.63^{a}	10.33	61.98	27.69
TX	6.50 ± 0.11^{ab}	0.43 ± 0.00^{c}	1.36 ± 0.03^{f}	69.35 ± 1.48^{d}	22.37 ± 1.4^{ab}	7.85	69.78	22.37
GU	6.87 ± 0.10^{ab}	0.66 ± 0.03^{ab}	3.73 ± 0.10^{a}	69.6 ± 0.74^{d}	19.15 ± 0.52^{bc}	10.60	70.26	19.14
NU	5.59 ± 0.20^{b}	0.41 ± 0.01^{c}	2.42 ± 0.08^{b}	77.63 ± 1.67^{a}	13.96 ± 1.39^{b}	8.00	78.04	13.96
BA	7.04 ± 0.06^{ab}	0.51 ± 0.04^{bc}	1.58 ± 0.08^{ef}	72.12 ± 2.08^{abcd}	18.76 ± 2.14^{bc}	8.62	72.63	18.76

$^{\delta}$ Ferragnes (FS), Ferraduel (FL), Marta (MA), Lauranne (LU), Texas (TX), Nonpareil (NL), Guara (GU), Yaltinskii (YA), Nurlu (NU), and Acıbadem (BA). PA: Palmitic acid, PLA: Palmitoleic acid, SA: Stearic acid, OA: Oleic acid, LA: Linoleic acid, SFA: Saturated fatty acid, MUSFA: Monounsaturated fatty acid, PUSFA: Polyunsaturated fatty acid. Different small letters in each column show the statistically significance ($p < 0.05$).

In a study performed by Karaat [1], the major fatty acids of two different almond genotypes (Ferragnes and Ferraduel) were reported to be oleic acid and linoleic acid. Oleic acid ranged between 78.9–82.4% and 75.8–82.4% for Ferragnes and Ferraduel almond variety, respectively, while the linoleic acid levels were in the range of 8.4–12.7% and 6.2–15.3% for the same samples, respectively. Palmitoleic acid was found to have the lowest level for all samples, as similar to the current study results. It was also informed that the oleic and linoleic acids were the most abundant unsaturated fatty acids in almonds, accounting for about 80–90%, while saturated fatty acids, such as palmitic and stearic fatty acids, are present in lower quantities (<10%) [9]. It was reported that the oleic acid and linoleic acid were reported to be the dominant fatty acids from different parts of the world such as Turkey, Spain, İtaly, Serbia, China, and California [23]. The main factor affecting the differences among the almond varieties were reported to be some growing conditions and year of cultivation. Some studies suggested that the poor water supply to the crop resulted a lower oleic/linoleic ratio indicating a significant effect of irrigation on almond fatty acid composition [24,25]. Kodad et al. [9] stated that the irrigation management and the environmental temperature levels of the growing area are the main factors affecting the total oil level and fatty acid profile of the almond varieties. They reported that the Spain almond varieties, which grow at lower temperature and abundant water supply, showed higher total oil content (58.65% vs. 55.58% (w/w)) and the percentage of oleic acid (71.1% vs. 68.6% (w/w)) compared to the ones obtained in samples grown in central Morocco. In a study conducted by Celik and Balta [25], it was reported that the USFA and SFA levels were in the range of 90.99 to 91.17% and 8.82 to 9.00%, respectively.

3.3. Mineral Contents of the Almond Samples

Major mineral compositions of all samples were characterized, and the results were given in Table 3. Nine different minerals (P, Mg, Ca, K, Zn, Mn, Cu, Fe, and Na) were determined as mineral elements, and it was monitored that the most abundant type of mineral was potassium (K); phosphorous (P) and magnesium (Mg) were the second and third major mineral for all almond samples. Cu and Mn were the elements showing the lowest concentration for all almond samples. Potassium levels of the almond samples were in the range of 6192.08–11,046.05 mg/kg sample. The highest K level was in the almond sample of NL and the lowest was for the sample of TX. Phosphorous (P) levels of the almond samples ranged between 4784.42–5527.03 mg/kg; the highest P levels were determined in the almond sample of FL, and the lowest P level was in the sample of YS. Mg level of the samples ranged between 2199–2657.19 mg/kg while the Ca levels of the samples were in the range of 459.34–1069.6 mg/kg. The highest Mg level was in the sample of NU (2657.19 mg/kg), while the lowest Mg was in the sample of FS. For the Ca levels, the sample of NU also showed the lowest Ca level, and for the sample of TX, Zn, Mn, Cu, Fe, and Na were the elements showing the lowest concentrations in the almond samples compared to K, P, Mg, and Ca, which, due to their levels, were trace for all almond samples. The lowest Zn level was observed in the sample of TX (33.78 mg/kg), and the highest Zn level was for the sample of FL (52.34 mg/kg). Mn levels of the samples were in the range of 12.95–24.18 mg/kg, and the sample of TX showed the highest Mn level while the sample of NU had the lowest Mn concentration (12.95 mg/kg). For the Cu levels of the samples, it was recorded that the sample of the NU showed the lowest Cu content (8.33 mg/kg) while the sample of the NL had the highest Cu concentration (15.28 mg/kg). Fe levels of the samples also ranged between 25.65–47.35 mg/kg while the levels of Na were in the range of 27.42–61.29% for all almond samples. The statistical analysis results revealed that there were significant differences among the samples in terms of the detected minerals ($p < 0.05$). Simsek and Kizmaz (2017) investigated the mineral profile of almond genotypes grown in Beyazsu district, and they reported that the potassium, phosphorous, magnesium, and calcium were in the range of 646.27–925.13 mg/100 g, 562.53–701.93 mg/100 g, 217.13–367.27 mg/100 g and 190.97–317.13 mg/100 g, respectively. It was reported that the almond kernel is accepted as a good source for the minerals [26–28]. For the almond kernels, the major minerals were reported as K, P, Ca, and Mg and minor elements were Na, Fe, Cu, Zn, and Mn [8]. As is known, the minerals present in the plant tissues are sourced from the soil, water, and fertilizers, the differences among the genotypes in terms of mineral profiles are also attributed to the geographical origin, which combines soil and weather conditions, and agricultural practices [28]. Additionally, it was informed that the ripening stage of the kernel was also the other factor affecting the mineral profile of the almonds. Beltrán Sanahuja et al. [23] stated that different almond genotypes could maturate in periods along the year and with different ripening period length, and thus it should be considered when comparing different cultivars. Drogoudi et al. [29] investigated the mineral profile of 72 different almond genotypes provided from different countries (France, Greece, and Italy), and they found that the major elements were K, Mg, and P as similar to the current research results. Şimsek et al. [30] characterized different almond varieties including Ferraduel and Ferragnes almond types, and they reported that the major elements were also K, P, and Mg for all samples. K and P levels of the Ferraduel and Ferragnes were 903.3 and 584.7 mg/100 g and 879.4 and 621.5 mg/100 g, respectively, and it was informed as very similar to the levels found in the present work for both almond varieties. Cu was the element found in the minor levels for all almond genotypes both in the present work and the work of Simsek et al. [30].

3.4. Volatile Composition of the Almond Samples

Table 4 shows the volatile composition of all almond samples. For the almond samples, 41 different volatile compounds were identified, and their levels showed a significant difference among the samples. As is seen, for all almond samples, toluene, 4-octanone, pinacol, and 2-methylpentanal were the dominant volatile compounds for all almond samples. The levels of toluene ranged between 64.3–201.05 μg/kg while the highest toluene level was in the almond sample of FL and the lowest level was for the sample of YS. It was determined that the differences among the samples in terms of toluene were significant statistically ($p < 0.05$). 4-octanone levels were in the range of 29.35–334.55 μg/kg for the samples. It was observed that the highest 4-octanone concentration was monitored for the sample of FS (334.55 μg/kg) and the lowest level of 4-octanone concentration was for the sample of TX (29.35 μg/kg).

The statistical analysis revealed that the differences were significant among the almond samples in terms of 4-octanone level. Pinacol was also the most common present compound in the almond samples, and its concentration ranged between 30.45–130.1 μg/kg. The highest level of the pinacol was detected in the sample of FL, and the lowest level was recorded for the sample of TX (30.45 μg/kg). For the other abundant volatile compound named as 2-methylpentanal levels, the sample of LU showed the highest concentration (368.25 μg/kg), and the sample of YS had the lowest level of 2-methylpentanal (94.05 μg/kg). Apart from these major volatile compounds, many different constituents were detected in almond samples, and their levels were also given in Table 4. When considering all compound types, a classification was carried out and the volatile compounds were grouped as esters, alcohols, aldehydes, ketones, acids, and terpenes, and their levels were tabulated in Table 4. As is seen from Table 4, it can be said clearly that the major volatile groups were aldehydes and ketones for all almond samples.

Table 3. Major mineral profile of almond samples (mg/kg).

Samples $^{\delta}$	K	Mg	Ca	P	Zn	Mn	Cu	Fe	Na
FS	8590.31 ± 15.41 ^{cd}	2199 ± 41.8 ^{c}	887.35 ± 44.68 ^{b}	5248.46 ± 75.27 ^{bc}	52.13 ± 1.46 ^{a}	19.09 ± 1.28 ^{bc}	14.27 ± 0.06 ^{ab}	45.99 ± 2.6 ^{a}	37.42 ± 3.67 ^{c}
FL	9196.78 ± 435.17 ^{bcd}	2303.85 ± 7.18 ^{bc}	883.81 ± 67.8 ^{b}	5527.03 ± 77.64 ^{ab}	52.34 ± 3.53 ^{a}	17.37 ± 0.69 ^{bcd}	14.34 ± 0.31 ^{ab}	47.35 ± 0.13 ^{a}	39.79 ± 1.21 ^{bc}
MA	9788.69 ± 231.49 ^{abc}	2280.6 ± 68.86 ^{bc}	928.35 ± 13.64 ^{ab}	4978.78 ± 94.58 ^{c}	41.91 ± 3.81 ^{abcd}	17.73 ± 1.32 ^{bcd}	11.33 ± 0.16 ^{c}	35.58 ± 0.87 ^{b}	42.55 ± 0.04 ^{bc}
LU	9775.08 ± 494.44 ^{abc}	2372.97 ± 37.04 ^{bc}	835.66 ± 3.97 ^{bc}	5252.74 ± 184.83 ^{bc}	46.16 ± 4.29 ^{abc}	18.05 ± 1.03 ^{bc}	11.88 ± 0.59 ^{c}	41.95 ± 1.5 ^{ab}	44.48 ± 4.04 ^{bc}
YS	8956.48 ± 16.5 ^{cd}	2284.41 ± 54.69 ^{bc}	581.35 ± 48.45 ^{de}	4784.42 ± 159.25 ^{c}	39.07 ± 0.81 ^{cd}	15.1 ± 0.04 ^{cd}	14.9 ± 0.01 ^{a}	25.65 ± 0.59 ^{c}	45.92 ± 2.41 ^{bc}
NL	11,046.05 ± 205.75 ^{a}	2444.33 ± 100.94 ^{abc}	719.11 ± 39.24 ^{cd}	4965.88 ± 185.29 ^{c}	50.04 ± 3.53 ^{ab}	19.04 ± 2.11 ^{bc}	15.28 ± 0.85 ^{a}	35.46 ± 0.37 ^{b}	43.11 ± 0.18 ^{bc}
TX	6192.08 ± 8.39 ^{e}	2251.05 ± 39.99 ^{bc}	1069.6 ± 30.51 ^{a}	4971.58 ± 41.13 ^{c}	33.78 ± 0.23 ^{d}	24.18 ± 0.04 ^{a}	12.66 ± 0.31 ^{bc}	36.05 ± 3.51 ^{b}	41.87 ± 0.79 ^{bc}
GU	10,364.24 ± 683.13 ^{ab}	2342.57 ± 163.79 ^{bc}	930.63 ± 31.93 ^{ab}	5151.56 ± 80.73 ^{bc}	49.56 ± 1.62 ^{ab}	21.4 ± 0.34 ^{ab}	13.72 ± 0.75 ^{ab}	36.73 ± 1.94 ^{b}	48.63 ± 3.63 ^{b}
NU	7945.02 ± 67.71 ^{d}	2657.19 ± 6.79 ^{a}	459.34 ± 26.28 ^{e}	5037.35 ± 72.41 ^{bc}	41.45 ± 2.45 ^{bcd}	12.95 ± 0.62 ^{d}	8.33 ± 0.35 ^{d}	46.05 ± 0.93 ^{a}	61.29 ± 1.68 ^{a}
BA	6232.97 ± 288.2 ^{e}	2527.92 ± 3.42 ^{ab}	662.72 ± 13.38 ^{d}	5173.36 ± 313.29 ^{a}	51.09 ± 0.53 ^{ab}	17.23 ± 2.2 ^{bcd}	14.12 ± 0.25 ^{ab}	41.88 ± 0.2 ^{ab}	43.8 ± 1.47 ^{ab}

$^{\delta}$ Ferragnes (FS), Ferraduel (FL), Marta (MA), Lauranne (LU), Texas (TX), Nonpareil (NL), Guara (GU), Yaltinskii (YA), Nurlu (NU), and Acıbadem (BA). Different small letters in each column show the statistically significance ($p < 0.05$).

Table 4. Volatile profile of almond samples $^{\delta}$.

No.	Compound	RI	FS	FL	MA	LU	YS	NL	TX	GU	NU	BA
1	Butanal	833	6.4 ± 0.42 ^{a}	3.15 ± 0.35 ^{b}	n.d	2.1 ± 0.42 ^{ab}	0.6 ± 0.28 ^{de}	n.d	n.d	1.1 ± 0.57 ^{cde}	1.3 ± 0.28 ^{bcd}	1.85 ± 0.35 ^{abc}
2	2-Methylbutanal	894	n.d	n.d	n.d	n.d	n.d	0.45 ± 0.07 ^{a}	0.3 ± 0.14 ^{a}	n.d	n.d	0.35 ± 0.07 ^{a}
3	3-Methylbutanal	920	n.d	n.d	n.d	n.d	n.d	0.15 ± 0.07 ^{b}	n.d	n.d	0.15 ± 0.07 ^{b}	0.45 ± 0.21 ^{a}
4	Ethanol	942	n.d	n.d	n.d	n.d	n.d	3.45 ± 0.92 ^{a}	2.15 ± 0.49 ^{ab}	n.d	0.9 ± 0.28 ^{bc}	n.d
5	3-Methyl-2-pentanone	1023	n.d	0.4 ± 0.14 ^{a}	0.65 ± 0.07 ^{a}	n.d	0.25 ± 0.07 ^{a}	n.d	n.d	n.d	n.d	n.d
6	α-pinene	1025	n.d	n.d	n.d	n.d	0.55 ± 0.2 ^{bc}	1.05 ± 0.21 ^{b}	2.15 ± 0.35 ^{a}	0.25 ± 0.07 ^{c}	1.1 ± 0.28 ^{b}	n.d
7	Toluene	1044	161.45 ± 9.55 ^{b}	201.05 ± 7.57 ^{a}	103.1 ± 7.92 ^{c}	145.6 ± 8.2 ^{b}	64.3 ± 5.52 ^{f}	85.9 ± 6.22 ^{cdef}	69.1 ± 2.4 ^{ef}	95.1 ± 7.5 ^{cde}	75.3 ± 5.37 ^{def}	101.4 ± 6.65 ^{cd}
8	Butyl acetate	1070	74.85 ± 3.18 ^{a}	87.05 ± 6.29 ^{a}	18.95 ± 1.63 ^{def}	57.2 ± 6.08 ^{b}	27.7 ± 2.26 ^{de}	15.9 ± 3.68 ^{ef}	11.6 ± 2.55 ^{f}	26.65 ± 3.89 ^{def}	45.85 ± 3.75 ^{bc}	31.9 ± 1.7 ^{cd}
9	Hexanal	1084	n.d	n.d	10.6 ± 1.13 ^{c}	n.d	n.d	18.05 ± 2.33 ^{b}	23.2 ± 2.26 ^{a}	8.3 ± 1.27 ^{c}	n.d	11.05 ± 0.35 ^{c}
10	2-Methyl-1-propanol	1088	6.4 ± 0.71 ^{a}	6.9 ± 0.85 ^{a}	n.d	2.15 ± 0.35 ^{b}	n.d	n.d	1.35 ± 0.21 ^{bc}	n.d	0.55 ± 0.21 ^{c}	n.d
11	6-hepten-3-one-4-methyl	1115	3.2 ± 0.14 ^{b}	n.d	n.d	n.d	1.55 ± 0.35 ^{d}	n.d	4.7 ± 0.28 ^{a}	n.d	n.d	2.45 ± 0.35 ^{c}
12	isoamyl acetate	1122	n.d	1.6 ± 0.28 ^{a}	1.15 ± 0.07 ^{b}	n.d	n.d	n.d	n.d	n.d	n.d	n.d
13	Ethylbenzene	1126	23.45 ± 1.06 ^{b}	22.1 ± 1.7 ^{bc}	17.6 ± 1.56 ^{cd}	28.9 ± 1.7 ^{a}	23.4 ± 1.27 ^{b}	n.d	n.d	10.25 ± 0.64 ^{e}	17.15 ± 1.48 ^{d}	8.6 ± 1.13 ^{e}
14	2.5-Dimethyl-3-hexanone	1144	24.55 ± 2.47 ^{a}	26.55 ± 1.63 ^{a}	n.d	n.d	n.d	16.1 ± 2.4 ^{b}	10.85 ± 0.49 ^{c}	n.d	6.55 ± 0.21 ^{c}	27.2 ± 0.57 ^{a}
15	butyl isobutyrate	1150	n.d	2.15 ± 0.49 ^{a}	n.d	1.15 ± 0.35 ^{b}	n.d	n.d	n.d	n.d	0.650.21 ^{bc}	n.d
16	butanol	1155	12.4 ± 0.99 ^{cd}	14.65 ± 0.64 ^{bc}	7.9 ± 0.57 ^{ef}	17.8 ± 1.56 ^{b}	5.45 ± 0.92 ^{f}	22.6 ± 1.13 ^{a}	15.45 ± 0.92 ^{bc}	10.05 ± 0.64 ^{de}	12.55 ± 0.92 ^{cd}	8.9 ± 0.42 ^{def}
17	3-heptanone	1158	7.35 ± 0.64 ^{b}	9.25 ± 0.78 ^{a}	n.d	10.05 ± 0.92 ^{a}	n.d	n.d	n.d	3.3 ± 0.28 ^{c}	n.d	2.6 ± 0.28 ^{c}
18	2-heptanone	1184	n.d	n.d	n.d	n.d	2.1 ± 0.28 ^{b}	n.d	1.6 ± 0.14 ^{b}	n.d	3.05 ± 0.35 ^{a}	n.d
19	Limonene	1195	n.d	n.d	3.85 ± 0.49 ^{a}	n.d	n.d	1.65 ± 0.21 ^{c}	n.d	2.6 ± 0.28 ^{b}	0.75 ± 0.21	n.d
20	3-Methyl-butanol	1195	1.05 ± 0.35 ^{c}	n.d	n.d	1.7 ± 0.14 ^{c}	5.4 ± 0.42 ^{a}	n.d	3.9 ± 0.28 ^{b}	n.d	n.d	n.d
21	butyl butyrate	1206	n.d	1.4 ± 0.28 ^{b}	n.d	2.25 ± 0.21 ^{a}	n.d	n.d	n.d	n.d	n.d	n.d
22	2-hexanol	1220	1.6 ± 0.28 ^{c}	n.d	2.25 ± 0.21 ^{ab}	n.d	0.7 ± 0.14 ^{d}	n.d	2.75 ± 0.21 ^{a}	0.8 ± 0.14 ^{d}	1.45 ± 0.21 ^{c}	1.85 ± 0.07 ^{bc}
23	4-octanone	1230	334.55 ± 8.27 ^{a}	318.5 ± 6.79 ^{a}	56.7 ± 3.68 ^{de}	129.8 ± 7.5 ^{c}	143.65 ± 5.44 ^{c}	40.45 ± 2.62 ^{ef}	29.35 ± 1.2 ^{f}	75.25 ± 5.59 ^{d}	83.45 ± 8.7 ^{d}	172.65 ± 12.09 ^{b}
24	3-octanone	1254	1.7 ± 0.28 ^{a}	2.05 ± 0.35 ^{a}	n.d	n.d	n.d	n.d	0.75 ± 0.21 ^{b}	0.55 ± 0.21 ^{bc}	n.d	n.d
25	m-Cymene	1262	n.d	2.75 ± 0.49 ^{b}	n.d	4.55 ± 0.21 ^{a}	0.5 ± 0.14 ^{cd}	n.d	1.05 ± 0.21 ^{c}	n.d	n.d	n.d

Table 4. *Cont.*

No.	Compound	RI	FS	FL	MA	LU	YS	NL	TX	GU	NU	BA
26	butyl valerate	1299	1.05 ± 0.21 ^{a}	0.85 ± 0.35 ^{a}	n.d	n.d	0.65 ± 0.21 ^{a}	n.d	n.d	n.d	n.d	n.d
27	2-heptenal	1332	n.d	n.d	2.4 ± 0.28 ^{a}	n.d	n.d	1.6 ± 0.28 ^{bc}	n.d	2.4 ± 0.14 ^{a}	1.2 ± 0.14 ^{c}	1.85 ± 0.21 ^{ab}
28	Pinacol	1336	120.65 ± 11.38 ^{ab}	130.1 ± 8.91 ^{a}	40.2 ± 2.69 ^{g}	87.1 ± 5.09 ^{cd}	73.4 ± 3.25 ^{de}	62.75 ± 2.05 ^{ef}	30.45 ± 0.92 ^{g}	50.55 ± 1.06 ^{fg}	103.15 ± 4.17 ^{bc}	40.6 ± 1.13 ^{g}
29	1-hexanol	1357	0.3 ± 0.14 ^{b}	n.d	n.d	n.d	0.25 ± 0.07 ^{b}	n.d	1.05 ± 0.35 ^{a}	n.d	n.d	n.d
30	2-nonanone	1396	11.45 ± 0.78 ^{bc}	14.1 ± 1.84 ^{b}	8.15 ± 0.35 ^{cde}	21.65 ± 2.05 ^{a}	n.d	7.05 ± 0.35 ^{de}	12.85 ± 0.64 ^{b}	10.85 ± 0.64 ^{bc}	4.7 ± 0.57 ^{e}	13.55 ± 0.78 ^{b}
31	2-methylpentanal	1451	296.55 ± 15.91 ^{bc}	327.6 ± 9.62 ^{ab}	175.8 ± 10.75 ^{cd}	368.25 ± 16.76 ^{a}	94.05 ± 6.15 ^{f}	131.15 ± 4.74 ^{f}	110.9 ± 10.32 ^{f}	202.05 ± 7.71 ^{c}	136.75 ± 4.74 ^{de}	262.8 ± 16.12 ^{c}
32	Benzaldehyde	1541	n.d	n.d	4.7 ± 0.28 ^{d}	1.55 ± 0.35 ^{ef}	3.25 ± 0.49 ^{de}	15.6 ± 0.71 ^{a}	12.55 ± 0.92 ^{b}	6.8 ± 0.42 ^{c}	n.d	7.15 ± 0.64 ^{c}
33	Isobutyric acid	1577	35.45 ± 3.32 ^{a}	37.6 ± 3.11	n.d	11.75 ± 2.33 ^{b}	n.d	n.d	n.d	n.d	n.d	n.d
34	γ-butyrolactone	1625	n.d	n.d	0.35 ± 0.07 ^{a}	n.d	n.d	n.d	n.d	n.d	0.4 ± 0.14 ^{a}	n.d
35	2-Decenal	1646	n.d	n.d	0.8 ± 0.14 ^{bc}	n.d	0.45 ± 0.07 ^{cd}	1.1 ± 0.14 ^{ab}	1.55 ± 0.21 ^{a}	1.25 ± 0.21 ^{ab}	n.d	1.3 ± 0.14 ^{a}
36	Phenylacetaldehyde	1668	n.d	n.d	4.05 ± 0.49 ^{b}	1.15 ± 0.21 ^{c}	n.d	5.8 ± 0.57 ^{a}	n.d	4.75 ± 0.64 ^{ab}	n.d	n.d
37	valeric acid	1719	2.05 ± 0.21 ^{a}	1.1 ± 0.28 ^{bc}	n.d	n.d	1.45 ± 0.21 ^{ab}	n.d	1.05 ± 0.35 ^{bc}	n.d	1.05 ± 0.21 ^{bc}	0.55 ± 0.07 ^{cd}
38	2-Phenyl-2-propanol	1768	10.15 ± 0.64 ^{b}	8.65 ± 0.64 ^{ab}	2.75 ± 0.21 ^{e}	13.7 ± 1.13 ^{a}	2.2 ± 0.42 ^{e}	3.6 ± 0.28 ^{de}	7.05 ± 0.49 ^{bc}	4.2 ± 0.42 ^{de}	5.35 ± 0.64 ^{cd}	8.45 ± 0.92 ^{ab}
39	γ-decalactone	2185	11.1 ± 0.85 ^{b}	n.d	15.55 ± 1.2 ^{a}	n.d	n.d	n.d	n.d	7.9 ± 0.42 ^{c}	1.65 ± 0.21 ^{de}	2.65 ± 0.35 ^{d}
40	γ-undecalactone	2246	2.85 ± 0.49 ^{b}	5.2 ± 0.42 ^{a}	n.d	2.6 ± 0.28 ^{b}	0.3 ± 0.14 ^{c}	n.d	0.5 ± 0.14 ^{c}	n.d	n.d	n.d
41	Diethyl Phthalate	2374	0.35 ± 0.07 ^{bc}	5.75 ± 0.64 ^{a}	1.1 ± 0.42 ^{b}	n.d	n.d	1.05 ± 0.21 ^{b}	n.d	n.d	1.05 ± 0.21 ^{b}	0.65 ± 0.07 ^{bc}
	Total (μg/kg)											
	Esters		76.25	98.8	21.2	60.6	28.35	16.95	11.6	26.65	47.55	32.55
	Alcohols		152.55	160.3	53.1	122.45	87.4	92.4	64.15	65.6	123.95	59.8
	Aldehydes		302.95	330.75	198.35	373.05	98.35	173.9	148.5	226.65	139.4	286.8
	Ketones		396.75	376.05	81.4	164.1	147.85	63.6	60.6	97.85	99.8	221.1
	Acids		37.5	38.7	0	11.75	1.45	0	1.05	0	1.05	0.55
	Terpenes		0	2.75	3.85	4.55	1.05	2.7	3.2	2.85	1.85	0
	Others		184.9	223.15	120.7	174.5	87.7	85.9	69.1	105.33	92.45	110

$^{\delta}$ Ferragnes (FS), Ferraduel (FL), Marta (MA), Lauranne (LU), Texas (TX), Nonpareil (NL), Guara (GU), Yaltinskii (YA), Nurlu (NU), and Acıbadem (BA). Different small letters in each column show the statistically significance ($p < 0.05$); n.d: not determined.

The minor chemical volatile group was also monitored as acids and terpenes (Table 4). The highest ester levels were determined in the sample of FL (98.8 μg/kg) while the lowest ester levels were recorded for the almond sample of TX (11.6 μg/kg). Alcohols ranged between 53.1–160.3 μg/kg while the sample of FL also showed the highest alcohol group and the sample of MA showed the lowest alcohol group levels. The almond samples of LU and FS showed the highest levels of aldehydes and ketones, respectively, while the samples of YS and TX had the lowest levels of aldehydes and ketones, respectively. For all almond samples, other compounds that do not place in the mentioned chemical class above were also characterized, and their levels were also tabulated in Table 4. For all identified volatile compounds, almond types showed a significant effect on the variation among the sample in terms of volatile compound concentration.

Xiao et al. [31] characterized volatiles in raw and dry-roasted almonds (*Prunus dulcis*); using NIST libraries and Kovats Index values, they identified 41 volatiles including 13 carbonyls, 1 pyrazine, 20 alcohols, and 7 additional volatiles. It was revealed that predominant volatile (2934.6 ± 272.5 ng/g) present in raw almonds was benzaldehyde, known as the breakdown product of amygdalin. In the present study, benzaldehyde was not the dominant compound, and it was detected in the almond samples except FS, FL, and NU while the sample of NL showed the highest benzaldehyde concentration. Mexis et al. [32] also reported the similar results to the results of Xiao et al. [31].

Karaat [1] investigated the volatile profile of two almond genotypes, and he reported that the detected volatile compounds were butanal, butyl acetate, ethylbenzene, p-chlorotoluene, and 4-octanone. In another study, Erten and Cadwallader [33] reported the identification of predominant aroma components of raw, dry roasted, and oil roasted almonds, and they revealed that the 1-octen-3-one and acetic acid were important aroma compounds in raw almonds. In most of the study, the authors reported the volatile profile of raw almonds revealed that the octanal, nonanal, acetic acid, and methional were the volatile constituents of raw almonds [31,34–36].

3.5. Principal Component and Correlation Analysis

To determine the relationship among the studied parameters, Pearson correlation was performed, and the results were illustrated in Figure 1. As is seen, there were generally no significant correlations among the proximate composition except fatty acid and mineral profile. It was determined that a significant and negative and positive correlation was observed between oleic acid and palmitoleic acid (r = −0.81) and linoleic acid and palmitic acid (r = 0.74), respectively. Additionally, for all almond varieties, the sample having the highest oleic acid content had the lowest linoleic acid level, and thus a negative correlation between oleic and linoleic acid level was monitored (r = −0.98). In addition to that, between the palmitoleic acid, zinc, and phosphorus contents and total phenolic levels of the almond samples, it was determined that there was a positive correlation. For the minerals, between Fe and P (r = 0.82), Na and Mg (0.78), Mn and Ca (0.87), and Mn and K (0.87), significant positive correlations were observed while Ca and Mg and K and Mg showed negative correlation.

For the volatile compounds, principal component (PC) analysis was performed and two main PCs explaining the total variation of the parameters as 55.2% was determined. Figure 2 top shows the factor loadings of the samples and volatile compounds, and it was determined that the almond samples showed differences in terms of the aroma. As is seen from the figure, TX, NL, BA, and YS were in the same class and showed significantly different volatile compounds compared to FL and LU. Additionally, FS differed in terms of aroma profile compared to MA, NU, and GU. BA from Turkey showed similar aroma with the sample of TX and NL from USA. Figure 2 bottom also showed the classification of the identified volatile compounds. As is seen, five main clusters were determined for the volatile profile of the samples, and it was also observed that the major compounds identified for all of the studied almond genotypes were aldehydes, ketones, and alcohols.

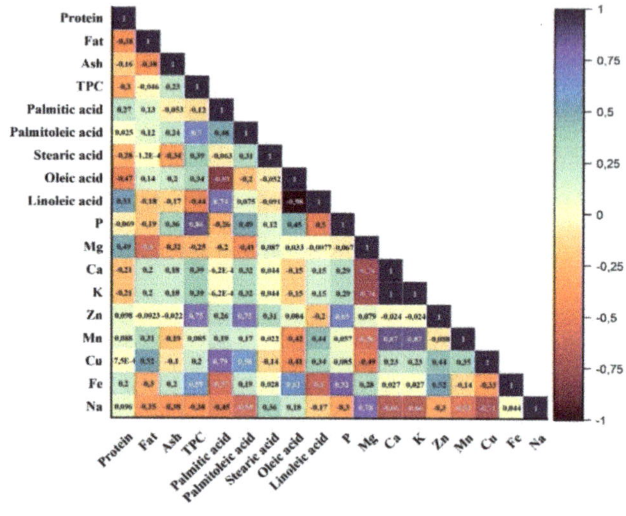

Figure 1. Pearson correlation matrix of almond characteristics.

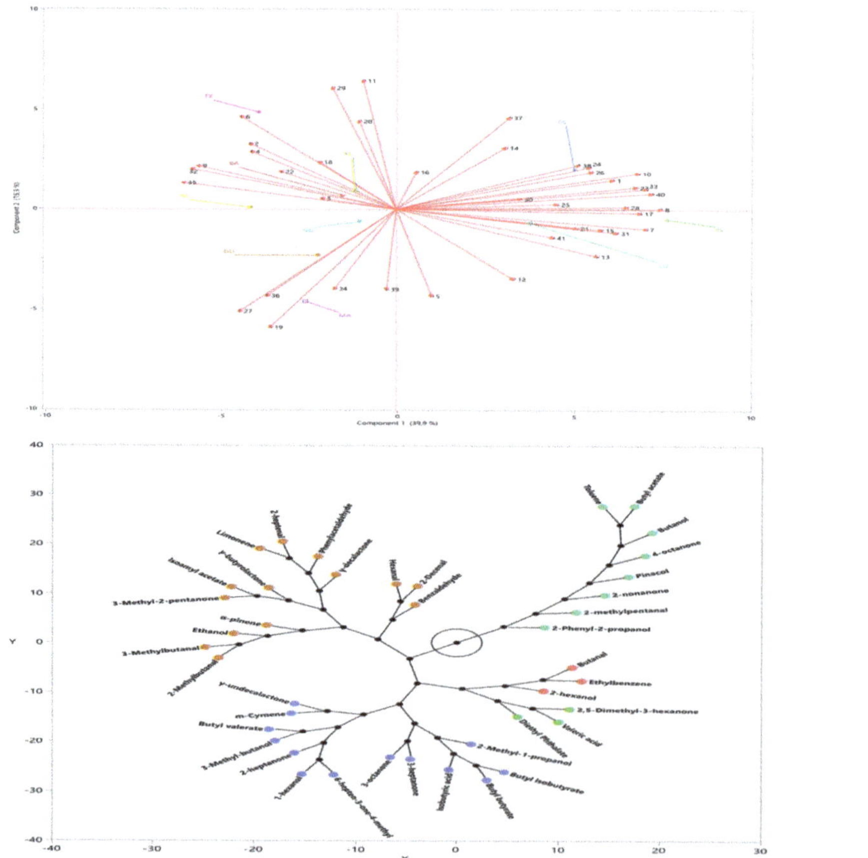

Figure 2. Factor loadings (**top**) and cluster (**bottom**) of the volatile compounds for almond samples.

3.6. Conclusions

It was observed that different almond genotypes from different countries showed a significant variation in terms of main proximate composition, fatty acid and mineral profile, and volatile composition. All almond samples were determined as rich in fat and protein. Additionally, they had phenolic compounds. It was revealed that the almond oils were composed of mainly unsaturated fatty acids such as oleic acid. For the mineral profile of the samples, all of them showed richness in terms of phosphorus, potassium, and magnesium. After volatile characterization, especially, aldehydes, ketones, and alcohols were the major dominant aroma compounds, and the almond varieties differed significantly in terms of the identified volatile component. It was concluded that the almonds showed significant difference according to the genotypes as well as the origin. This comparative research presented the main proximate composition and values compounds (fatty acid, mineral, and aroma) concentrations of the different almond genotypes to see the differences in the nutritive values of the studied varieties.

Funding: This research received no external funding.

Institutional Review Board Statement: Not applicable.

Informed Consent Statement: Not applicable.

Data Availability Statement: Not applicable.

Acknowledgments: I would like to thank Fırat EGE (Adıyaman University, Faculty of Agriculture, and Department of Horticulture) for the help to provide the almond samples.

Conflicts of Interest: The author declares no conflict of interest.

References

1. Karaat, F. Organic vs conventional almond: Market quality, fatty acid composition and volatile aroma compounds. *Appl. Ecol. Environ. Res.* **2019**, *17*, 7793. [CrossRef]
2. Simsek, M. Chemical, Mineral, and Fatty Acid Compositions of Various Types of Walnut (*Juglans regia* L.) in Turkey. *Bulg. Chem. Commun.* **2016**, *48*, 6670.
3. Jenkins, D.J.; Kendall, C.W.; Marchie, A.; Parker, T.L.; Connelly, P.W.; Qian, W.; Haight, J.S.; Faulkner, D.; Vidgen, E.; Lapsley, K.G.; et al. Dose Response of Almonds on Coronary Heart Disease Risk Factors: Blood Lipids, Oxidized Low-Density Lipoproteins, Lipoprotein(a), Homocysteine, and Pulmonary Nitric Oxide. *Circulation* **2002**, *106*, 1327–1332. [CrossRef] [PubMed]
4. Barreca, D.; Nabavi, S.M.; Sureda, A.; Rasekhian, M.; Raciti, R.; Silva, A.S.; Annunziata, G.; Arnone, A.; Tenore, G.C.; Süntar, İ.; et al. Almonds (*Prunus dulcis* Mill. DA web): A Source of Nutrients and Health-Promoting Compounds. *Nutrients* **2020**, *12*, 672. [CrossRef]
5. Barreira, J.C.M.; Casal, S.; Ferreira, I.C.F.R.; Peres, A.M.; Pereira, J.A.; Oliveira, M.B.P.P. Supervised Chemical Pattern Recognition in Almond (*Prunus dulcis*) Portuguese PDO Cultivars: PCA- and LDA-Based Triennial Study. *J. Agric. Food Chem.* **2012**, *60*, 9697–9704. [CrossRef]
6. Beyhan, Ö.; Aktaş, M.; Yilmaz, N.; Şimşek, N.; Gerçekçioğlu, R. Determination of Fatty Acid Compositions of Some Important Almond (*Prunus amygdalus* L.) Varieties Selected from Tokat Province and Eagean Region of Turkey. *J. Med. Plant Res.* **2011**, *5*, 4907–4911.
7. Askin, M.A.; Balta, M.F.; Tekintas, F.E.; Kazankaya, A.; Balta, F. Fatty Acid Composition Affected by Kernel Weight in Almond [*Prunus dulcis* (Mill.) D. A. Webb.] Genetic Resources. *J. Food Compos. Anal.* **2007**, *20*, 7–12. [CrossRef]
8. Yada, S.; Huang, G.; Lapsley, K. Natural Variability in the Nutrient Composition of California-Grown Almonds. *J. Food Compos. Anal.* **2013**, *30*, 80–85. [CrossRef]
9. Kodad, O.; Company, R.S. Variability of Oil Content and of Major Fatty Acid Composition in Almond (*Prunus amygdalus* Batsch) and Its Relationship with Kernel Quality. *J. Agric. Food Chem.* **2008**, *56*, 4096–4101. [CrossRef]
10. Nizamlıoğlu, N.M.; Nas, S. Physical and Chemicial Properties of a Type of Almond Called" Akbadem" Grown in the Aegean Region in Turkey. *Int. J. Second Metabol.* **2017**, *4*, 134–141. [CrossRef]
11. Yıldırım, A.; Mavi, A.; Kara, A.A. Determination of Antioxidant and Antimicrobial Activities of *Rumex crispus* L. Extracts. *J. Agric. Food Chem.* **2001**, *49*, 4083–4089. [CrossRef] [PubMed]
12. Durmaz, G.; Alpaslan, M. Antioxidant properties of roasted apricot (*Prunus armeniaca* L.) kernel. *Food Chem.* **2007**, *100*, 1177–1181. [CrossRef]
13. Mertens, D. AOAC official method 922.02. Plants preparation of laboratuary sample. In *Official Methods of Analysis*, 18th ed.; Horwitz, W., Latimer, G.W., Eds.; AOAC Intl.: Gaitherburg, MA, USA, 2005; Chapter 3, pp. 1–2.

14. Mertens, D. AOAC official method 975.03. Metal in plants and pet foods. In *Official Methods of Analysis*, 18th ed.; Horwitz, W., Latimer, G.W., Eds.; Chapter 3; AOAC Intl.: Gaitherburg, MA, USA, 2005; pp. 3–4.
15. Doguer, C.; Yıkmış, S.; Levent, O.; Turkol, M. Anticancer Effects of Enrichment in the Bioactive Components of the Functional Beverage of Turkish Gastronomy by Supplementation with Purple Basil (*Ocimum basilicum* L.) and the Ultrasound Treatment. *J. Food Process. Preserv.* **2021**, *45*, e15436. [CrossRef]
16. Gradziel, T.M. *Almond Quality: A Breeding Perspective Horticultural Reviews*; Janick, J., Ed.; John Wiley & Sons, Inc.: Zarogaza, Spain, 2008; Volume 34.
17. García-López, C.; Grané-Teruel, N.; Berenguer-Navarro, V.; García-García, J.E.; Martín-Carratalá, M.L. Major Fatty Acid Composition of 19 Almond Cultivars of Different Origins. A Chemometric Approach. *J. Agric. Food Chem.* **1996**, *44*, 1751–1755. [CrossRef]
18. Ruggeri, S.; Cappelloni, M.; Gambelli, L.; Carnovale, E. Chemical Composition and Nutritive Value of Nuts Grown in Italy. *Ital. J. Food Sci.* **1998**, *10*, 243–252.
19. Nizamlioğlu, N.M. Kavurma ve Depolama Koşullarının Bademin Bazı Fiziksel, Kimyasal ve Duyusal Özellikleri Üzerine Etkisi. Ph.D. Thesis, Pamukkale University, Pamukkale, Turkey, 2015.
20. Milbury, P.E.; Chen, C.-Y.; Dolnikowski, A.G.G.; Blumberg, J.B. Determination of Flavonoids and Phenolics and Their Distribution in Almonds. *J. Agric. Food Chem.* **2006**, *54*, 5027–5033. [CrossRef]
21. Esfahlan, A.J.; Jamei, R.; Esfahlan, R.J. The importance of almond (*Prunus amygdalus* L.) and its by-products. *Food Chem.* **2010**, *120*, 349–360. [CrossRef]
22. Pinelo, M.; Rubilar, M.; Sineiro, J.; Núñez, M. Extraction of antioxidant phenolics from almond hulls (*Prunus amygdalus*) and pine sawdust (Pinus pinaster). *Food Chem.* **2004**, *85*, 267–273. [CrossRef]
23. Beltrán Sanahuja, A.; Maestre Pérez, S.E.; Grané Teruel, N.; Valdés García, A.; Prats Moya, M.S. Variability of Chemical Profile in Almonds (*Prunus dulcis*) of Different Cultivars and Origins. *Foods* **2021**, *10*, 153. [CrossRef]
24. Zhu, Y.; Taylor, C.; Sommer, K.; Wilkinson, K.; Wirthensohn, M. Influence of deficit irrigation strategies on fatty acid and tocopherol concentration of almond (*Prunus dulcis*). *Food Chem.* **2015**, *173*, 821–826. [CrossRef]
25. Celik, F.; Balta, M.F. Kernel Fatty Acid Composition of Turkish Almond (*Prunus dulcis* L.) Gen-otypes: A Regional Comparison. *J. Food Agric. Environ.* **2011**, *9*, 171–174.
26. Simşek, M.; Kizmaz, V. Determination of Chemical and Mineral Compositions of Promising Almond (*Prunus amygdalus* L.) Genotypes from Beyazsu (Mardin) Region. *Int. J. Agric. Wild Sci.* **2017**, *3*, 6–11.
27. Roncero, J.M.; Álvarez-Ortí, M.; Pardo-Giménez, A.; Rabadán, A.; Pardo, J.E. Review about Non-Lipid Components and Minor Fat-Soluble Bioactive Compounds of Almond Kernel. *Foods* **2020**, *9*, 1646. [CrossRef] [PubMed]
28. Piscopo, A.; Romeo, F.; Petrovicova, B.; Poiana, M. Effect of the harvest time on kernel quality of several almond varieties (*Prunus dulcis* (Mill.) D.A. Webb). *Sci. Hortic.* **2010**, *125*, 41–46. [CrossRef]
29. Drogoudi, P.D.; Pantelidis, G.; Bacchetta, L.; De Giorgio, D.; Duval, H.; Metzidakis, I.; Spera, D. Protein and Mineral Nutrient Contents in Kernels from 72 Sweet Almond Cultivars and Acces-sions Grown in France, Greece and Italy. *Int. J. Food Sci. Nutr.* **2013**, *64*, 202–209. [CrossRef]
30. Simsek, M.; Gulsoy, E.; Yavic, A.; Arikan, B.; Yildirim, Y.; Olmez, N.; Erdogmus, B.; Boguc, F. Fatty acid, Mineral and Proximate Compositions of Various Genotypes and Commercial Cultivars of Sweet Almond from the Same Ecological Conditions. *Appl. Ecol. Environ. Res.* **2018**, *16*, 2957–2971. [CrossRef]
31. Xiao, L.; Lee, J.; Zhang, G.; Ebeler, S.E.; Wickramasinghe, N.; Seiber, J.; Mitchell, A.E. HS-SPME GC/MS characterization of volatiles in raw and dry-roasted almonds (*Prunus dulcis*). *Food Chem.* **2013**, *151*, 31–39. [CrossRef]
32. Mexis, S.F.; Badeka, A.V.; Chouliara, E.; Riganakos, K.A.; Kontominas, M.G. Effect of C-Irradiation on the Physicochemical and Sensory Properties of Raw Unpeeled Almond Kernels (*Prunus dulcis*). *Innov. Food Sci. Emerg. Technol.* **2009**, *10*, 87–92. [CrossRef]
33. Erten, E.S.; Cadwallader, K.R. Identification of predominant aroma components of raw, dry roasted and oil roasted almonds. *Food Chem.* **2017**, *217*, 244–253. [CrossRef]
34. Agila, A.; Barringer, S. Effect of Roasting Conditions on Color and Volatile Profile Including HMF Level in Sweet Almonds (*Prunus dulcis*). *J. Food Sci.* **2012**, *77*, C461–C468. [CrossRef]
35. Beck, J.J.; Mahoney, N.E.; Cook, D.; Gee, W.S. Volatile analysis of ground almonds contami-nated with naturally occurring fungi. *J. Agric. Food Chem.* **2011**, *59*, 6180–6187. [CrossRef] [PubMed]
36. Valdés, A.; Beltrán, A.; Karabagias, I.; Badeka, A.; Kontominas, M.G.; Garrigós, M.C. Monitoring the oxidative stability and volatiles in blanched, roasted and fried almonds under normal and accelerated storage conditions by DSC, thermogravimetric analysis and ATR-FTIR. *Eur. J. Lipid Sci. Technol.* **2015**, *117*, 1199–1213. [CrossRef]

horticulturae

MDPI

Article

Seasonal Variations in the Starch Properties of Sweet Potato Cultivars

Thaís Paes Rodrigues dos Santos [1], Magali Leonel [1,*], Luciana Alves de Oliveira [2], Adalton Mazetti Fernandes [1], Sarita Leonel [1,3] and Jason Geter da Silva Nunes [1]

1 Center for Tropical Roots and Starches (CERAT), São Paulo State University (UNESP), Botucatu 18610307, Brazil
2 Embrapa Cassava and Fruits, Cruz das Almas 44380000, Brazil
3 School of Agriculture (FCA), São Paulo State University (UNESP), Botucatu 18610307, Brazil
* Correspondence: magali.leonel@unesp.br; Tel.: +55-14-3880-7612

Abstract: Starch is widely used in the food and non-food industries, and this is related to its physicochemical characteristics. In the coming years, climate changes will become unpredictable, and these conditions may affect the process of starch biosynthesis and polymer properties. The sweet potato starch market has grown substantially in recent years and understanding the environmental impacts on starch characteristics will contribute to advances for the sector. Herein, the effects of the growing season on the structural, morphological, and physicochemical properties of sweet potato starches were evaluated. Sweet potato trials with two Brazilian cultivars (Canadense and Uruguaiana) were installed in the dry season (planting in March and harvesting in July) and rainy season (planting in October and harvesting in March). Regardless of the cultivar, starches isolated from plants grown in the rainy season have a more ordered structure, with higher gelatinization temperatures, thermal stability, and resistant starch content. Starches from plants grown in the dry season have a higher percentage of small granules with lower crystallinity and lower gelatinization temperatures. These findings can be useful as early knowledge of these changes can help the supply chain to better plan and target suitable markets for naturally modified sweet potato starches.

Keywords: *Ipomoea batatas* (L.) Lam; growing season; native starch; physicochemical properties

Citation: dos Santos, T.P.R.; Leonel, M.; de Oliveira, L.A.; Fernandes, A.M.; Leonel, S.; da Silva Nunes, J.G. Seasonal Variations in the Starch Properties of Sweet Potato Cultivars. *Horticulturae* **2023**, *9*, 303. https://doi.org/10.3390/horticulturae9030303

Academic Editors: Alessandra Durazzo and Massimo Lucarini

Received: 21 January 2023
Revised: 16 February 2023
Accepted: 17 February 2023
Published: 23 February 2023

1. Introduction

Sweet potato is an important crop for global food security and has been considered an emerging source of starch. In 2021, the world produced about 88.87 million tons (Mt) of sweet potato roots, with Asia standing out as the largest producer (61.45%), followed by Africa (33.72%) and the Americas (3.78%), with Brazil producing 824,680 tons [1–4].

Starch accounts for 17.5% of the fresh mass and 40–50% of the dry mass in sweet potato roots [5,6]. This polymer is used in a wide range of foods for various purposes, including thickening, gelling, adding stability, and replacing or extending the more expensive ingredients [7].

In the last decade, studies have reported an increase in the use of sweet potato starch in the food industry. The global sweet potato starch market is expected to grow by 3.9% over the next years, from USD 560 million in 2019 to USD 710 million in 2024 [8,9].

The sweet potato growth cycle (4 to 6 months) is characterized by an initial phase (adventitious root growth), an intermediate phase (root tuberization) and a final phase (accumulation of photo assimilates in the tuberous roots). Although sweet potato is a moderately drought tolerant crop, water stress affects plant development by limiting photosynthetic activity, and affecting storage root development, volume, and yield [10,11]. Favorable environmental conditions can lead to an early onset of tuberization, prolonging the period of reserve accumulation in the roots, with increased starch accumulation rate and improved productivity [5]. The process of starch biosynthesis in plants involves isoforms of several enzymes. In addition to genotypic interference and climate variations from year to year, changes in planting or growing seasons at various locations can have a strong influence on the action of synthesis enzymes and thus on starch functionality [12–14].

Due to the different applicability requirements of starch, chemical modification has been applied. However, currently a greater number of consumers are concerned about their health and have avoided the consumption of products that contain modified starch as an additive on their labels, which has increased the demand for natural starches, which are considered as ingredients [15–18].

The effects of climate change on starch synthesis in plants, structure and functional properties of starches are still poorly explored. In this study, starches isolated from sweet potato plants grown in the dry and rainy season were analyzed for morphological, structural, and physicochemical properties. The results will help broaden the understanding of the impacts of climatic conditions on starch properties, encouraging farmer and industry integration to add value to naturally modified sweet potato starches.

2. Materials and Methods

2.1. Cultivars, Experimental Area, and Treatments

In this study, the cultivars Canandese and Uruguaina were evaluated. These are the main cultivars planted in the state of São Paulo, Brazil.

Experimental trials were installed at São Manuel Experimental Farm of the São Paulo State University, São Manuel city, SP, Brazil. (22°44′28″ S, 48°34′37″ W, at an altitude of 740 m a.s.l). The region has a Cwa climate (tropical with a dry winter and a hot, rainy summer) according to the Köppen classification system. The soil in the experimental area was classified as a sandy textured dystroferric Red Latosol, corresponding to a dystrophic Typic Hapludox. Prior to the installation of the experiments, soil samples were collected at a depth of 0–20 cm and the chemical characteristics of the soils were determined: pH in $CaCl_2$, 4.8; Soil organic matter, 13 g dm^{-3}; P_{resin}, 12 mg dm^{-3}; H + Al, 26 mmol dm^{-3}; K, 2.9 mmol dm^{-3}; Ca, 11 mmol dm^{-3}; Mg, 4 mmol dm^{-3}; ECEC, 43 mmol dm^{-3}.

The first planting period was in March and the harvest was in July 2018 (dry season—DS), and the second planting period was in October 2018 and the harvest was in March 2019 (rainy season—RS). Climatic variables were monitored and presented in Figure 1.

At planting, branches of healthy plants with a length of 40 cm containing about eight internodes were selected. The branches were planted at a depth of 10 to 12 cm. The experimental plots consisted of six lines of 5 m in length (0.80 m spacing between lines and 0.3 m between plants). The four central lines were considered as the useful area of the plot, disregarding 0.5 m at the ends of each line. Soil preparation, fertilization for planting and coverage, and phytosanitary management were carried out in accordance with the recommendations for the crop [19] (Figure 2). The plants were harvested at 165 days after planting.

2.2. Starch Isolation

The sweet potato roots were washed, peeled, and disintegrated in an industrial stainless-steel blender. The suspension passed through 80-mesh (0.177 mm) and 150-mesh (0.105 mm) aperture sieves to rinse the fibrous residue. The residue retained on the 80-mesh sieve passed again through the same procedure to rinse the residual starch. The recovered starch suspension was mixed into the first suspension and kept in a cold chamber for decantation at 5 °C for 12 h. Afterwards, the recovered starch was siphoned, rinsed in distilled water, recovered by centrifugation, and dried in an oven with air circulation for 12 h at 40 °C [20].

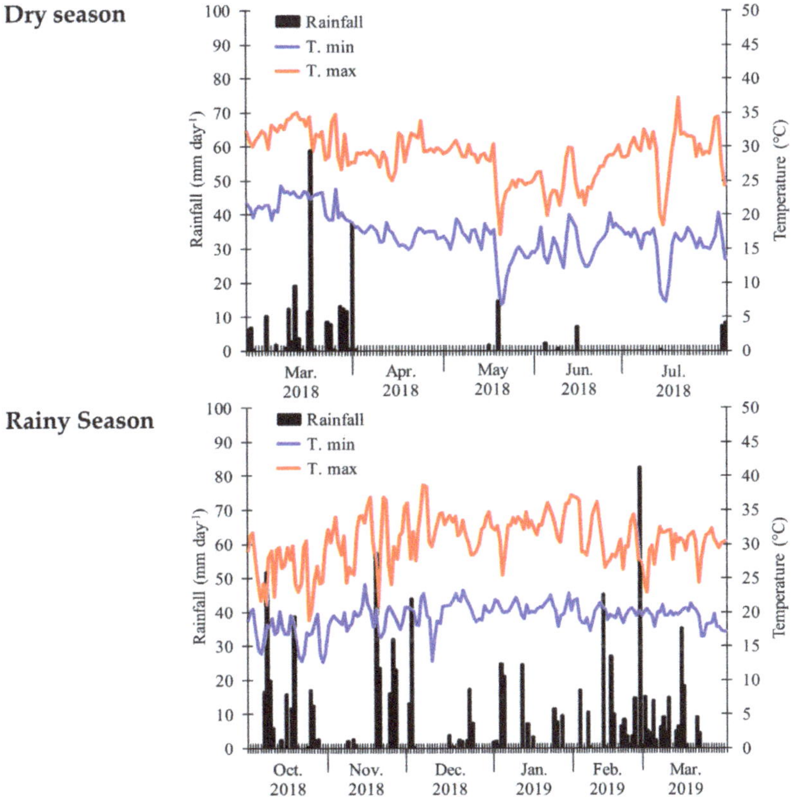

Figure 1. Rainfall, maximum and minimum air temperatures recorded in experimental area.

Figure 2. Images of the experimental area and roots of sweet potato cultivars.

2.3. Starch Analysis

2.3.1. Morphology and Granule Size

The starch granule morphologies were evaluated using scanning electron microscopy (model EVO LS15, Carl Zeiss, Oberkochen, Germany). Starch samples were applied to an aluminum stub with double-sided tape and covered with a thin gold layer (20 nm) in a metallizer for 220 s (Sputter Coater SCD 050- Balzers). The images were obtained using 2000× magnification in high vacuum (10^{-3} Pa) and recorded through the Finepix digital camera and smart SEM software (Carl Zeiss, Oberkochen, Germany).

Starch granule size and size distribution were determined through laser diffraction analysis by using a Helium–Neon laser (Mastersizer 2000, Laser Scattering Spectrometer, Model MAM 5005-Instruments Ltd., Worcestershire, UK). Starch samples were dispersed in distilled water until an obscuration of 5.5% was reached. The refractive indexes of starch samples and solvent were 1.500 and 1.360, respectively. Surface-weighted diameter (D[3,2]), volume-weighted diameter (D[4,3]), median particle size (D[0.5]), and size distribution of the particles were obtained and chosen as granule size through the manufacturer's software (Malvern Application version 5.6, Malvern Instruments Ltd., Worcestershire, UK) [20].

2.3.2. X-ray Diffraction Pattern (XRD) and Relative Crystallinity (RC)

Starch samples were incubated in a desiccator containing saturated $BaCl_2$ solution (25 °C, aw = 0.9) for 10 days to reach humidity equilibrium (90%) and improve the quality of the diffraction diagram. X-ray patterns were examined using the goniometer system Rigaku MiniFlex 600 powder X-ray diffractometer (Cu Kα radiation, λ = 0.1542 nm) (Rigaku, Tokyo, Japan). The scanning speed was 5 °θ min^{-1} and the irradiation was performed at 40 kV and 15 mA. The relative crystallinity was calculated based on the relation between the peak and the total area of the diffractogram [21].

2.3.3. Amylose and Resistant Starch

The amylose content of the starch was determined using the method described by Williams et al. [22]. A starch sample (20 mg) was taken, and 10 mL of 0.5 N KOH was added to it. The suspension was thoroughly mixed. The dispersed sample was transferred to a 100 mL volumetric flask and diluted to the mark with distilled water. An aliquot of this solution (10 mL) was pipetted into a 50 mL volumetric flask and 5 mL of 0.1 N HCl was added followed by 0.5 mL of iodine reagent. The volume was diluted to 50 mL and the absorbance was measured at 625 nm.

Resistant starch content was determined according to Goñi et al. [23]. The samples were subjected to: incubation (40 °C, 60 min, pH 1.5) with pepsin (0.1 mL (10 mg/mL), Sigma P-7012) for protein removal; incubation (37 °C, 16 h, pH 6.9) with α-amylase (1 mL (40 mg/mL), Sigma A-3176) to hydrolyze digestible starch; residue treatment with 2 M KOH for solubilization of resistant starch; incubation (60 °C, 45 min, pH 4.75) with amyloglucosidase (80 mL (140 U/mL), Sigma A-7255) to hydrolyze the resistant starch solubilized; and the glucose contents in the mixture were measured using glucose oxidase and peroxidase assay kits (GAGO-20, Sigma–Aldrich Company, Saint Louis, MO, USA).

2.3.4. Swelling Power (SP) and Solubility (SS)

The starch samples (0.2 g, wet basis) were placed in tubes and 20 g of distilled water was added based on the initial moisture content. The suspension tubes were immersed in a water bath under constant agitation for 30 min at 95 °C. All tubes were covered with plastic to prevent water loss. Each sample was then centrifuged at 2000× *g* for 15 min; an aliquot (mL) of the supernatant was then collected and left to dry in an oven at 105 °C until constant weight was reached (Ws). The precipitated paste was separated from supernatant and weighed (Wp) [24].

$$SS\ (\%) = [Ws/\ \text{sample weight (dry basis)}] \times 100$$
$$SP\ (g\ g^{-1}) = [Wp \times 100]\ /\ [\text{sample weight (dry basis)} \times (100 - \%\ S)]$$

2.3.5. Pasting and Thermal Properties

The pasting properties of sweet potato starches were analyzed using a Rapid Visco Analyzer (RVA), RVA-4500, Newport Scientific Pty. Ltd., Warriewood, Australia), using Thermocline for Windows, version 3.0. For the analysis, 3 g of each sample were weighed according to their respective moistures, adding approximately 25 g of water to reach a concentration of 10% starch, and were placed in the sample holder of the equipment. For approximately 10 s, the mixture was stirred at 960 rpm (160 rpm during the test). The temperature program used was STD 1. The samples were held at 50 °C for 1 min, followed by heating from 50 °C to 95 °C at a rate of 6 °C minutes^{-1}; holding at 95 °C for 5 min, and cooling at 50, at 6 °C min^{-1}. The equipment generated viscosity in Rapid Visco Units (RVU), where one unit is equivalent to 12 cP [25].

Thermal properties were evaluated using a differential scanning calorimeter (Perkin Elmer DSC-8500, Norwalk, CT, USA). Starch samples (2.0 mg, dry starch) were mixed with distilled water (6.0 μL) in aluminum pans. The pans were sealed and kept for 2 h at room temperature until balanced; an empty sealed pan was used as reference. The scanning temperature range was 25 °C to 100 °C and the heating rate was 10 °C min^{-1}. The equipment was calibrated with indium. The thermal parameters including the temperature of the onset (To), peak (Tp), conclusion (Tc), and the enthalpy change (ΔH) were obtained using the software Pyris 1 (Perkin Elmer, EUA) [20].

2.4. Data Analysis

Analysis of variance (ANOVA) was performed with a significance level of 5% and differences between means were determined by Tukey's test, using the Sisvar program (Lavras, MG, Brazil). All measurements were performed in quadruplicate and data are presented as mean and standard deviation.

3. Results and Discussion

3.1. Morphology and Granule Size

Starch granule morphology is related to amyloplast biochemistry and plant source. For both cultivars grown in dry season (DS), the microscopic analysis showed granules with smooth surfaces, concave and convex polygonal shapes with curved sides with some depression points and a number of small granules (Figures 3 and 4). In the rainy season (RS), the shape of the starch granules was similar to that of the DS, but with a predominance of the rounded shape and a reduced number of small granules (Figure 3). Some authors have already observed similar shapes of sweet potato starch granules [8,26–28].

Starches isolated from plants grown in the rainy season were larger, indicating the interference of water availability on root tuberization and distribution of granule sizes (Table 1). Teerawanichpan et al. [29] compared cassava plants grown in different climates and observed that starch granules were larger in plants grown under higher rainfall than in plants under dry season. In the dry season, there are fewer hours of light; nighttime temperatures are low, in addition to low precipitation, interfering with the rate of photosynthesis and root tuberization. Consequently, there is a higher percentage of granules that are not fully formed.

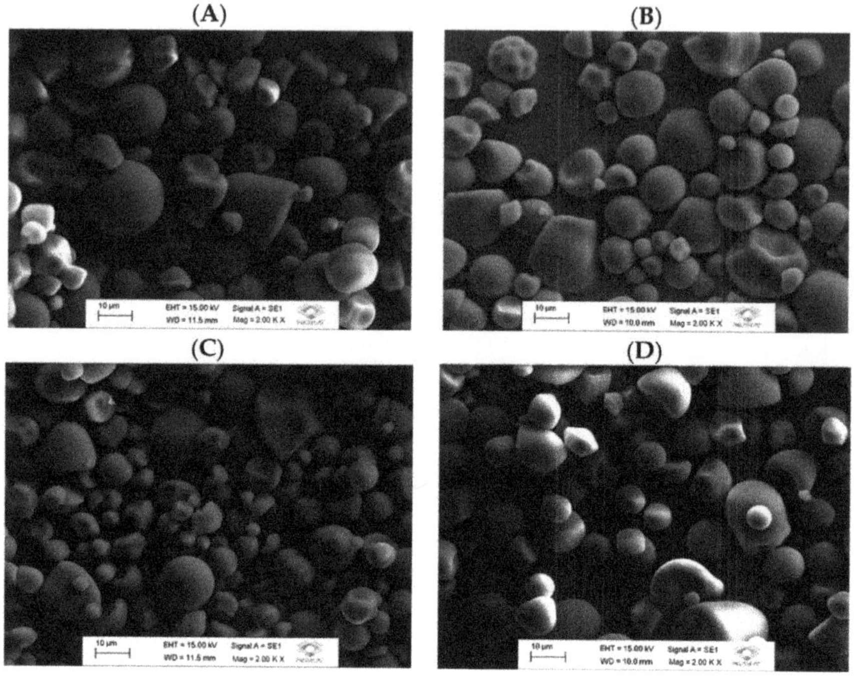

Figure 3. Microphotograph of starch granules from sweet potato cultivars. (**A**) Starch from cultivar Canadense (Dry season); (**B**) Starch from cultivar Canadense (Rainy season); (**C**) Starch from cultivar Uruguaiana (Dry season); (**D**) Starch from cultivar Uruguaiana (Rainy season).

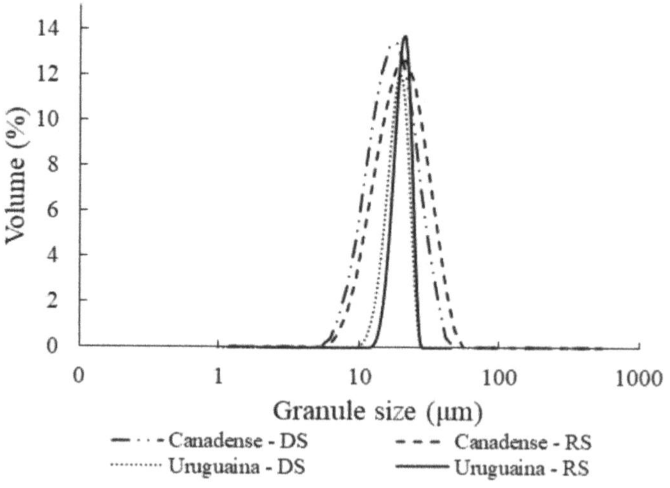

Figure 4. Granule size distribution of sweet potato starches. DS = dry season; RS = rainy season.

Table 1. Effect of growing season on sweet potato starches in terms of particle size parameters, amylose content, relative crystallinity, resistant starch, swelling power and solubility, and analysis of variance.

	Canadense		Uruguaiana		ANOVA Source of Variation		
	Dry Season	Rainy Season	Dry Season	Rainy Season	C	GS	CxGS
D[4,3] (μm)	19.29 ± 0.12 bA	22.61 ± 0.02 aA	14.68 ± 0.24 bB	17.14 ± 0.22 aB	***	***	**
D[3,2] (μm)	16.72 ± 0.14 bA	19.28 ± 0.02 aA	12.23 ± 0.11 bB	14.93 ± 0.19 aB	***	***	ns
D(0.5) (μm)	18.11 ± 0.12 bA	21.13 ± 0.02 aA	13.49 ± 0.16 bB	16.12 ± 0.20 aB	***	***	*
Relative crystalinity (%)	25.67 ± 0.30 bB	29.92 ± 0.77 aB	26.96 ± 0.32 bA	31.13 ± 0.40 aA	**	***	ns
Amylose (%)	25.49 ± 5.6 aB	25.35 ± 1.1 aB	26.29 ± 2.0 aA	25.61 ± 1.3 bA	*	ns	*
Resistant Starch (%)	59.16 ± 9.7 bA	64.34 ± 6.2 aA	54.88 ± 9.8 bB	69.22 ± 2.2 aB	***	***	ns
Swelling Power (g g^{-1})	40.09 ± 1.26 aA	27.18 ± 0.15 bB	39.07 ± 1.31 aA	30.56 ± 0.13 bA	*	***	**
Solubility (%)	33.38 ± 0.30 aA	18.48 ± 0.05 bB	32.15 ± 1.37 aA	20.39 ± 0.33 bA	**	***	ns

The same lower-case letter in line, in each cultivar, indicates that the results do not differ statistically between growing season; and the same upper-case letter in line, in each season, indicates that the results do not differ statistically between cultivars using the Tukey's HSD test ($p < 0.05$). D[3,2], surface-weight diameter; D[4,3], volume-weighted diameter; D(0.5), median particle size. The means are based on four repetitions. C = cultivar; GS = growing season. ns = Non-significant at the 0.05 probability level; * = Significant at the 0.05 probability level; ** = Significant at the 0.01 probability level; *** = Significant at the 0.001 probability level.

Regardless of the growing season, starches isolated from 'Canadense' had a higher average size, with a wider size distribution. Starches from 'Uruguaiana' had a uniform distribution of granule sizes. These differences may be related to the root system of the cultivar, which interferes with nutrient uptake, storage root formation and tuberization, and adaptation to different climatic conditions, interfering with starch biosynthesis and granule size.

The size of sweet potato starch granules is quite variable. Guo et al. [30] analyzing the granule size distribution of starches extracted from sweet potatoes with white, yellow, and purple pulp observed that the granule sizes ranged from 12.33 to 18.09 μm. Wang et al. [31] after analyzing starches obtained from colored sweet potato varieties observed that of the eight varieties analyzed, three starches showed a bimodal size distribution, with small granules (1–4 μm) and large granules (5–84 μm). The other starches showed unimodal distribution with a granule size ranging from 4.5 to 84 μm. The average size of the starch granules ranged from 16.10 μm to 23.94 μm.

The relationship between the size of granules and the applicability of starches is important. Smaller granules have been valued in edible products such as sauces and dairy desserts, which require a soft mouth feel. They can also be used as fat substitutes due to their similar size to lipid mycelia. Other applications where granule size is important are the production of biodegradable plastic films, paper coatings, and cosmetic products.

3.2. X-ray Diffraction Pattern and Granule Size

Starch granules have crystal structures with specific X-ray diffraction patterns, called A, B and C, due to the packing of the double helices of amylopectin. The crystal structure of sweet potato starch is variable and may present patterns of types A, C or C_A [31–33]. Starch isolated from 'Canadense' grown in rainy season showed an A-type diffraction pattern with the most intense peaks at 15, 17, 18, and 23° 2θ. The same cultivar presented a C_A-type in dry season with the most intense peaks at 15, 17, and 23° 2θ and a shoulder peak at approximately 18° 2θ, that is, an indicative of great similarity to A-type polymorph. Starch from 'Uruguaiana' had a C_A-type diffraction pattern in both seasons (Figure 5).

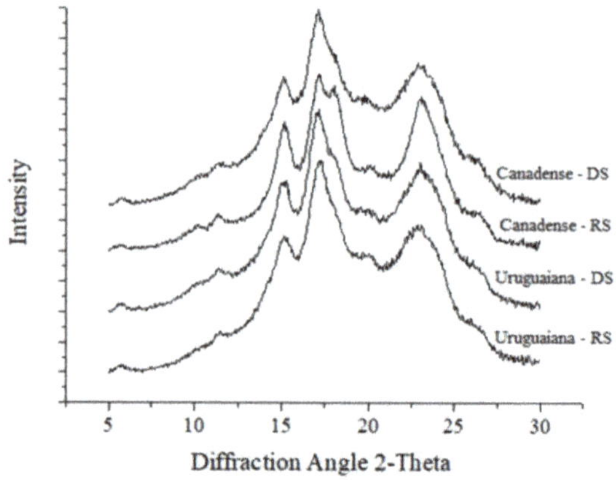

Figure 5. X-ray diffraction patterns of sweet potato starches. DS = dry season; RS = rainy season.

Genkina et al. [34] reported the impact of sweet potato crop soil temperature on starch properties. The starches isolated from plants grown at 33 °C had an A-type polymorph, while the isolate from plants grown at 15 °C had a C-type polymorph. These results contribute to the understanding of the pattern variation observed in the starches of the 'Canadense', since in the rainy season daytime and nighttime temperatures are higher than in the dry season, which influences the soil temperature.

Starch is a semi-crystalline material, and the degree and type of crystallinity depends mainly on the structural characteristics of amylopectin. Starches with A-type have a high proportion of short chains in amylopectin. Starches with B and C-type polymorph amylopectin are highly branched, forming long chains linked to amylose molecules. Starches from 'Uruguaiana' showed higher relative crystallinity than those from 'Canadense'. The crystallinities of starches isolated from plants grown in the rainy season were about 4% higher than those of starches from plants grown in the dry season (Table 1). Starches from plants grown in the dry season showed percentages of crystallinity within the ranges reported in other studies [27,28]. The changes resulting from the growing seasons may be related to the actions of starch synthesizing enzymes [28]. Starch crystallinity affects the physical, mechanical, and technological properties of various starchy products, and is therefore important for product development, quality, and process control [25].

3.3. Amylose and Resistant Starch

The amylose content and the characteristics of the particles and microstructure of the granules determine whether the starch can be used as a stabilizer, gelling agent or thickener in industries. The amylose content of sweet potato starch, according to several studies ranges from 15.3 to 28.8% [20,26,28,33,35–37].

The amylose content of sweet potato starches varied between cultivars in the dry season, with the highest content observed for starch from 'Uruguaiana' (Table 1). Teerawanichpan et al. [29] observed that the change in amylose content with the growing season was specifically related to the cultivar.

Starches with a high content of amylose are capable of forming inclusion complexes with food ingredients such as essential oils, fatty acids and flavoring molecules, acting as an encapsulant that contributes to an increase in the shelf life of products. In addition, high-amylose starches have interesting nutritional properties, since high amylose is linked to high levels of resistant starch in processed starchy foods [38]. Starches isolated from plants of the same cultivar grown in the rainy season showed higher levels of resistant starch. The same response was observed for starch granule sizes showing that larger granules were more resistant to hydrolysis ($r = 0.95$, $p < 0.001$). Furthermore, it was possible to verify that starches with higher crystallinity obtained in the rainy season also had a higher resistant starch content, with a positive correlation between the two characteristics (r = 0.73, $p < 0.01$). Starches isolated from 'Canadense' showed higher levels of resistant starch than those isolated from 'Uruguaiana', which improved its functional properties.

Higher levels of resistant starch can increase the commercial value of natural sweet potato starch. Research indicates that the resistant starch market was valued at USD 10.5 billion in 2022 and is expected to reach USD 19.9 billion by 2032 (CAGR of 6.6%). This market has been driven by growing awareness of the health benefits of resistant starch and increasing use of resistant starch in bakery products, confectionery products, dairy products, breakfast cereals, beverages, among others [39].

3.4. Swelling Power (SP) and Solubility (SS)

In gelatinization, the starch structure breaks down, leading to the weakening of hydrogen bonds and the interaction of water molecules with the hydroxyl groups of amylose and amylopectin, causing swelling and partial solubilization of the starch.

Starches isolated from plants grown in the dry season had higher SP and SS than those isolated from plants grown in the rainy season. Starch from the 'Uruguaiana' had higher SP and SS in the rainy season (Table 1). The values observed in this study for starches isolated in the dry season were close to those observed in other studies conducted in Brazil with sweet potato starches [27,40], and those obtained in the rainy season were similar to those observed in studies conducted in China [30,37].

The effects of the dry season on SP and SS are consistent with the smallest granule size (r = −0.99, $p = 0.01$), higher amylose (r = 0.79, $p < 0.05$), and lower resistant starch (r = −0.81, $p < 0.05$) observed in starches isolated from plants grown in these climatic conditions. The relationship between these physicochemical parameters, and the water absorption capacity and solubilization of starches has already been reported by Guo et al. [30].

3.5. Pasting and Thermal Properties

Data analysis of pasting properties of sweet potato starches showed a greater effect of the growing seasons on the viscosity parameters (Table 2, Figure 6). Starches from 'Canadense' have higher viscosity peaks and breakdown and starches from 'Uruguaina' have higher retrogradation tendencies.

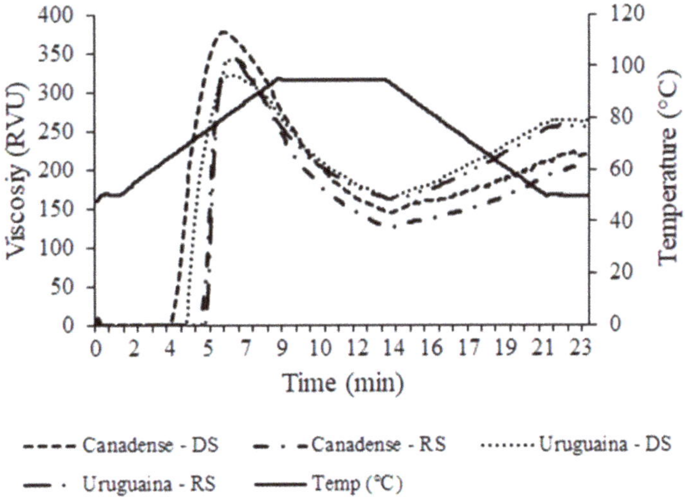

Figure 6. Rapid Visco Analyzer (RVA) profiles of sweet potato starches. DS = dry season; RS = rainy season.

Table 2. Effect of growing season on sweet potato starches in terms of paste properties, and analysis of variance.

	Canadense		Uruguaiana		ANOVA Source of Variation		
	Dry Season	Rainy Season	Dry Season	Rainy Season	C	GS	CxGS
Peak Viscosity (RVU)	382.08 ± 13.92 aA	315.42 ± 0.34 bB	338.47 ± 5.10 aB	324.25 ± 2.22 bA	**	**	***
Breakdown (RVU)	234.22 ± 10.24 aA	193.63 ± 1.54 bA	177.55 ± 3.30 aB	156.97 ± 5.17 bB	***	**	***
Final Viscosity (RVU)	222.39 ± 9.94 aB	198.71 ± 2.13 bB	251.08 ± 4.63 bA	271.92 ± 8.81 aA	***	**	ns
Setback (RVU)	74.53 ± 1.08 bB	76.92 ± 0.25 aB	90.17 ± 1.28 bA	107.75 ± 0.82 aA	***	*	**
Pasting Temperature (°C)	66.23 ± 1.18 bB	75.63 ± 0.03 aA	69.57 ± 0.45 bA	75.38 ± 0.42 aA	*	***	**

The same lower-case letter in line, in each cultivar, indicates that the results do not differ statistically between growing seasons; and the same upper-case letter in line, in each season, indicates that the results do not differ statistically between cultivars using the Tukey's HSD test ($p < 0.05$). Rapid Visco Unit (1 RVU = 12 cP). The means are based on four repetitions. C = cultivar; GS = growing season. ns = Non-significant at the 0.05 probability level; * = Significant at the 0.05 probability level; ** = Significant at the 0.01 probability level; *** = Significant at the 0.001 probability level.

The resistance to mechanical action during the period of constant temperature maintenance in the RVA analysis indicates that the starch has strong intermolecular bonds. Granules with low swelling power are more resistant to prolonged heating and/or mechanical agitation, therefore, less susceptible to granule rupture, which is related to viscosity stability. Starches isolated from plants grown in the rainy season had higher crystallinity, lower swelling power and lower viscosity breakdown. This type of starch is preferred as a thickener in foods that require a long heat treatment time under agitation, for example, in processes that involve treatment with high temperatures and pressure, such as autoclaving, and manufacture of soups and canned products [41].

All starches showed an increase in viscosity on cooling, with the highest setback values observed for starches isolated from the 'Uruguaiana' in the two growing seasons of the plants. Regardless of the cultivar, the setback was positively correlated with the amylose content in the two growing seasons (Dry season, $r = 0.73$, $p < 0.05$; Rainy season, $r = 0.88$, $p < 0.05$). Setback viscosity is an indirect measure of starch retrogradation tendency. Retrogradation is the process of crystallization of starch chains, particularly amylose molecules, which occurs after the gelatinized starch paste has cooled, forming a cohesive three-dimensional network. The higher setback indicates lower stability of the starch paste in cold and this parameter allows the estimation of the stability of the starch gel during storage at low temperatures, considering that starches with lower tendencies to retrogradation are more desirable by the food industry.

The gelatinization process depends mainly on the dissociation of the helical structure within the starch chains, and the energy required to dissociate this structure varies with different starch sources. The results observed in the analysis of the thermal properties showed that both cultivars had higher gelatinization temperatures and enthalpy range in the rainy season (Table 3, Figure 7). Starches isolated from 'Uruguaiana' had higher gelatinization temperatures and enthalpy range in the dry season, differing in all parameters from those isolated from 'Canadense'. In the rainy season, the cultivars differed in the initial and peak temperatures.

Table 3. Effect of growing seasons on sweet potato starches in terms of thermal properties, and analysis of variance.

	Canadense		Uruguaiana		ANOVA Source of Variation		
	Dry Season	Rainy Season	Dry Season	Rainy Season	C	GS	CxGS
T_{onset} (°C)	57.47 ± 0.19 bB	70.32 ± 0.20 aB	58.10 ± 0.17 bA	71.72 ± 0.05 aA	***	***	**
T_{peak} (°C)	62.07 ± 0.07 bB	74.69 ± 0.20 aB	63.30 ± 0.19 bA	75.31 ± 0.14 aA	***	***	**
$T_{conclusion}$ (°C)	68.23 ± 0.09 bB	79.28 ± 0.10 aA	70.50 ± 0.41 bA	79.34 ± 0.23 aA	***	***	***
$T_{conclusion}$-T_{onset} (°C)	10.77 ± 0.27 aB	8.96 ± 0.15 bA	12.40 ± 0.41 aA	7.62 ± 0.19 bB	***	***	***
ΔH (J g^{-1})	12.51 ± 0.29 bB	14.84 ± 0.19 aA	13.94 ± 0.26 bA	14.85 ± 0.30 aA	*	***	**

T, temperature; $T_{conclusion}$-T_{onset}, temperature range (ΔT); ΔH, enthalpy change. C = cultivar; GS = growing season. The means are based on four repetitions. ns = Non-significant at the 0.05 probability level; * = Significant at the 0.05 probability level; ** = Significant at the 0.01 probability level; *** = Significant at the 0.001 probability level.

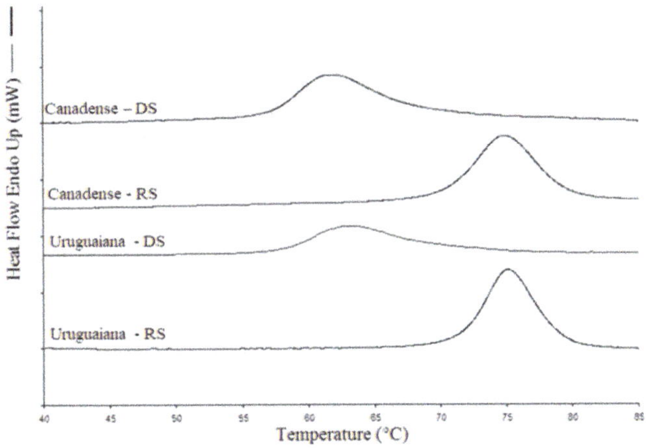

Figure 7. Differential scanning calorimetry (DSC) curves of sweet potato starches. DS = dry season; RS = rainy season.

Tsakama et al. [42] analyzed the pasting properties of eleven sweet potato genotypes and observed that the pasting temperature ranged from 73.4 to 75.88 °C, the peak viscosity ranged from 1947 to 2596 cP, the hot viscosity ranged from 1496 to 2049 cP, the breakdown ranged from 221 to 889 cP, the cold viscosity ranged from 2304 to 2762 and the setback ranged from 1.51 to 1.71 cP.

Gelatinization temperature provides a measure of granule crystallinity [43], and in this study, differences of approximately 12 °C in gelatinization temperatures were observed between the growing seasons for the two cultivars, indicating a possible presence of a larger area of crystallinity in starches isolated from rainy season plants in relation to the dry season. The higher gelatinization temperatures and enthalpy range can be explained by the higher relative crystallinity of the starch granules, which provides a higher structural stability [44].

Campanha and Franco [45] reported gelatinization temperatures of sweet potato starch ranging from 62.9 to 77.9 °C, with the peak at 70.6 °C, and an enthalpy range of 12.9 J/g. The gelatinization temperatures of sweet potato starch were higher than those observed for cassava (59.0 to 71.2 °C) and potato (61.9 to 69.9 °C) starches, suggesting stronger crystalline structures and a higher molecular order of sweet potato starch.

The onset, peak, and conclusion temperatures of gelatinization observed in this study were similar to those reported in other studies [8,30,41]. The enthalpy of gelatinization (ΔH) provides a general measure of crystallinity and is an indicator of the loss of molecular order within the granule during gelatinization. In both growing seasons, the enthalpy of gelatinization of sweet potato starches was positively correlated with the crystallinity of the starches (DS, r = 0.92, $p < 0.001$; RS, r = 0.88, $p < 0.01$). Higher ΔH as observed for starches isolated from the 'Uruguaiana' and those from plants grown in the rainy season suggests a greater degree of organization or greater stability of the crystals.

4. Conclusions

Different plant-growing seasons led to the production of naturally modified sweet potato starches, regardless of cultivar. Starches isolated from plants grown in the rainy season have a more ordered structure with higher resistance to thermal processes. Starches isolated from plants grown in the dry season showed lower resistance to heat and agitation, tendency to retrogradation, paste temperature and enthalpy of gelatinization. In addition, dry season starches had a lower content of resistant starch. These differences point to different potential uses of naturally modified starch as an ingredient for food products. With the global sweet potato starch market growing, our findings are also important for defining industrial planning strategies.

Author Contributions: Conceptualization, M.L., A.M.F., T.P.R.d.S. and S.L.; methodology, validation, formal analysis, and investigation, T.P.R.d.S., L.A.d.O. and J.G.d.S.N.; data curation, writing—original draft preparation, writing—review and editing, T.P.R.d.S., M.L. and S.L.; supervision, project administration, funding acquisition, M.L. All authors have read and agreed to the published version of the manuscript.

Funding: This work was partially supported by the Brazilian National Council for Scientific and Technological Development (CNPq), grant numbers 302611/2021-5 and 302848/2021-5 and Coordination for the Improvement of Higher Education Personnel (CAPES—PNPD scholarship).

Data Availability Statement: Data are contained within the article.

Conflicts of Interest: The authors declare no conflict of interest.

References

1. FAO—Food and Agriculture Organization of the United Nations. FAOSTAT: Production-Crops. 2022. Available online: http://www.fao.org/faostat/en/#data/QC (accessed on 5 January 2022).
2. Galvao, A.C.; Nicoletto, C.; Zanin, G.; Vargas, P.F.; Sambo, P. Nutraceutical content and daily value contribution of sweet potato accessions for the European market. *Horticulturae* **2021**, *7*, 23. [CrossRef]
3. Soto-Caro, A.; Luo, T.; Wu, F.; Guan, Z. The U.S. sweet potato market: Price response and impact of supply shocks. *Horticulturae* **2022**, *8*, 856. [CrossRef]
4. Adu-Kwarteng, E.; Baafi, E.; Amoa-Owusu, A.; Okyere, F.; Carey, E. Expanding industrial uses of sweet potato for food security and poverty alleviation. *Open Agric.* **2021**, *6*, 382–391. [CrossRef]
5. Somasundaram, K.; Santhosh Mithra, V.S. Madhuram: A simulation model for sweet potato growth. *World J. Agric. Sci.* **2008**, *4*, 241–254.
6. Erpen, L.; Streck, N.A.; Uhlmann, L.O.; De Freitas, C.P.O.; Andriolo, J.A. Tuberization and yield of sweet potato as affected by planting date in a subtropical climate. *Bragantia* **2013**, *72*, 396–402. [CrossRef]
7. Leonel, M.; Sarita Leonel, S.; Santos, T.P.R.; Souza, J.M.A.; Martins, R.C.; Silva, M.S.C. Agronomic yield and starch properties of banana cultivars. *Pesq. Agropec. Bras.* **2021**, *56*, e02491. [CrossRef]
8. Zhu, F.; Wang, S. Physicochemical properties, molecular structure, and uses of sweet potato starch. *Trends Food Sci. Technol.* **2014**, *36*, 68–78. [CrossRef]
9. Marketwatch. Sweet Potato Starch Market Size 2020: Top Countries Data, Definition, SWOT Analysis, Applications, Trends and Forecast to 2024. Available online: https://www.marketwatch.com/press-release/sweet-potato-starch-market-size-2020-top-countries-data-defination-swot-analysis-applications-trends-and-forecast-to-2024-2020-02-14. (accessed on 15 May 2022).
10. Lewthwaite, S.L.; Triggs, C.M. Sweet potato cultivar response to prolonged drought. *Agron. N. Z.* **2012**, *42*, 1–10.
11. Yooyongwech, S.; Theerawitaya, C.; Samphumphuang, T.; Cha-um, S. Water-deficit tolerant identification in sweet potato genotypes (*Ipomoea batatas* (L.) Lam.) in vegetative developmental stage using multivariate physiological indices. *Sci. Hortic* **2013**, *162*, 242–251. [CrossRef]
12. Tester, R.F.; Karkalas, J. The effects of environmental conditions on the structural features and physico-chemical properties of starches. *Starch-Stärke* **2001**, *53*, 513–519. [CrossRef]
13. Noda, T.; Kobayashi, T.; Suda, I. Effect of soil temperature on starch properties of sweet potatoes. *Carbohydr. Polym.* **2001**, *44*, 239–246. [CrossRef]

14. Beckles, D.M.; Thitisaksakul, M. How environmental stress affects starch composition and functionality in cereal endosperm. *Starch-Stärke* **2014**, *66*, 58–71. [CrossRef]
15. Nascimento, K.O.; Dias, S.N.P.; Augusta, I.M. A review 'clean labeling': Applications of natural ingredients in bakery products. *J. Food Nutr. Res.* **2018**, *6*, 285–294. [CrossRef]
16. Aschemann-Witzel, J.; Varela, P.; Peschel, A.O. Consumers' categorization of food ingredients: Do consumers perceive them as 'clean label' producers expect? An exploration with projective mapping. *Food Qual. Prefer.* **2019**, *71*, 117–128. [CrossRef]
17. Noguerol, A.T.; Pagán, M.J.; García-Segovia, P.; Varela, P. Green or clean? Perception of clean label plant-based products by omnivorous, vegan, vegetarian and flexitarian consumers. *Food Res. Int.* **2021**, *149*, 110652. [CrossRef]
18. Park, S.; Kim, Y.-R. Clean label starch: Production, physicochemical characteristics, and industrial applications. *Food Sci. Biotechnol.* **2021**, *30*, 1–17. [CrossRef] [PubMed]
19. Feltran, J.C.; Peressin, V.A.; Granja, N.P.; Silva Filho, H.M.; Lorenzi, J.O.; Fernandes, A.M.; Soratto, R.P.; Factor, T.L.; Rós, A.B.; Aguiar, E.B. Raízes e Tubérculos. In *Boletim 100: Recomendações de Adubação e Calagem Para o Estado de São Paulo*; Cantarella, H., Quaggio, J.A., Mattos, D., Jr., Boaretto, R.M., Van Raij, B., Eds.; Instituto Agronômico (IAC): Campinas, Brasil, 2022; Volume 1, pp. 314–338.
20. Santos, T.P.R.; Franco, C.M.L.; Leonel, M. Gelatinized sweet potato starches obtained at different preheating temperatures in a spray dryer. *Int. J. Biol. Macromol.* **2020**, *149*, 1339–1346. [CrossRef]
21. Nara, S.; Komiya, T. Studies on the relationship between water-satured state and crystallinity by the diffraction method for moistened potato starch. *Starch* **1983**, *35*, 407–410. [CrossRef]
22. Williams, P.C.; Kuzina, F.D.; Hlynka, L. A rapid calorimetric procedure for estimating the amylose content of starches and flour. *Cereal Chem.* **1970**, *47*, 411–420.
23. Goñi, I.; García-Diz, I.; Mañas, E.; Saura-Calixto, F. Analysis of resistant starch: A method for foods and food products. *Food Chem.* **1996**, *56*, 445–449. [CrossRef]
24. Schoch, T.J. Swelling power and solubility of granular starches. In *Methods in Carbohydrate Chemistry*, 4th ed.; Whistler, R.L., Ed.; Academic Press: New York, NY, USA, 1964; pp. 106–109.
25. Lossolli, N.A.B.; Leonel, M.; Leonel, S.; Izidoro, M.; Paula, G.V.d.; Santos, T.P.R.; Oliveira, L.A.d. Cultivars and fruit part as differentiating factors of physicochemical characteristics of mango starches. *Horticulturae* **2023**, *9*, 69. [CrossRef]
26. Zhu, F.; Corke, H.; Bertoft, E. Amylopectin internal molecular structure in relation to physical properties of sweet potato starch. *Carbohydr. Polym.* **2011**, *84*, 907–918. [CrossRef]
27. Santos, T.P.R.; Franco, C.M.L.; Mischan, M.M.; Leonel, M. Behavior of sweet potato starch after spray-drying under different pretreatment conditions. *Starch-Stärke* **2019**, *71*, 1800245. [CrossRef]
28. Gou, M.; Wu, H.; Saleh, A.S.M.; Jing, L.; Liu, Y.; Zhao, K.; Su, C.; Zhang, B.; Jiang, H.; Li, W. Effects of repeated and continuous dry heat treatments on properties of sweet potato starch. *Int. J. Biol. Macromol.* **2019**, *129*, 869–877. [CrossRef]
29. Teerawanichpan, P.; Lertpanyasampatha, M.; Netrphan, S.; Varavinit, S.; Boonseng, O.; Narangajavana, J. Influence of cassava storage root development and environmental conditions on starch granule size distribution. *Starch-Stärke* **2008**, *60*, 696–705. [CrossRef]
30. Guo, K.; Liu, T.; Xu, A.; Zhang, L.; Bian, X.; Wei, C. Structural and functional properties of starches from root tubers of white, yellow, and purple sweet potatoes. *Food Hydrocoll.* **2019**, *89*, 829–836. [CrossRef]
31. Wang, H.; Yang, Q.; Ferdinand, U.; Gong, X.; Qu, Y.; Gao, W.; Ivanistau, A.; Feng, B.; Liu, M. Isolation and characterization of starch from light yellow, orange, and purple sweet potatoes. *Int. J. Biol. Macromol.* **2020**, *160*, 660–668. [CrossRef]
32. Lee, B.H.; Lee, Y.T. Physicochemical and structural properties of different colored sweet potato starches. *Starch-Stärke* **2017**, *69*, 3–4. [CrossRef]
33. Yong, H.; Wang, X.; Sun, J.; Fang, Y.; Liu, J.; Jin, C. Comparison of the structural characterization and physicochemical properties of starches from seven purple sweet potato varieties cultivated in China. *Int. J. Biol. Macromol.* **2018**, *120*, 1632–1638. [CrossRef]
34. Genkina, N.K.; Wasserman, L.A.; Noda, T.; Tester, R.F.; Yuryev, V.P. Effects of annealing on the polymorphic structure of starches from sweet potatoes (Ayamurasaki and Sunnyred cultivars) grown at various soil temperatures. *Carbohydr. Res.* **2004**, *339*, 1093–1098. [CrossRef]
35. Kitahara, K.; Fukunaga, S.; Katayama, K.; Takahata, Y.; Nakazawa, Y.; Yoshinaga, M.; Suganuma, T. Physicochemical properties of sweet potato starches with different gelatinization temperatures. *Starch-Stärke* **2005**, *57*, 473–479. [CrossRef]
36. Aina, A.J.; Falade, K.O.; Akingbala, J.O.; Titus, P. Physicochemical properties of caribbean sweet potato (*Ipomoea batatas* (L) Lam) starches. *Food Bioprocess Technol.* **2012**, *5*, 576–583. [CrossRef]
37. Zhang, L.; Zhao, L.; Bian, X.; Guo, K.; Zhou, L.; Wei, C. Characterization and comparative study of starches from seven purple sweet potatoes. *Food Hydrocoll.* **2018**, *80*, 168–176. [CrossRef]
38. Zhao, X.; Andersson, M.; Andersson, R. Resistant starch and other dietary fiber components in tubers from a high-amylose potato. *Food Chem.* **2018**, *251*, 58–63. [CrossRef]
39. Future Market Information. Resistant Starch Market Outlook (2022–2032). Available online: https://www.futuremarketinsights.com/reports/resistant-starch-market (accessed on 12 October 2022).
40. Peroni, F.H.G.; Rocha, T.S.; Franco, C.M.L. Some structural and physicochemical characteristics of tuber and root starches. *Food Sci. Technol. Int.* **2006**, *12*, 505–513. [CrossRef]

41. Abegunde, O.K.; Mu, T.; Chen, J.; Deng, F. Physicochemical characterization of sweet potato starches popularly used in Chinese starch industry. *Food Hydrocoll.* **2013**, *33*, 169–177. [CrossRef]
42. Tsakama, M.; Mwangwela, A.M.; Manani, T.A.; Mahungu, N.M. Physicochemical and pasting properties of starch extracted from eleven sweet potato varieties. *Afr. J. Food Sci.* **2010**, *1*, 90–98.
43. Almeida, V.O.; Batista, K.A.; Di-Medeirosa, M.C.B.; Moraes, M.G.; Fernandes, K.F. Effect of drought stress on the morphological and physicochemical properties of starches from *Trimezia juncifolia*. *Carbohydr. Polym.* **2019**, *212*, 304–311. [CrossRef] [PubMed]
44. Singh, N.; Kaur, A.; Shevkani, K.; Ezekiel, R.; Kaur, P.; Isono, N.; Noda, T. Structural, morphological, thermal, and pasting properties of starches from diverse Indian potato cultivars. *Starch-Stärke* **2018**, *70*, 1700130. [CrossRef]
45. Campanha, R.B.; Franco, C.M.L. Gelatinization properties of native starches and their Näegeli dextrins. *J. Therm. Anal. Calorim.* **2011**, *106*, 799–804. [CrossRef]

horticulturae

Article

Variation in Yield, Berry Distribution and Chemical Attributes of *Coffea arabica* Beans among the Canopy Strata of Four Genotypes Cultivated under Contrasted Water Regimes

Miroslava Rakocevic [1,2,3,*], Maria Brigida dos Santos Scholz [2], Ricardo Antônio Almeida Pazianotto [3], Fabio Takeshi Matsunaga [2,4] and José Cochicho Ramalho [5,6,*]

1 Plant Physiology Laboratory (LMGV), State University of North Fluminense (UENF), Campos dos Goytacazes 28013-602, RJ, Brazil
2 Agronomic Institute of Paraná, IAPAR, Londrina 86047-902, PR, Brazil
3 Embrapa Environment, Campinas 13820-000, SP, Brazil
4 Department of Computing, National Service for Industrial Training (SENAI), Londrina 86026-040, PR, Brazil
5 Plant Stress and Biodiversity Lab, Forest Research Center (CEF), School of Agriculture, University of Lisbon (ISA/ULisboa), 2784-505 Oeiras, Portugal
6 GeoBioSciences, GeoTechnologies and GeoEngineering (GeoBioTec), Faculdade de Ciências e Tecnologia, Universidade NOVA de Lisboa (FCT/UNL), 2829-516 Caparica, Portugal
* Correspondence: mima.rakocevic61@gmail.com (M.R.); cochichor@mail.telepac.pt (J.C.R.)

Abstract: Water availability, light, management practices, and harvest time impacts on *Coffea arabica* L. yield and bean quality remain uncertain. It was hypothesized that the soil water and light availability could impact berry distribution, yield, and bean chemical attributes within the plant canopy. Therefore, it was aimed to study berry yield, berry distribution, and bean chemical traits along the canopy strata of four coffee genotypes (Iapar 59, Catuaí 99 and two Ethiopian wild accessions, 'E083' and 'E027'), cultivated with (IRR) and without irrigation (NI) in the two initial harvest years. The maximum height of berry occurrence was lower in NI than in IRR plants in both harvest years. In the 2nd harvest year, higher leaf-to-fruit ratio was found under NI than under IRR for all genotypes, except for Catuaí 99, while the most regular berry distribution among canopy strata was obtained in IRR 'E083', the highest bean yield in IRR 'E083' and Iapar 59, and the highest percent of useful bean biomass in NI Catuaí 99. The reduced lipid content under IRR was more important in the 1st (all genotypes) than in the 2nd harvest year (Iapar 59 and 'E027'). As a novelty, chemical bean composition was additionally impacted by light availability along the canopy strata. Proteins declined from bottom (shaded) to upper (highly light exposed) strata, regardless of genotype and harvest year. Similar stratification was observed in caffeine in the 2nd year. Although some traits were somewhat changed among strata, no substantial quality changes occurred, thus allowing that harvest might include the entire plant and not only some specific strata. Iapar 59 and 'E083' showed chemical composition usually associated with high bean quality, with the highest lipid, sucrose, and soluble sugar contents, and the lowest caffeine, chlorogenic acids, and phenolic components among four genotypes, but Iapar 59 plants were less affected in their yield under NI. Based on additional responses from space occupation and yield only under IRR, the wild accession 'E083' must be considered in future breeding programs as promising material for intensive input conditions. High bean quality and the less variated yield under lower soil water availability qualified the Iapar 59 as the most prominent among the four genotypes.

Keywords: biomass; caffeine; chlorogenic acids; Ethiopian accessions; irrigation; leaf-to-fruit ratio; lipids; proteins; sucrose

Citation: Rakocevic, M.; dos Santos Scholz, M.B.; Pazianotto, R.A.A.; Matsunaga, F.T.; Ramalho, J.C. Variation in Yield, Berry Distribution and Chemical Attributes of *Coffea arabica* Beans among the Canopy Strata of Four Genotypes Cultivated under Contrasted Water Regimes. *Horticulturae* **2023**, *9*, 215. https://doi.org/10.3390/horticulturae9020215

Academic Editors: Alessandra Durazzo and Massimo Lucarini

Received: 29 December 2022
Revised: 31 January 2023
Accepted: 3 February 2023
Published: 6 February 2023

1. Introduction

Among the 130 species from the *Coffea* genus [1], *C. arabica* L. and *C. canephora* Pierre ex Froehner commercially dominate the coffee trade [2]. Coffee plays a relevant role for the subsistence of nearly 25 million coffee-farming families from about 80 developing countries of Asia, Africa, and Latin America [3]. *C. arabica* contributes today to about 60% of the world's coffee consumption [4], and *C. canephora* to the rest.

Coffee trees build their architecture following the Roux model, considering a continuous growth and dimorphism of branches—orthotropic (1st order) and plagiotropic [5]. In *C. arabica*, the plagiotropic axes develop from the 2nd to the 5th orders [6]. Coffee metamer (basic architectural segment in tree construction) is characterized by internode, two leaves, and two serial buds (up to 5–6 buds formed in each leaf axil). Buds may differ in either inflorescence (up to 4 flowers from each bud in *C. arabica*), or more plagiotropic branches [7]. Therefore, the maximum possible flower/fruit number produced by one metamer is 40. All plagiotropic orders bear fruits, especially in late production years, but the 5th order axes are very rare [6,8].

The alternate vegetative and reproductive phases, six in total [9], occur in the coffee plant over a two-year cycle, with the coexistence of two distinct phenophases within the meta-population of axes and metamers in a same moment of the biennial cycle [6]. When the number of daylight hours begins to decrease, such photoperiod changes can induce the differentiation of the reproductive buds [10]. These buds grow and enter in dormancy, matching with the dry season in most growing regions [7]. After the first rainfalls, the break of bud dormancy initiates anthesis, or floral opening. In non-equatorial regions, encompassing most Brazilian coffee production areas, blossom occurs at different periods (from August to November) in two or more unsynchronized flushes [11], with the latter promoting the simultaneous presence of mature and immature berries at harvest [6,12]. Expression of coffee florigen suggests a continuum of floral induction that allows different starting points for floral activation, explaining developmental asynchrony and prolonged anthesis events in coffee [13]. After fertilization, fruits develop along ca. 6–9 months, with successive divisions and elongation of the perisperm and endosperm tissues [14].

In the mature coffee fruit (drupe, also called coffee cherry or berry), the exocarp is red or yellow. The berry consists of pericarp and seeds or beans (usually two in *C. arabica*). To obtain the commercial coffee beans, the pericarp outer skin, pulp, pectic adhesive layer, and parchment (usually together with bean silverskin) are removed, through either dry or wet processing [15]. The remaining part of the coffee beans (processed beans) are then roasted using dry heat at temperatures usually between 200 and 240 °C, with constant stirring to ensure even heat distribution [16].

To obtain a good fruit development, a certain leaf area expressed in leaf-to-fruit ratio is required [17]. Consequently, plant investment in leaf area development must increase with increased leaf area index, to sustain the growth and maintenance of berries. This requirement increases with plant density [6], considering that leaf photosynthesis decreases in the lower canopy layers with greater self-shading [18].

Arabica coffee beans contain up to 15% nitrogenous compounds (10–11% proteins, 0.9–1.3% caffeine, 0.6–2.0% trigonelline, 0.5% of free amino acids), 15–18.5% lipids, 50–60% carbohydrates (6–9% sucrose, 0.1% reducing sugars, 33–44% polysaccharides, 3% lignin, 2% pectin), 3–4.2% minerals, and 4.1–7.9% chlorogenic acids [19]. Under different environmental conditions, the same coffee species, or even genotype, can produce coffee beans with a wide chemical composition [16,20], flavors, and aromas or sensory attributes [21,22]. The estimation is that chemical bean attributes and coffee cup quality are affected by 40% pre-harvest, 40% post-harvest, and 20% export handling [23]. In fact, plant performance, bean yield, and quality can be altered by: (1) genotype and its geographical origin [16,24], (2) environmental conditions, such as soil, topography, altitude and climate [2,25], shade density under agroforestry [24,26,27], and (3) cultural management practices, such as planting density [28], mineral fertilization [29], irrigation [30,31], pruning [32], or even "CO_2 fertilization" [18,31,33]. The chemical composition is less impacted by biennial cycle [34], or year of production [22], but the year impact can be significant in some experiments [28].

To ensure large, high-quality seed yields, abundant water availability is crucial during the period of rapid berry expansion [35]. In drought-impacted regions, coffee yield increases with irrigation [36,37], but contradictory results were obtained about the impacts of this management practice on coffee bean and/or cup coffee quality. Partial root zone drying and normal deficit irrigation seemed to preserve cup coffee quality [36], although only minor changes of bean quality were found under irrigation [38]. The 5-caffeilquinic acid (5-CQA), the main chlorogenic acid (CGA) isomer found in coffee seeds, and the lipid content do not vary with irrigation, while coffee bean sucrose and caffeine contents are found to increase in non-irrigated coffee plants grown in warmer regions [38]. Other reports pointed to greater caffeine and CGA contents in beans of irrigated than in non-irrigated plants [30]. Additionally, low shading (ca. 30% irradiance reduction) can enhance dry bean yield, and total sugar and CGA contents in coffee beans [37]. Some evidence suggests that increased altitude (with a more prolonged maturation due to lower temperatures), as well as shade, might improve the sensory attributes of coffee [39]. Contradictory reports about environment and management practice impacts on coffee quality that might result from distinct *C. arabica* cultivars used in the studies. They intrinsically respond differently to shade/altitude, from yield to quality aspects [26,27], further interacting with management conditions that strongly influence the outcome, or with adverse thermal/light and water availabilities that can greatly reduce the potential bean yield and quality [30,40]. This can interact with harvest year, for example, protein and caffeine contents increase and lipids decrease in inferior plant layers (self-shading) only in latter harvest years [28].

One of the research gaps is related to shifts of coffee yield and quality (increase, decrease, or non-linear of principal primary and secondary metabolites) and their dependency on water deficit and/or light conditions in different genotypes. It was hypothesized that, depending on genotypes, the additional soil water availability and light availability along the canopy strata, could impact on berry yield and bean chemical attributes, from the initial harvest years. To examine this hypothesis, the variations in chemical traits of coffee beans and productivity were evaluated along the canopy strata of plants cultivated under natural rainfall or with supplementary irrigation, in four cropped genotypes (Iapar 59, Catuaí 99, 'E027' and 'E083') of different genetic origins, in two subsequent harvest years.

2. Materials and Methods

2.1. Plant Material and Experimental Conditions

Seedlings of *Coffea arabica* L. from the approximately 100 Ethiopian wild accessions and the two test cultivars (Iapar 59 and Catuaí 99) were established in 2009 in nursery. They were planted in 2010 in the experimental fields of IAPAR, Londrina (23°18′ S and 51°17′ W, altitude 620 m.a.s.l.), Paraná state, Brazil. Coffee rows were oriented east–west, with 2.5 m distance among them, and 0.5 m between plants in the row (planting density of 8000 plants ha^{-1}), with four repetitions (plants) of each genotype in each of two water regimes. Plants were randomly distributed in the experimental field. Two Ethiopian wild accessions ('E027' and 'E083') were chosen because of their outstanding architectural characters shown in spring 2011. The 'E027' presented visually large and long leaf blades, branched structure, and very few flowers, whilst the 'E083' had smaller elongated leaves, a high number of flowers, and quick vegetative space occupation by higher order branching. Iapar 59 is originated from the cross between the cultivar Villa Sarchi CIFC 971/10 and hybrid of Timor CIFC 832/2, representing *C. canephora* introgression by spontaneous specific cross with *C. arabica* [41], while Catuaí 99 is a highly productive *C. arabica* cultivar characterized by high cup quality [42].

The soil was dusky-red dystrophic latosol, characterized by 790.02 g clay, 160.31 g silt, and 49.67 g sand per kg of soil particle-size in 0 to 0.20 m depth layer [43]. Climate is subtropical, Köpen–Geiger climate type Cfa, with average annual precipitation of about 1585 mm, ranging from 55 mm in the driest month (August) to 245 mm in wettest one (January). The limiting factors for *C. arabica* growth in Cfa climate are defined by low autumn and winter temperatures [44], such that almost all plants from this experiment died after the strong frost occurred after harvest in 2013, precluding the experimental continuation. The data of 2012–2013 daily rains were obtained from the IAPAR local meteorological station.

Coffee plants were cultivated under rainfed (not irrigated—NI) and with additional irrigation (irrigated—IRR) water regimes. For the latter, drip irrigation was implemented, being triggered based on a soil water balance method, aiming at supplying the difference of rain and soil storage [45]. The irrigation intensity was 3.5 L h^{-1} by dripper, placed near to the trunk of each coffee plant. Fertilization NPK (20:5:15) was added at 1000 kg ha^{-1} $year^{-1}$, split in four times during the most demanding phenophases of the culture.

Light irradiance, measured as photosynthetic photon flux density (PPFD), was obtained from stational sensors distributed along the plant canopy strata, starting from 20 cm from the soil (average height of coffee trunks), and positioned at every subsequent 40 cm of the plant height (60 cm, 100 cm). They were installed in one representative plant of each genotype and each water regime. PPFD was measured before the harvests, during one representative cloudy day, in June of 2012 and 2013. Stational photodiode sensors (Hamamatsu G1118, Japan) used in the experiment were calibrated with sensors (LI-COR 190R, Lincoln, NE, USA). For data acquisition and storage, a datalogger (CR21X, Campbell Scientific, Logan, UT, USA) and a multiplexer (AM416, Campbell Scientific, Logan, UT, USA) were used, allowing the simultaneous data collection of up to 32 irradiance sensors. This included Hamamatsu sensors positioned at the base of each existing stratum along the plant canopy strata of representative plants, while one LI-COR 190R was installed in the middle of the field, as a PPFD reference, at 2 m height (thus, above the plant top). Data were collected every 60 s and were presented as mean values for each 15 min interval.

2.2. Plant Coding, Computational Processing, Berry Harvests, and Yield

Coffee plant coding was performed in the two harvesting periods, in June of 2012 and 2013. Harvest was effectuated in two or three passages, two weeks apart, always collecting only red (cherry) berries under adequate state of maturation. For berry collection, plant canopy was divided in four strata (S): S1: 20–60 cm (or <60 cm); S2: 61–100 cm; S3: 101–140 cm; S4 > 140 cm (Figure 1).

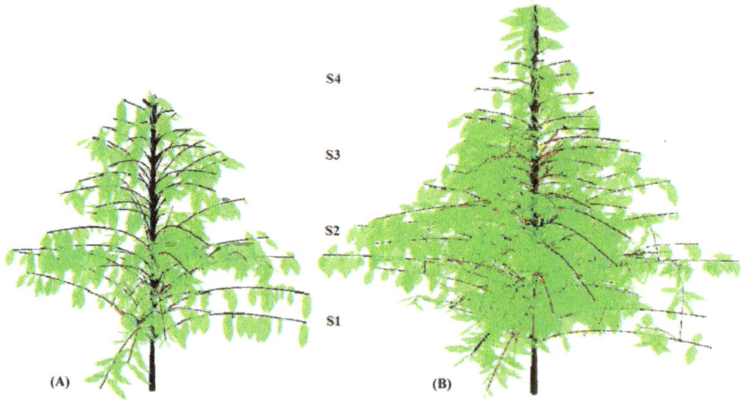

Figure 1. Snapshots of 3D reconstruction of one 'E083' irrigated plant in the (**A**) 1st (2012) and (**B**) 2nd (2013) harvest year, illustrating the strata along plant canopy: S1: 20–60 cm (or <60 cm); S2: 61–100 cm; S3: 101–140 cm; S4 > 140 cm.

Topological and geometric codification of coffee trees was performed in three botanical scales—metamers, branches, and plants [8] in multiscale tree graphs [46]. The orthotropic axes were always described at metamer scale, collecting a maximum number of variables, including length of each metamer, leaf position, size and elevation, orientation, and total length of all plagiotropic branches inserted in the orthotropic axis. Four representative 2nd order plagiotropic axes were sampled (one for each cardinal point) in each stratum above the trunk [42]. The sampled plagiotropic axes were described in detail following the same logic as orthotropic axes. In addition, the 3rd to 5th order plagiotropic lateral branches that belonged to the sampled and decomposed 2nd order branches were also described in detail. All other 2nd order plagiotropic axes were described according to their positions along the orthotropic axis, considering elevation and orientation cardinal points, total live length of the axes, mortality/vivacity of 2nd order plagiotropic axes terminal apex, supported by total number of berries [47]. The plants were reconstructed using the VPlants modeling platform [48]. The reconstructions followed the specific proposed modules use, such as AmostraCafe3D, VirtualCafe3D, and Cafe3D and procedures [6,47]. Reconstructed plants (Figure 1) permitted the calculation of leaf and branch area. In this paper, only the information about the total berry number, leaf/branch area per plant, together with berry position relative to the trunk distance, or over the strata were estimated from the virtual 3D plants (Figure 1). This information was used to calculate the ratio of average leaf-to-fruit and branch-to-fruit ratios (cm^2 $berry^{-1}$) of plants in the two harvest years.

The berry yield of coffee cherries (fresh berry mass, FM) was measured separately for each plant and each plant stratum. The collected berries were dried in the sun at a concrete yard until 12.5% of moisture (dry berry mass, DM). Afterwards, the pulp and parchment were removed, and only beans without visual defects (processed bean mass, BM) were selected for further analysis. The initial berry moisture was calculated from the ratio of DM to FM. The dry and processed bean mass performances were calculated as percent of BM or DM to FM, respectively.

2.3. Chemical Attributes of the Coffee Beans

The processed beans were stored in a dry local on paper bags. The coffee beans from each genotype, stratum, water availability, and harvest year were frozen with liquid nitrogen at −196 °C for chemical analyses, ground in a laboratory disk mill (Perten 3600, Kungens Kurva, Sweden) to a particle size of 0.5 mm, packed in plastic bottles, and kept at −18 °C until analysis. The determination of protein (PRO), caffeine (CAF), lipids (LIP), sucrose (SUC), total soluble sugars (TS), total chlorogenic acids (CGA), and total phenolic components (PC) in coffee beans was performed using a near infrared spectroscopy (NIR, SYSTEM 6500 spectrophotometer, Foss-Perstorp employing ISIscan software, Foss, Silver Spring, MD, USA). The NIR reflectance spectra were collected at 2 nm intervals from 1100 to 2500 nm using a rectangular cell containing 6 g of ground coffee beans, and data were saved as the average of 32 scans. The two replicates of the NIR spectra were collected for each coffee sample. The blank spectrum was used from the ceramic plate supplied with the instrument. The ISIscan software package was used to control the recorder, collect the spectra, import, and analyze the data. The concentration of each compound was calculated using the prediction models for coffee beans, developed by Scholz et al. [49].

2.4. Statistical Analyses

The 'R' software [50] was used for all statistical analyses. The experimental design was completely randomized, with plant or stratum as statistical units, and four repetitions. Data were subjected to the two-way analysis of variance (ANOVA), after testing the hypothesis of variance homogeneity. ANOVA considered a mixed linear model ('nlme' package) and maximum likelihood to test the significance of differences between two water regimes (IRR and NI), and four genotypes ('E083', 'E027', Iapar 59, and Catuaí 99). If no significant interaction was found, the model reduction was applied and fitted. The significance of local strata conditions (three in 2012 and four in 2013) was estimated for each genotype, not including the water regime as a 3rd factor, to simplify the model and explanations. In comparison among the averages estimated by the ANOVA models, the Tukey HSD test with the significance of 0.05 was used, supported by 'lsmeans', and 'multcompView' packages. In the 1st harvest year (2012), the 'E027' showed extremely low frequency or even lack of berries in plants/strata, which was insufficient for any kind of further chemical analyses.

To analyze the radial and vertical distribution of berries in coffee trees, the resulting 3D reconstructions were explored, using exact berry positions in the Cartesian coordinate system (x, y, z). The cumulative empirical distribution weighted by the inverse of the total number of berries per plant (function 'Ecdf', 'Hmisc' package) were estimated. The weighted Kolmogorov–Smirnov two-sample test was used to compare water regime levels for each genotype (function 'ks_test', 'Ecume' package) separately for each harvest year. Finally, the measured and calculated fruit architectural, yield, and chemical parameters along the plant canopy strata were correlated separately for each harvest year, and graphically presented using the 'Hmisc' and 'corrplot'.

3. Results

3.1. Environmental Conditions—Rainfall and Light Distribution along the Plant Canopy Strata

The experimental fields were localized in Northwest of Paraná state, Brazil (Figure 2A). During the experimental period, one not usual summer drought occurred in the beginning of 2012, in the phenophase of leaf area and berry expansion (Figure 2B), the critical phase for coffee yield and quality. The winter of 2012 was very dry, preceding the blossom. The early autumn dry period in 2013 matched with the phenophases of berry maturation [9], and the last dry period (winter of 2013) was used for berry harvest. The experiment finished due to strong frost that killed almost all plants in the experimental area, with exception of four IRR 'E083' plants.

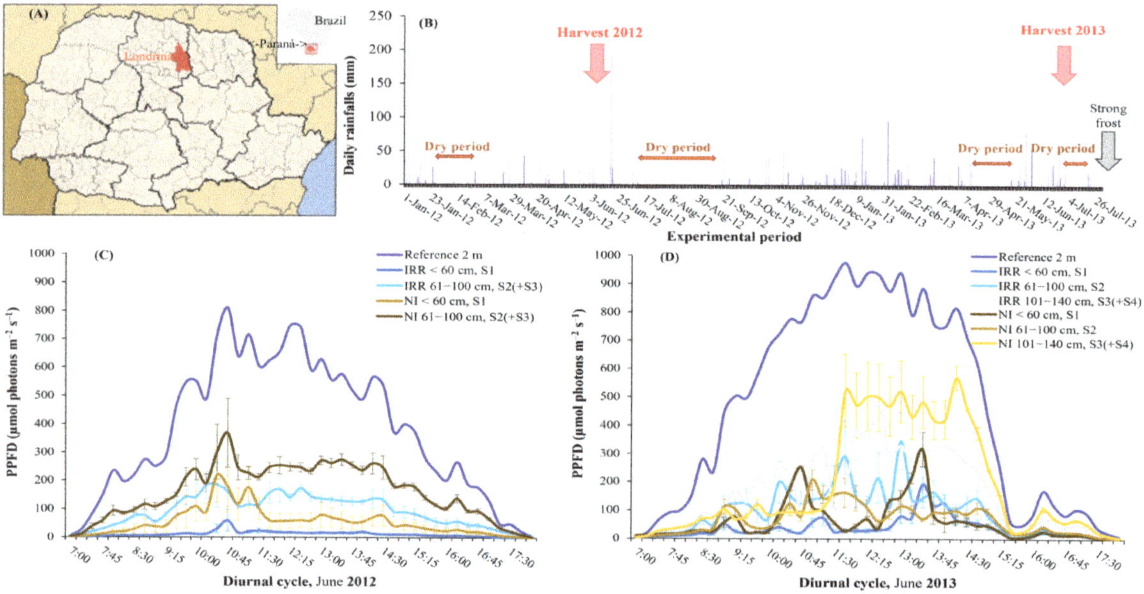

Figure 2. Experimental site localization and the environmental factors during experiment. (**A**) Londrina localization is tagged in red at Northwest of Paraná state, Brazil; (**B**) Daily rainfalls during the experimental period, periods of 2012 and 2013 harvests, dry periods, and the strong frost that ended the experiment are tagged. Photosynthetic photon flux density (PPFD, µmol photons m^{-2} s^{-1}), values (means ± standard errors, n = 4), registered along the plant canopy strata and in the reference (2 m above soil) in (**C**) June 2012, and (**D**) June 2013.

Irradiance strongly and progressively decreased from the top to the bottom layers of plant canopy (Figure 2C,D), even in the 1st harvest—2012 (Figure 2C), with quite small plants. In fact, PPFD values registered at plant base (<60 cm) showed that a large part was intercepted by leaves, since from the ~700 µmol photons m^{-2} s^{-1} measured at 2 m height at noon (cloudy sky), only ~10 µmol photons m^{-2} s^{-1} and 55 µmol photons m^{-2} s^{-1}

reached S1 in IRR or NI plants, respectively. Most irradiance was intercepted by S2 and S3, as inferred from the difference among PPFD of reference and at the bottom of S2(+S3), in IRR and NI plants in 2012. In this year, the IRR plants intercepted more than NI plants, as shown by the PPFD at S2(+S3), thus pointing to a greater leaf area of IRR than NI plants.

A more complex light distribution along canopy strata occurred in the 2nd harvest year (2013, Figure 2D), due to an additional leaf stratum (S4, Figure 1B) as compared with 2012 (Figure 1A), which increased light interception (Figure 2D). The S3 + S4 strata would have lower leaf area to intercept the incident light in NI than in IRR plants, as inferred from the higher light measured at the bottom of S3 in NI plants, when compared to their IRR counterparts (Figure 2D), in line with the 3D reconstruction (Figure 1B).

3.2. Plant Scale—Berry Production and Distribution, Leaf-to Fruit, and Branch-to Fruit Dependency on Water Regime and Genotype

Total berry number per plant in 2012 greatly differed among the four genotypes, between ca. 530 ('E083' = Iapar 59) and ca. 50 berries ('E027'), but with no significant impact of water regime, despite a common tendency to somewhat lower values in NI plants of all genotypes except 'E027' (Figure 3A). In the 2nd harvest year, the 'E027' also showed the lowest berry number per plant under IRR, whilst the other three genotypes showed similar values, close to 3000 (Figure 3B). Contrasting with the 1st harvest year, significant reductions of berry number per plant were observed under NI in all genotypes except in Catuaí 99, as compared with their respective IRR plants.

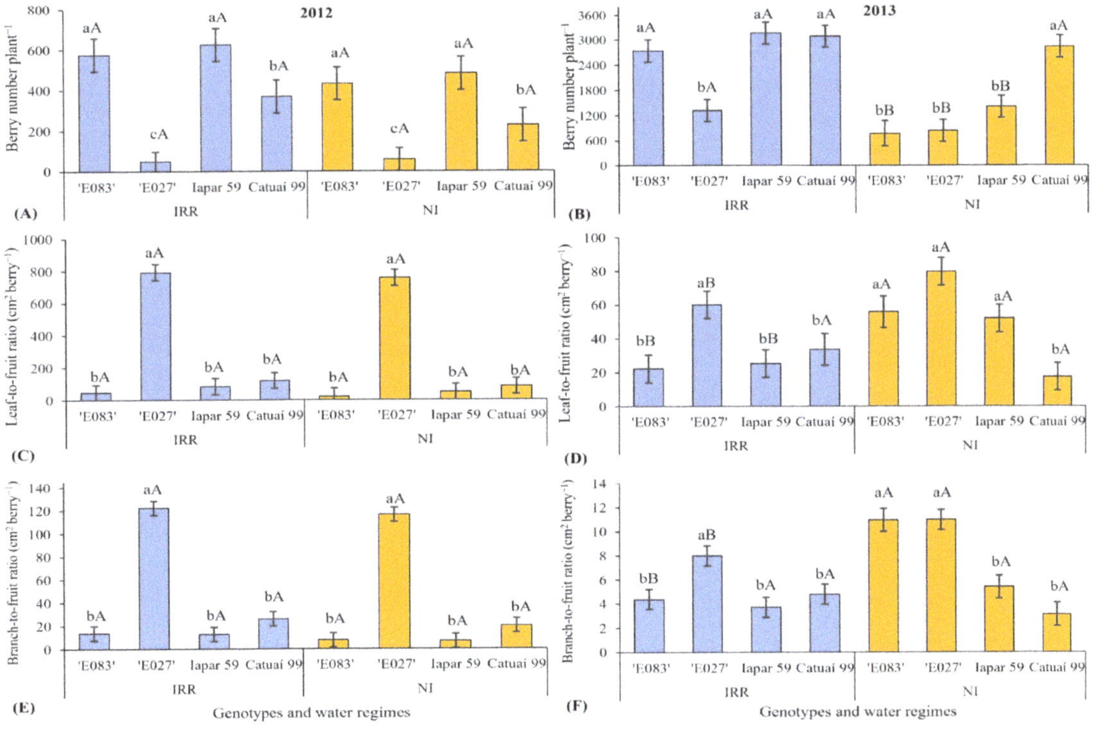

Figure 3. Reproductive and vegetative investments at plant scale: (**A**) Berry number in 2012 and (**B**) 2013; Leaf-to-fruit ratio (cm^2 berry^{-1}) in (**C**) 2012 and (**D**) 2013; Branch-to-fruit ratio (cm^2 berry^{-1}) in (**E**) 2012 and (**F**) 2013) of four genotypes of *Coffea arabica* cultivated under two water regimes (irrigated-IRR; not irrigated-NI). Mean values ± standard errors (n = 4) are shown. Lower-case letters compare genotype effects within each water regime; upper-case letters compare water regime effects within each genotype, always separately for each harvest year.

In 2012, the greatest leaf-to-fruit ratio was found in 'E027' (Figure 3C), and with quite low values in the other three genotypes, but without impact of soil water conditions in all genotypes. In 2013, this parameter still presented maximal values in 'E027', but in all genotypes presented much lower values than in 2012. Additionally, the response to soil water availability was different in 2013, when greater leaf-to-fruit ratio values were found under NI than under IRR for all genotypes, except Catuaí 99 (Figure 3D, *p*-value of Genotype × Water regime = 0.0330).

In the 1st harvest year, the branch-to-fruit ratio (Figure 3E) showed a similar pattern to the leaf-to-fruit ratio, with the highest values found in 'E027', quite low values in the other three genotypes, and without impact of soil water conditions in all genotypes (Figure 3E). In the 2nd harvest year, 'E027' showed higher values of branch-to-fruit ratio than the other three genotypes under IRR (Figure 3F). At the same time under NI, two wild Ethiopian accessions showed higher values of this parameter than two cultivars (significant interaction of Genotype × Water regime, *p*-value = 0.0012). In 2013, the branch-to-fruit ratio had higher values under NI than under IRR in two Ethiopian accessions, but without soil water regime impact in two cultivars.

In both harvest years, the distribution of berries along the height of the plant canopy differed between the two water regimes, in all four studied genotypes (Figure 4A). Higher density of berries in NI than in IRR was observed in lower plant strata. The berry distribution reached lower height in NI than in IRR plants of all genotypes, in both harvest years. In 2013, the IRR 'E083' showed a regular berry distribution along the canopy height that tended to linearity, attaining 200 cm of height, the highest among four genotypes.

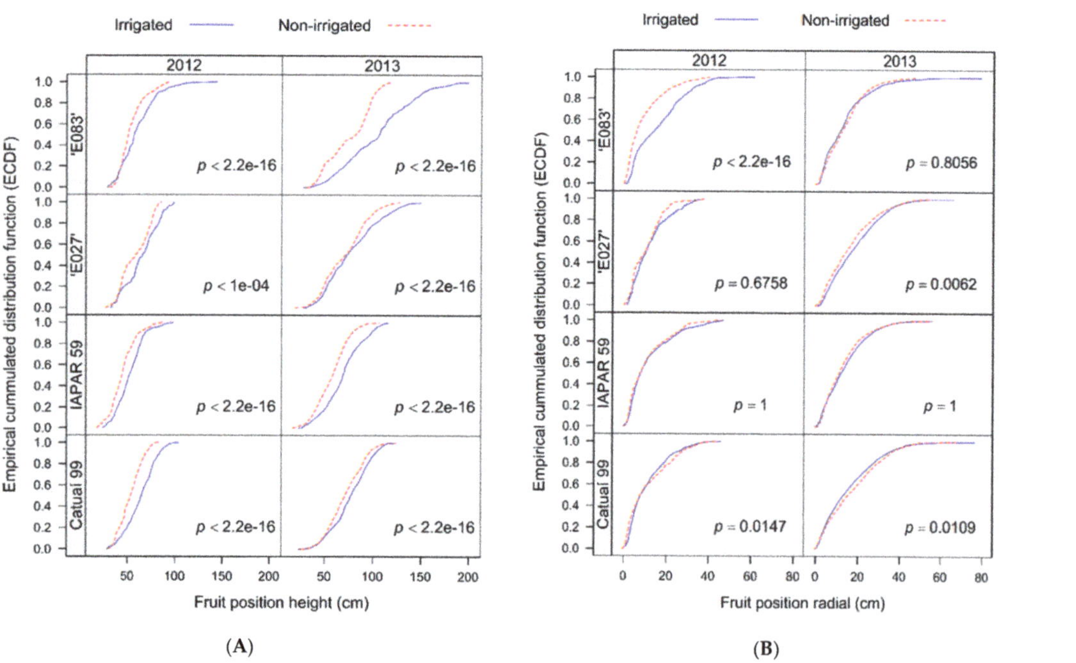

Figure 4. Empirical cumulative distribution function (ECDF) for berries along the (**A**) height of the canopy and (**B**) radial plant profile analyzed for four genotypes of *Coffea arabica* cultivated under two water regimes (irrigated and not irrigated). The *p*-values are shown ($n = 4$), comparing water regimes, separately for each genotype and harvest year (2012 and 2013).

Additionally, the radial berry distribution (horizontal distribution over x- and y-axes) was also assessed, showing slightly delayed berry accumulation over the plagiotropic branches in NI compared to IRR in Catuaí 99 plants, in both harvest years (Figure 4B). The opposite, i.e., the delayed berry accumulation over the plagiotropic branches in IRR than in NI plants was calculated for 'E083 in the 1st harvest year and for 'E027' the 2nd harvest year. The radial berry occupation zone referent to trunk attained the maximum from ca. 40 cm ('E027') to 60 cm ('E083') in the 1st harvest year. In the 2nd harvest year, the radial zone of berry occupation attained the maximum from ca. 60 cm (Iapar 59) to 80 cm ('E083' and Catuaí 99).

3.3. Components of the Berry and Bean Yields at Plant Scale Dependent on Water Regime, Genotype, and Local Light Availability

In the 1st harvest year (2012) the yield components (fresh dry mass (FM), dry mass (DM), processed bean mass (BM)), and mass performances were positively impacted by irrigation in the three analyzed genotypes (Table 1). In this year, greater initial berry moisture (indicator of delayed fruit maturation), and lower processed bean mass performance were found in Catuaí 99 than in 'E083', while Iapar 59 showed intermediary values.

Table 1. Components of the berry and bean yields: fresh berry mass (FM), dry berry mass (DM), and processed bean mass (BM) (g plant^{-1}), initial berry moisture (%), DM and BM performances (%) of two harvests (2012 and 2013) analyzed in four genotypes of *Coffea arabica* cultivated under two water regimes (irrigated—IRR; not irrigated—NI). Estimated means and *p*-values are shown (n = 4), where "-" indicates that the model reduction and fitting were applied. When significant, *p*-value is marked in bold. Lower-case letters compare genotype effects within each water regime; upper-case letters compare water regime effects within each genotype, always separately for each harvest year.

Harvest Year	Water Regime	Genotype	FM	DM	BM	Initial Berry Moisture	DM Performance	BM Performance
2012	IRR	'E083'	573.1 aA	240.8 aA	117.4 aA	54.8 bA	47.4 aA	20.6 aA
		Iapar 59	701.0 aA	254.8 aA	118.5aA	64.3 abA	48.1 aA	17.6 abA
		Catuaí 99	469.1 aA	168.3 aA	75.3 aA	66.8 aA	43.6 aA	15.0 bA
	NI	'E083'	271.4 aB	117.0 aB	52.8 aB	55.3 bA	40.6 aB	17.7 aB
		Iapar 59	398.9 aB	131.0 aB	53.9 aB	63.8 abA	41.4 aB	14.6 abB
		Catuaí 99	166.9 aB	44.6 aB	44.6 aB	66.3 aA	36.8 aB	12.1 bB
	p-value	Genotype	0.2535	0.1609	0.0876	**0.0465**	0.1085	**0.0024**
		Water regime	**0.0242**	**0.0088**	**0.0031**	0.8881	**0.0033**	**0.0161**
		Gen × Water	-	-	-	-	-	-
2013	IRR	'E083'	3943 aA	1287 aA	599 aA	65.3 aA	38.6 bB	17.4 cB
		'E027'	1464 bA	503 bA	255 bA	66.9 aA	33.3 dA	15.9 dA
		Iapar 59	3596 aA	1300 aA	637 aA	65.9 aA	37.6 cA	18.9 bA
		Catuaí 99	1816 bA	724 bA	354 bA	57.9 bA	40.8 aB	20.4 aB
	NI	'E083'	645 cB	248 bB	109 bB	61.3 aA	40.1 bA	20.7 bA
		'E027'	640 cB	230 bB	107 bB	62.8 aA	31.3 dB	14.9 dB
		Iapar 59	1830 aB	654 aB	311 aB	61.9 aA	32.9 cB	16.3 cB
		Catuaí 99	1212 bB	622 aB	318 bB	53.8 bA	49.1 aA	24.3 aA
	p-value	Genotype	**0.0422**	**0.0036**	**0.0049**	**0.0454**	**<0.0001**	**<0.0001**
		Water regime	**<0.0001**	**0.0001**	**0.0001**	0.0990	**<0.0001**	**<0.0001**
		Gen × Water	**0.0248**	**0.0186**	**0.0216**	-	**<0.0001**	**<0.0001**

In the 2nd harvest year (2013), important increments in berry FM, DM, and bean BM were obtained due to supplementary irrigation, as compared to NI values (Table 1). Those increments were higher for two Ethiopia accessions than for two cultivars, although 'E083' and Iapar 59 did not differ as regards their absolute values under IRR conditions. The mass performances were positively impacted by irrigation for 'E027' and Iapar 59, while the opposite pattern was found for 'E083' and Catuaí 99 in the 2nd harvest year. Among the four genotypes in 2013, the DM and BM performances showed maximum values in Catuaí 99, followed by 'E083', both of which with greater values than Iapar 59 and 'E027', with the latter showing the smallest values. In 2013, berries of Catuaí 99 had the lowest initial berry moisture among the four studied genotypes, meaning the most advanced maturity of Catuaí 99.

The variation in yield parameters was additionally studied among the canopy strata of studied genotypes, irrespective of water regime. In 2012, no significant differences in FM, DM, BM, or initial berry moisture were observed among three canopy strata in three studied genotypes, despite the variation in estimated average values (Table 2). The DM and BM performances showed lower values in Catuaí 99 than in other genotypes, in all canopy strata, in 2012.

Table 2. Components of the berry and bean yields: fresh berry mass (FM), dry berry mass (DM), and processed bean mass (BM) (g stratum^{-1} initial berry moisture (%), DM and BM performances (%) of two harvests (2012 and 2013) analyzed in four genotypes of *Coffea arabica* along the canopy strata (S1: 20–60 cm; S2: 61–100 cm; S3: 101–140 cm; S4: > 40 cm). Estimated means and *p*-values are shown (n = 2–4), where "-" indicates that the model reduction and fitting were applied. When significant, *p*-value is marked in bold. Lower-case letters compare genotype effects in each stratum, whilst the upper-case letters compare stratum effects in each genotype, always separately for each harvest year.

Harvest Year	Stratum	Genotype	FM	DM	BM	Initial Berry Moisture	DM Performance	BM Performance
2012	S3	'E083'	178.2 aA	85.2 aA	44.2 aA	57.1 aA	47.4 aA	20.5 aA
		Iapar 59	303.0 aA	108.4 aA	52.3 aA	62.2 aA	45.8 aA	18.8 aA
		Catuaí 99	187.1 aA	65.2 aA	30.5 aA	65.0 aA	42.3 bA	15.4 bA
	S2	'E083'	220.4 aA	98.1 aA	46.3 aA	53.1 aA	45.2 aA	20.9 aA
		Iapar 59	336.1 aA	121.3 aA	54.4 aA	58.3 aA	48.1 aA	19.2 aA
		Catuaí 99	211.4 aA	78.1 aA	32.6 aA	61.0 aA	40.1 bA	15.8 bA
	S1	'E083'	151.2 aA	70.0 aA	35.1 aA	55.3 aA	46.0 aA	20.5 aA
		Iapar 59	268.1 aA	93.2 aA	43.2 aA	60.4 aA	46.6 aA	18.7 aA
		Catuaí 99	142.3 aA	49.9 aA	21.4 aA	63.2 aA	40.8 bA	15.3 bA
	p-value	Genotype	0.3217	0.4274	0.3739	0.2044	**0.0298**	**0.0245**
		Stratum	0.7031	0.5813	0.6561	0.6534	0.7322	0.9427
		Gen × Stratum	-	-	-	-	-	-
2013	S4	'E083'	488.7 abA	152.1 abAB	65.3 abAB	61.4 abA	43.2 bA	18.6 bA
		'E027'	131.6 bA	37.1 bAB	14.4 bAB	64.8 aA	50.5 bAB	19.4 bA
		Iapar 59	752.1 aA	263.7 aAB	124.4 aAB	63.7 aA	44.5 bA	18.3 bA
		Catuaí 99	283.0 abA	143.3 abAB	68.0 abAB	55.7 bA	51.4 aA	24.8 aA
	S3	'E083'	799.0 abA	290.7 abA	135.9 abA	60.8 abA	42.5 bA	17.3 bA
		'E027'	441.9 bA	175.6 bA	85.1 bA	64.2 aA	49.6 aAB	18.1 bA
		Iapar 59	1062.4 aA	402.2 aA	195.0 aA	63.1 aA	46.7 abA	17.0 bA
		Catuaí 99	593.3 abA	281.9 abA	138.6 abA	55.1 bA	48.7 aB	23.5 aA
	S2	'E083'	777.9 abA	252.2 abAB	116.2 abAB	60.8 abA	43.3 bA	17.2 bA
		'E027'	420.8 bA	137.1 bAB	65.3 bAB	64.2 aA	51.3 aA	18.0 bA
		Iapar 59	1041.3 aA	363.8 aAB	175.3 aAB	63.1 aA	48.9 aA	16.9 bA
		Catuaí 99	572.2 abA	243.4 abAB	118.9 abAB	55.1 bA	46.5 abB	23.4 aA
	S1	'E083'	408.9 abA	122.7 abB	52.2 abB	62.2 abA	44.8 abA	16.1 bA
		'E027'	51.8 bA	7.7 bB	1.33 bB	65.6 aA	44.6 abB	16.9 bA
		Iapar 59	672.2 aA	234.3 aB	111.3 aB	64.4 aA	49.3 aA	15.8 bA
		Catuaí 99	203.1 abA	114.0 abB	54.9 abB	56.5 bA	41.0 bB	22.3 aA
	p-value	Genotype	**0.0358**	**<0.0001**	**0.0353**	**0.0213**	**<0.0001**	**<0.0001**
		Stratum	0.1581	**0.0429**	**0.0473**	0.9569	**0.0428**	0.2244
		Gen × Stratum	-	-	-	-	**0.0005**	-

Notably, in 2013, the highest FM, DM, and BM values were obtained in Iapar 59, and the lowest in 'E027' consistently in all canopy strata (Table 2). The highest and lowest values of DM and BM were found in S3 and S1, respectively, for all genotypes, showing greater berry production in the stratum that received high PPFD. In 2013, the initial berry moisture did not differ among strata, but was the lowest in Catuaí 99, opposite to BM performance that was the highest in this genotype. The DM performance showed the interaction of stratification with genotype: it was stable among the strata in 'E083' and Iapar 59; had the highest values in S2 and the lowest in S1 (two self-shaded strata) in 'E027'; and had the highest values in the well-lighted S4 in Catuaí 99. Among the four genotypes, Catuaí 99 showed the lowest DM performance in the S1, intermediate in S2, but the highest in

S3 and S4, which was related to low initial berry moisture in this genotype. This genotype also showed the highest BM performance in all strata among the four genotypes, suggesting its lowest losses in bean processing.

3.4. Chemical Attributes of the Coffee Dependency on Water Regime, Genotype, and Irradiance Availability

High water availability (IRR) significantly altered only PRO and LIP contents, which were reduced although not in great extent (between 5 and 6%) in the 1st harvest year in all genotypes (Table 3). Furthermore, these compounds were mostly unresponsive to water conditions in the 2nd year, with a marginal (but significant) decline of LIP content only in 'E027' and Iapar 59 under IRR. On the other hand, the PRO, LIP, CAF, and CGA varied among studied genotypes in the 1st harvest year, and in all studied components in the 2nd harvest year in both water conditions. The highest PRO and CAF contents were found in Catuaí 99 and Iapar 59 in the 1st harvest year, and in Catuaí 99 in the 2nd year, when this genotype also showed the greatest contents of CGA and PC in both water regime. By contrast, Catuaí 99 showed the lowest values of LIP, SUC, and TS in the 2nd harvest year, with the highest contents of those compounds being found in 'E083' and Iapar 59 under both water regimes.

Additionally, the variation in chemical composition of coffee beans was analyzed along the canopy strata in studied genotypes (Figure 5). In the 1st harvest year, only PRO content showed a stratification, being the lowest in the upper S3 in all three analyzed genotypes, as compared to their respective S1 and S2 (Figure 5A). In the 2nd harvest year, additionally to PRO, bean CAF content of all four genotypes also varied among the canopy strata, with greater values found in S1 in all genotypes, and usually gradually decreasing at higher (S3 or S4) strata, that is, with greater irradiance (Figure 5B).

Table 3. Chemical contents (% of dry matter): PRO (proteins), LIP (lipids), CAF (caffeine), SUC (sucrose), TS (total soluble sugars), CGA (total chlorogenic acids), PC (phenolic compounds) of two harvests (2012 and 2013) analyzed in four genotypes of *Coffea arabica* cultivated under two water regimes (irrigated—IRR; not irrigated—NI). Estimated means and *p*-values are shown ($n = 4$), where "-" indicates that the model reduction and fitting were applied. When significant, *p*-value is marked in bold. Lower-case letters compare genotype effects within each water regime, whilst the upper-case letters compare water regime effects within each genotype, always separately for each harvest year.

Harvest Year	Water Regime	Genotype	PRO	LIP	CAF	SUC	TS	CGA	PC
2012	IRR	'E083'	15.0 bB	15.4 aB	1.32 bA	7.41 aA	7.82 aA	5.31 bA	6.80 aA
		Iapar 59	16.3 aB	14.0 bB	1.58 aA	7.01 aA	7.43 aA	5.75 aA	7.28 aA
		Catuaí 99	16.4 aB	13.5 bB	1.56 aA	6.43 aA	6.75 aA	5.10 bA	7.60 aA
	NI	'E083'	15.8 bA	16.2 aA	1.25 bA	6.47 aA	6.94 aA	5.22 bA	7.27 aA
		Iapar 59	17.1 aA	14.8 bA	1.51 aA	6.07 aA	6.55 aA	5.66 aA	7.76 aA
		Catuaí 99	17.2 aA	14.3 bA	1.49 aA	5.49 aA	5.87 aA	5.01 bA	8.08 aA
	p-value	Genotype	**<0.0001**	**0.0004**	**0.0004**	0.1401	0.1053	**0.0036**	0.0855
		Water regime	**0.0003**	**0.0258**	0.1776	0.0703	0.0848	0.5799	0.1759
		Gen × Water	-	-	-	-	-	-	-
2013	IRR	'E083'	14.6 cA	14.2 aA	1.34 dA	6.74 aA	6.99 aA	4.96 cA	5.88 bA
		'E027'	16.3 bA	12.3 cB	1.80 bA	4.47 bA	4.59 bA	5.65 bA	6.57 bA
		Iapar 59	15.4 cA	13.2 bB	1.61 cA	6.13 aA	6.37 aA	5.52 bA	6.08 bA
		Catuaí 99	18.3 aA	10.9 dA	2.03 aA	3.28 bA	3.48 cA	6.84 aA	7.93 aA
	NI	'E083'	14.2 cA	14.0 aA	1.30 dA	6.87 aA	7.19 aA	5.12 cA	6.23 bA
		'E027'	15.9 bA	13.1 bA	1.76 bA	4.60 bA	4.79 bA	5.81 bA	6.93 bA
		Iapar 59	15.0 cA	14.4 aA	1.57 cA	6.26 aA	6.57 aA	5.67 bA	6.43 bA
		Catuaí 99	18.0 aA	10.9 cA	1.98 aA	3.42 bA	3.59 cA	7.00 aA	8.28 aA
	p-value	Genotype	**<0.0001**	**<0.0001**	**<0.0001**	**<0.0001**	**<0.0001**	**<0.0001**	**<0.0001**
		Water regime	0.1634	**0.0019**	0.2509	0.6737	0.5294	0.2812	0.1308
		Gen × Water	-	**0.0001**	-	-	-	-	-

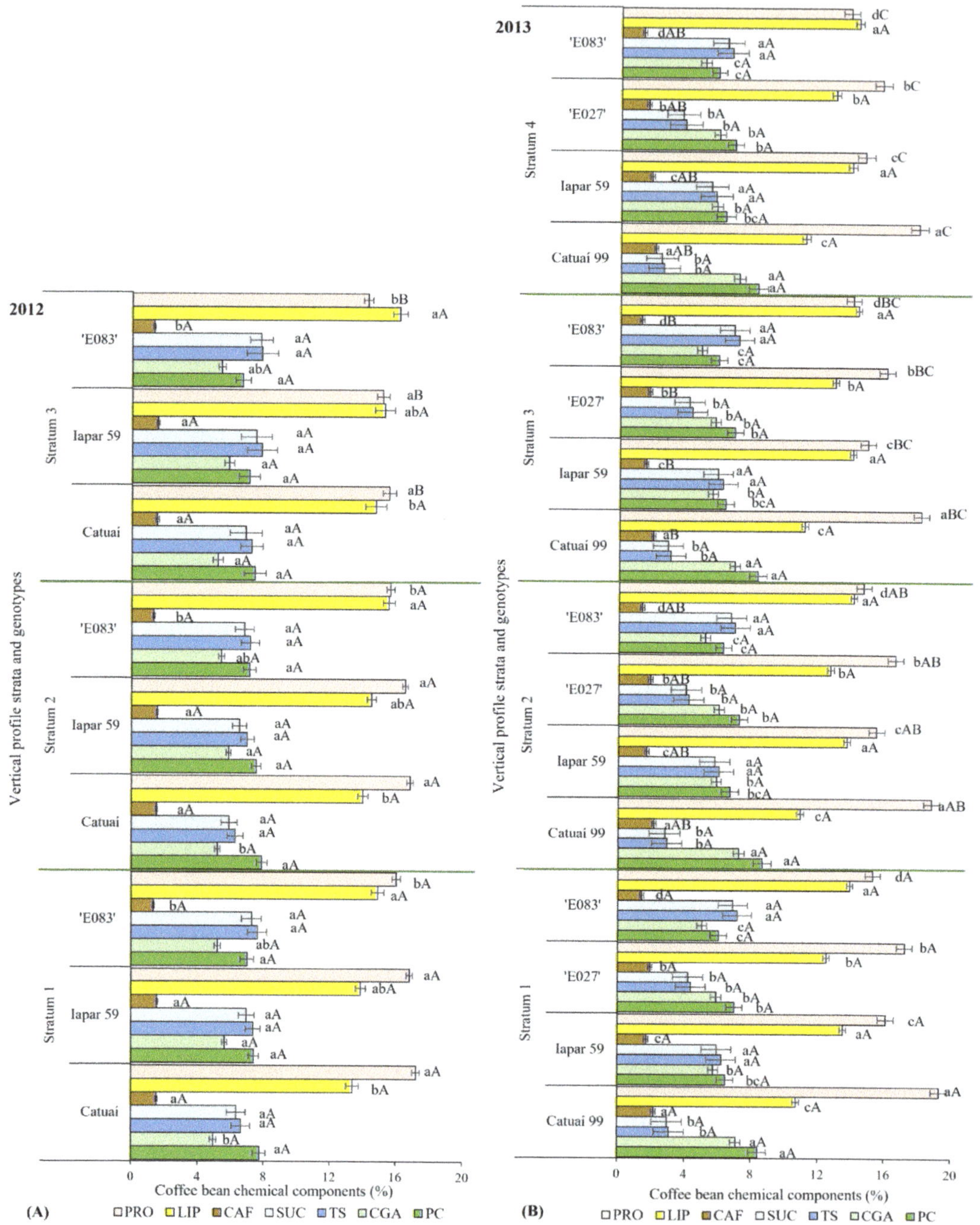

Figure 5. Chemical contents (% of dry matter) of coffee grains: PRO (proteins), LIP (lipids), CAF (caffeine), SUC (sucrose), TS (total soluble sugars), CGA (total chlorogenic acids), PC (phenolic compounds) of two harvests, (**A**) 2012 and (**B**) 2013, analyzed in four genotypes of *Coffea arabica* among the plant canopy strata (S1: 20–60 cm; S2: 61–100 cm; S3: 101–140 cm; S4: >140 cm). Estimated means ± standard error (n = 2–4) are shown. Lower-case letters compare genotype effects in each stratum, whilst the upper-case letters compare stratum effects in each genotype.

3.5. Correlations among the Berry Distribution per Strata, Yield, and Chemical Attributes

To synthesize the previous results, a correlation analysis was performed (Figure 6) considering berry distribution along the strata (sequential growing numbers for 'Stratum' in correlations), yield, and chemical coffee bean characteristics. In 2012, a decline of berry number was observed from S1 to S3 (Figure 6A), as a general response not observed when genotype and soil water availability impacts were analyzed (Table 2). Positive correlation of FM, DM, and BM was observed with increased canopy strata in 2013 (Figure 6B). Increased stratum height was associated with a decline of protein content in both harvest years (Figure 6A,B), decreased content in CAF in 2013 (Figure 6B), and increased in LIP in 2012 (Figure 6A). The FM over the plant canopy strata was positively associated to DM, BM in both years (Figure 6), and to increased initial berry moisture in 2012 (Figure 6A). The last suggested that in the first harvest year, the higher strata produced berries that that were induced lately (from latter flowering), gaining the red color rapidly, but still retaining more moisture than lower strata.

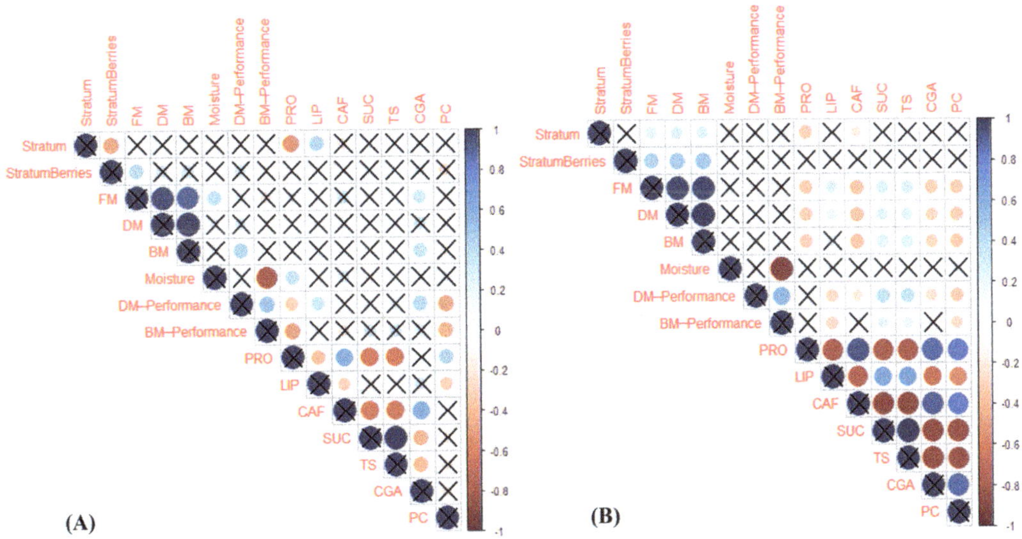

Figure 6. Graphical presentation of coefficients (values corresponding to circle size and color intensities) and p-values < 0.05 (not crossed circles) for general correlations among berry distribution along plant canopy strata (Stratum (S1–S4), StratumBerries (number of berries stratum^{-1})), yield [FM and DM (fresh and dry berry mass stratum^{-1} respectively), BM (processed bean mass stratum^{-1}), initial berry moisture, DM and BM performances, and chemical parameters (PRO (proteins), LIP (lipids), CAF (caffeine), SUC (sucrose), TS (total soluble sugars), CGA (total chlorogenic acids), PC (phenolic compounds)) of four *Coffea arabica* genotypes in (**A**) 2012, and (**B**) 2013.

Interestingly, greater FM and DM of berries per stratum was associated with greater LIP, SUC, and TS bean contents, and lower PRO, CAF, CGA, and PC values in 2013 (Figure 6B). The initial berry moisture was negatively correlated to BM performance in both harvest years (Figure 6), and positively to PRO grain content in 2012 (Figure 6A). The DM and BM performances were negatively associated with LIP, CAF, CGA and PC contents and positively to sugar contents in 2012 (Figure 6B).

The general patterns in bean chemical composition in both harvest years were negative correlations of PRO to LIP, SUC, and TS contents, and positive to CAF, TC, and PC, coherent with the opposite situation observed in LIP (negative correlations to CAF, TC, and PC, and positive to SUC and TS contents), especially in 2013 (Figure 6B). TS and SUC were negatively correlated to CGA and PC compounds.

4. Discussion

Coffee bean quality results from complex interactions between the relation of genetics and environment, which influences the presence/content of more than 1000 bean components [21]. The ongoing climate changes and the predicted future conditions demand evidence-based insights as regards coffee yield and quality, to guarantee the sustainability and promote resilience of the coffee sector, from field to cup [39], namely through plant breeding and adequate crop management. Here, we provided some novel data regarding yield and bean quality traits in the two initial harvest years, associated with water management, genotype, and microenvironment irradiance from the bottom (shaded) to the top (highly lighted) strata of plant canopy.

Additional irrigation significantly increased the berry number plant^{-1} in the 2nd harvest year (except Catuaí 99) (Figure 3) and on FM, DM, and BM in both harvest years (Table 1). This denoted a positive impact of irrigation on plant yield potential, in line with the previously reported gains in accumulation of biomass per fruit in *C. canephora* cv. Conilon [51]. In 2013, IRR plants showed strong increases of FM, DM, and BM in all genotypes, with 'E083' having the strongest (up to ca. 500%) and Catuaí 99 moderate yield improvements (ca. 50%). Conversely, the extent of the yield reduction (especially of BM) under NI conditions in the 2nd harvest year, was lower in Iapar 59 and Catuaí 99 than in the two Ethiopian wild accessions (Table 1). The studied coffee plants under NI experienced some dry periods (Figure 2B). The molecular mechanisms of drought-resistance in *C. arabica* genotypes involve ABA signaling, together with predominance of protective genes expression, associated with antioxidant activities, including genes involved in water deprivation and desiccation [52]. That could have been the case in Catuaí 99 and Iapar 59 plants under NI, as judged by their elevated berry/bean yielding (Figure 3, Table 1). Such lower sensitivity to lower water availability in Iapar 59 than in other genotypes could be related to its *C. canephora* introgression genetics [41].

The initial berry moisture is about 55–65%, which after drying falls to ca. 12% [53]. Our results of initial berry moisture (54–67%) were close to the usual values [53], although they additionally differed among the genotypes in the 2nd harvest year, the lowest being in Catuaí 99, indicating its advanced fruit maturation [54], as compared to the other three genotypes (Tables 1 and 2). As the exocarp is more attached in immature berries, which will induce more breaks and losses during the dry processing than in mature berries [15], consequently diminishing the bean mass. The bean mass performance was 12.1–20.6% in 2012 and 14.9–24.3% in 2013. In 2012, this trait was higher under IRR than under NI in all genotypes, but values differed among the genotypes, the highest in irrigated being 'E083' and the lowest in not irrigated Catuaí 99 (Table 1). Those trends were strongly modified in 2013, showing the greatest values in NI Catuaí 99, followed by 'E083', and with the lowest values in 'E027'. This suggested that Catuaí 99 (and 'E083') plants, with increased yield in the 2nd compared to the 1st harvest year, especially under IRR, could have higher flowering and maturation uniformity and earlier maturation, which impacted on higher yield performances than other studied genotypes. The higher flowering uniformity is normally expected from IRR than from NI coffees [55], but the Catuaí 99 and 'E083' responses in the 2nd year suggested their advanced maturation, and lower bean defect contents under lower water availability, even if a much greater yield was obtained under IRR conditions.

The leaf-to-fruit and branch-to-fruit ratios differed among the genotypes in both harvest years, always being the highest in 'E027' (Figure 3C–F), which resulted mainly from the lowest berry yield among these genotypes, and a concomitant elevated investment in vegetative structures (Tables 1 and 2). Leaf-to-fruit ratio is an expression of source–sink ratios considering growing sinks and maintenance of sources [56]. Previous reports estimated that ca. 20 cm^2 of leaf area is needed to support each coffee berry development [17]. Still, this value can be somewhat smaller in heavily bearing coffee trees [57], in line with our smallest values (ca. 17 cm^2 berry^{-1}). Leaf-to-fruit ratios below the 20 cm^2 berry^{-1} were registered in NI plants of 'E083' in the 1st harvest year, while in the 2nd year, those values were higher than 20 cm^2 $berry^{-1}$, and greater in NI than IRR plants (except in Catuaí 99). These high values of leaf-to-fruit ratio, although being positive for bean development, also suggest that the berry production was relatively more impacted by low soil water availability (Table 1) than the vegetative growth [40]. The elevated branch-to-fruit ratios under NI in 'E027' and 'E083' in 2013 showed relatively higher structural than reproductive investments of Ethiopian accessions than of test cultivars, indicated a higher sensitivity of former to lower water availability.

Greater concentration of some secondary metabolites of coffee beans is usually associated with increased sensory attributes (e.g., trigonelline), whereas an increase of others (e.g., feruloylquinic acids) is often associated with a decline in sensory attributes, although with a high degree of uncertainty resulting from their specific thresholds, impacts, and cross interaction [21,22,39]. In fact, some compounds, such as chlorogenic acids, are associated to both positive and negative impacts in cup quality, depending on their absolute values and their interactions with other chemical compounds in each specific coffee [2,58]. More than 30 chlorogenic acid isomers are detected in coffee beans, as bioactive compounds with antioxidant activity against free radicals and metal ions [59]. They accumulate in the beans as the berries mature, and greatly contribute to the final acidity, aroma, flavor, bitterness, and astringency of the coffee beverage [2]. The presence/content of primary metabolites (proteins, lipids, and carbohydrates) are also related to cup quality [22,60]. In both harvest years, the highest PRO and CAF and the lowest LIP contents, together with the highest CGA and PC contents in the 2nd harvest year were registered in Catuaí 99. In addition, the highest LIP, SUC, and TS content was found 'E083', irrespective of harvest year or water regime (Table 3). On the other hand, the most elevated CGA and PC content were found in Iapar 59 in the 1st harvest year, showing strong genetic impact on bean chemical quality. Phenolic compounds are secondary metabolites that can be involved in leaves during the plant adaptation to environmental stress conditions, among them the chlorogenic acids that are the main components of the phenolic fraction of coffee beans [61]. Chlorogenic acid impacts on the beverage are more dependent on roasting level than on coffee species [58], but we found here that CGA can be genotype dependent.

Iapar 59 has more genetic variation, more uniform maturation, and earlier yielding than Catuaí 99 [62]. In our experiment, the berries were collected when they were visually mature in all plants, excluding maturity as chemical composition factor. Later berry harvest performed in the same year can increase LIP and reducing sugar and decrease CAF contents [63], but maturity states of red berries do not show differences in organic acids, free fatty acids, lipids, total chlorogenic acids, proteins, alkaloids, or sucrose [54]. Catuaí 99 is seen as a highly productive cultivar, of high bean quality for the final coffee cup [42], but here it showed elevated PRO, CAF, TCA, and PC contents, generally associated with lower cup quality [22]. 'E083', usually together with Iapar 59, seemed to present the best chemical components composition, with the highest LIP, SUC, and TS and the lowest CAF, CGA, and PC contents among genotypes, irrespective of harvest year or water regime (Table 3).

An interesting finding was that irrigation promoted a decline on LIP and PRO bean contents of all genotypes harvested in 1st harvest year, and on LIP of Iapar 59 and 'E027' in the 2nd harvest year (Table 3). This can be an interesting issue since, lipids of *C. arabica* beans are responsible for flavor carriers, texture, and mouthfeel in the beverage [19,64–66]. Additionally, sugars are also greatly present and intended components of coffee beans, with SUC being the most represented TS irrespective of the environmental conditions [2], in line with the absence of changes in SUC and TS between water conditions (Table 3). Shade is reported to either reduce sucrose content [67,68], or increase reducing sugars in coffee beans [69], contrary to our findings that, in turn, agreed with absence of effect of light level on the content of sucrose, glucose, fructose, arabinose, and total soluble sugars [2]. In this regard, the greater values LIP, SUC, and TS could have a potential positive impact in cup quality [70], thus pointing to a greater potential bean quality in Iapar 59 and 'E083', despite the relevant sensitivity to low water availability in the latter, as reflected in the greatest yield decline from IRR to NI conditions.

The lower levels of CAF, CGA, PC in 'E083' in both harvest years, irrespective of water availability and in Iapar 59 in the 2nd harvest year (Table 3) could have positive implications on coffee quality [22], although with a certain degree of uncertainty. It is usually accepted that CAF and CGA contribute to coffee bitterness, and their significant rise under higher temperature conditions can result in poor bean quality [67,69]. Chlorogenic acids are greatly represented by monocaffeoylquinic acids (CQAs), among them by 5-CQA. The increase of CQAs level has an inverse association with cup quality, particularly concerning 5-CQA [21,71]. 5-CQA constitutes the major substrate for polyphenol oxidase, producing ortho-quinones that in turn will cause darkening of the beans and a worse coffee quality [72]. The 5-CQA rise is inconsistent with first grade coffee beans [73], as it is also characterized as low acidic, with a small amount of bitterness [58]. Furthermore, the presence of feruloylquinic acids and diCQAs isomers (also CGA) can have detrimental impact on the sensory properties of espresso coffee beverage, associated to greater bitterness, together with a metallic taste and astringency [74]. However, the impacts on cup quality of increased presence of chlorogenic acids and phenolic compounds would strongly depend on the specific compounds in each category and, mostly, with the interaction with other compounds [58]. Low soil water availability can promote the increase of CGA in the bean [75], which was not observed in our experiment regardless of genotype.

Altogether, lower water availability resulted in important yield decline but the direct impact on bean quality reported by others [30] was not likely to occur in our experimental conditions and genotypes. In fact, in 2013, with exception of LIP content in Iapar 59 and 'E027', no significant changes were depicted among the studied compounds, irrespective of genotype (Table 3).

The 'E083' presented the most expansive vertical and horizontal berry spatial occupations under irrigation (Figure 4). Despite Iapar 59 had less expansive space occupation than 'E083', the berry distribution tended to occupy the whole canopy strata, continuing with berry production in self-shaded strata, and producing the most in both 'E083' and Iapar 59 (Table 2, Figure 3). Judging by the produced fresh berry mass related to berry number, 'E083' produced the berries of smaller size than Iapar 59. In the 2nd harvest year, the highest biomass was produced in the 3rd stratum, and the lowest in the 1st and 4th strata, irrespective of genotype (Table 2). Among the studied chemical components considering both harvest years, only the PRO varied between the canopy strata, showing a pattern of increase from the top to the bottom in all genotypes, particularly in 2013 (Figures 5 and 6). Under high irradiance, the common stress responses in green leaves are usually associated with efficient dissipation of excess of excitation energy, and antioxidant mechanisms, an increased abundance of proteins associated with the photosynthetic apparatus, together with relevant dynamics of the lipid profile of cell membranes [76]. As regards the bean, LIP fraction did not vary among the coffee canopy strata, which were submitted to contrasting irradiances (Figures 5 and 6), thus showing a different response than that found in leaves. In the 2nd harvest year, CAF showed differences between S1 and S3, with lower values in the latter. In fact, under greater self-shading (S1), the beans tended to have greater CAF content and lower production, whereas moderately lighted S3 (Figure 2D) showed the opposite yield (Table 2) and CAF content (Figures 5 and 6), confirming earlier findings [18].

Recently, various review studies tried to synthetize results about coffee responses to stresses associated with coffee quality and chemical bean composition, but many data are still contradictory [2,23,39,59]. Yield and chemical traits differed more regarding the strata in the 2nd than in the 1st harvest year, indicating that those interannual variations were apparently more related to overall increased variation in microenvironment light conditions among canopy strata due to tree growth, than to soil water availability. This rationale is associated to one dry period in the 1st harvest year that occurred in extremely dry summer, during leaf and berry expansion, the crucial phenophase for fruit development and quality, while in the 2nd harvest year one dry period occurred during berry maturation, when its impact was less detrimental. The greater was the grain biomass per strata, greater were the LIP, SUC, and TS, and lower the PRO, CAF, CGA, and PC bean contents in the 2nd harvest year (Figure 6), suggesting strong light/temperature impacts on berry content, and/or costs of investments of the primary and secondary metabolism products. The expensive investments in the bean in biosynthesis of secondary metabolites that play role in protection and acclimation to stress conditions in the leaves (PC, CGA, CAF) including the synchronization with primary metabolism (PRO, TS, and LIP fractions) [77], might indicate that will be prevalent under climate challenging conditions, which will continue to be a focus of world agriculture interest.

5. Conclusions

Our hypothesis was that, depending on genotype, additional soil water availability on one side, and light availability along the canopy strata on the other, could impact coffee berry yield and bean chemical traits, studied in the two initial harvest years. Additional irrigation strongly increased the berry number and yield, modifying the berry distribution, but revealed to influence only some chemical bean attributes (reducing LIP and PRO contents). On the other hand, the genotype factor impacted practically all observed variables. The microenvironment light variation impacted on berry distribution and only on PRO and CAF bean contents among canopy strata. All metrics were variable between the two the harvest years, namely, berry yield and distribution, leaf-to-fruit ratio, and chemical composition of the bean. Yield and chemical traits differed more regarding the strata in the 2nd than in the 1st harvest year, indicating that those interannual variations were apparently more related to overall increased variation in microenvironment light conditions among canopy strata due to tree growth, than to soil water availability.

Iapar 59 plants were the less influenced by lower water availability among the four genotypes, showing the lowest yield reduction from irrigated to not irrigated plants at the 2nd harvest year. CGA was shown as genotype dependent, being the highest in Iapar 59 and Catuaí 99, irrespective of water regime of strata. Among the studied chemical components, PRO presented the greatest variation among the canopy strata, with a clear trend to decline from the shaded (bottom) to the well-lighted (upper) strata, in all genotypes and both harvest years. Similar stratification (now between S1 and S3) was observed for CAF, but only in the 2nd harvest year. In this harvest year, the most regular berry distribution along the plant canopy strata was obtained in irrigated 'E083', the highest bean yield in irrigated 'E083' and Iapar 59, and the highest DM and BM performances in not irrigated Catuaí 99. Iapar 59 and the Ethiopian wild accession 'E083' showed chemical composition associated with high cup quality, with the highest LIP, SUC, and TS and the lowest CAF, CGA, and PC contents among studied genotypes. Overall, these genotypes showed intrinsic differences on yield potential, which was impacted by water availability and self-shading.

Although some chemical bean traits had varied among strata (PRO), no substantial quality changes occurred. These suggest that the entire plant could be harvested, not only some specific strata, without an important deleterious impact on quality. High bean quality and the less reduced yield under reduced soil water availability qualified Iapar 59 as the most prominent among the four genotypes. Based on berry vertical and radial space occupation, yield, bean quality, the Ethiopian wild accession 'E083' could also be an interesting genetic material for the future breeding programs, although to be used with irrigation input, together with high fertilization condition, as applied in this experiment.

Author Contributions: Conceptualization, M.R.; methodology, M.R. and M.B.d.S.S.; software, F.T.M.; validation, M.R. and J.C.R.; formal analysis, R.A.A.P. and M.R.; investigation, M.R. and M.B.d.S.S.; resources, M.R.; data curation, R.A.A.P. and M.R.; writing—original draft preparation, M.R.; writing—review and editing, J.C.R.; visualization, M.R. and J.C.R.; supervision, M.R.; project administration, M.R.; funding acquisition, M.R. and M.B.d.S.S. All authors have read and agreed to the published version of the manuscript.

Funding: The research work was carried out with the support of Consórcio Pesquisa Café (Brazil) projects (02.09.20.008.00.00 and 02.13.02.042.00.00). The authors acknowledge the Conselho Nacional de Desenvolvimento Científico e Tecnológico (CNPq), Brazil, for Invited Researcher fellowships awarded to M.R. (350509/2020–4) and the Fundação para a Ciência e a Tecnologia I.P., Portugal, awarded to J.C.R., through the units UID/04129/2020 (CEF), and UIDP/04035/2020 (GeoBioTec).

Institutional Review Board Statement: Not applicable.

Informed Consent Statement: Not applicable.

Data Availability Statement: Publicly available datasets were analyzed in this study. The authors can provide the experimental data for all interested researchers.

Conflicts of Interest: The authors declare no conflict of interest.

References

1. Davis, A.P.; Rakotonasolo, F. Six new species of coffee (*Coffea*) from northern Madagascar. *Kew. Bull.* **2021**, *76*, 497–511. [CrossRef]
2. Cassamo, C.T.; Mangueze, A.V.; Leitão, A.E.; Pais, I.P.; Moreira, R.; Campa, C.; Chiulele, R.; Reis, F.O.; Marques, I.; Scotti-Campos, P.; et al. Shade and altitude implications on the physical and chemical attributes of green coffee beans from Gorongosa Mountain, Mozambique. *Agronomy* **2022**, *12*, 2540. [CrossRef]
3. FAO. Available online: https://www.fao.org/markets-and-trade/commodities/coffee/en/ (accessed on 18 October 2022).
4. Coffee Industry Statistics. Available online: https://coffeeaffection.com/coffee-industry-statistics/ (accessed on 4 November 2022).
5. Hallé, F.; Oldeman, R.A.A.; Tomlinson, P.B. *Tropical Trees and Forests—An Architectural Analysis*; Springer: Berlin, Germany, 1978; p. 441.
6. Rakocevic, M.; Matsunaga, F.T.; Baroni, D.F.; Campostrini, E.; Costes, E. Multiscale analyses of growth and berry distributions along four branching orders and vertical profile of *Coffea arabica* L. cultivated under high-density planting systems. *Sci. Hortic.* **2021**, *281*, 109934. [CrossRef]
7. Ferreira, T.; Shuler, J.; Guimarães, R.; Farah, A. Introduction to coffee plant and genetics. In *Coffee: Production, Quality and Chemistry*; Royal Society of Chemistry: Cambridge, UK, 2019; pp. 1–25. [CrossRef]
8. Rakocevic, M.; Androcioli-Filho, A. Morphophysiological characteristics of *Coffea arabica* L. in different arrangements: Lessons from a 3D virtual plant approach. *Coffee Sci.* **2010**, *5*, 54–166.
9. Camargo, A.P.; Camargo, M.B.P. Definição e esquematização das fases fenológicas do cafeeiro arábica nas condições tropicais do Brasil. *Bragantia* **2001**, *60*, 65–68. [CrossRef]
10. Quiñones, A.J.P.; Builes, V.H.R.; Robledo, A.J.; Sáenz, J.R.R.; Pulgarín, J.A. Effects of daylength and soil humidity on the flowering of coffee *Coffea arabica* L. in Colombia. *Rev. Fac. Nac. Agron. Medellin* **2011**, *64*, 5745–5754.
11. Rena, A.B.; Barros, R.S. Aspectos críticos no estudo da floraçãao do café. In *Efeitos da Irrigaçãao Sobre a Qualidade e Produtividade do Café*; Zambolim, L., Ed.; Universidade Federal de Viçosa: Viçosa, Brazil, 2004; pp. 149–172.
12. Rakocevic, M.; Braga, K.S.M.; Batista, E.R.; Maia, A.H.N.; Scholz, M.B.S.; Filizola, H.F. The vegetative growth assists to reproductive responses of Arabic coffee trees in a long-term FACE experiment. *Plant Growth Regul.* **2020**, *91*, 305–316. [CrossRef]
13. Cardon, C.H.; de Oliveira, R.R.; Lesy, V.; Ribeiro, T.H.; Fust, C.; Pereira, L.P.; Colasanti, J.; Chalfun-Junior, A. Expression of coffee florigen CaFT1 reveals a sustained floral induction window associated with asynchronous flowering in tropical perennials. *Plant Sci.* **2022**, *325*, 111479. [CrossRef]
14. Morais, H.; Caramori, P.H.; Koguishi, M.S.; Ribeiro, A.M.A. Detailed phenological scale of the reproductive phase of *Coffea arabica*. *Bragantia* **2008**, *67*, 257–260. [CrossRef]
15. Kitzberger, C.S.G.; Pot, D.; Marraccini, P.; Pereira, L.F.P.; Scholz, M.B.S. Flavor precursors and sensory attributes of coffee submitted to different post-harvest processing. *AIMS Agric. Food* **2020**, *5*, 700–714. [CrossRef]
16. Alnsour, L.; Issa, R.; Awwad, S.; Albals, D.; Al-Momani, I. Quantification of total phenols and antioxidants in coffee samples of different origins and evaluation of the effect of degree of roasting on their levels. *Molecules* **2022**, *27*, 1591. [CrossRef] [PubMed]
17. Cannell, M.G. Physiology of the coffee crop. In *Coffee—Botany, Biochemistry and Production of Beans and Beverage*; Clifford, M.N., Willson, K.C., Eds.; Crom Helm: London, UK, 1985; pp. 108–134.
18. Rakocevic, M.; Batista, E.R.; Pazianotto, R.A.A.; Scholz, M.B.S.; Souza, G.A.R.; Campostrini, E.; Ramalho, J.C. Leaf gas exchange and bean quality fluctuations over the whole canopy vertical profile of Arabic coffee cultivated under elevated CO_2. *Funct. Plant Biol.* **2021**, *48*, 469–482. [CrossRef] [PubMed]
19. Farah, A. Coffee constituents. In *Coffee: Emerging Health Effects and Disease Prevention*; Chu, Y.-F., Ed.; Blackwell Publishing Ltd.: Hoboken, NJ, USA, 2012; pp. 21–58.
20. Brige, F.A.A.; Celestino, S.M.C.; Amabile, R.F.; Fagioli, M.; Delvico, F.M.S.; Montalvão, A.P.L.; Sala, P.I.A.L. Genetic variability in conilon coffee related to grain attributes in an irrigated crop in the Cerrado. *Pesqui. Agropecu Bras.* **2019**, *54*, e00358. [CrossRef]
21. Farah, A.; Monteiro, M.C.; Calado, V.; Franca, A.S.; Trugo, L.C. Food chemistry correlation between cup quality and chemical attributes of Brazilian coffee. *Food Chem.* **2006**, *98*, 373–380. [CrossRef]
22. Scholz, M.B.S.; Kitzberger, C.S.G.; Durand, N.; Rakocevic, M. From the field to coffee cup: Impact of planting design on chlorogenic acid isomers and other compounds in coffee beans and sensory attributes of coffee beverage. *Eur. Food. Res. Technol.* **2018**, *244*, 793–1802. [CrossRef]

23. Velásquez, S.; Banchón, C. Influence of pre-and post-harvest factors on the organoleptic and physicochemical quality of coffee: A short review. *J. Food Sci. Technol.* **2022**, *15*, 1–13. [CrossRef]
24. Lambot, C.; Herrera, J.C.; Bertrand, B.; Sadeghian, S.; Benavides, P.; Gaitán, A. Cultivating coffee quality—Terroir and agro-ecosystem. In *The Craft and Science of Coffee*; Folmer, B., Ed.; Academic Press: Cambridge, MA, USA, 2017; pp. 17–49. [CrossRef]
25. Ginbo, T. Heterogeneous impacts of climate change on crop yields across altitudes in Ethiopia. *Clim. Chang.* **2022**, *170*, 12. [CrossRef]
26. Koutouleas, A.; Sarzynski, T.; Bertrand, B.; Bordeaux, M.; Bosselmann, A.S.; Campa, C.; Etienne, H.; Turreira-García, N.; Léran, S.; Markussen, B.; et al. Shade effects on yield across different Coffea arabica cultivars-how much is too much? A meta-analysis. *Agron. Sustain. Dev.* **2022**, *42*, 55. [CrossRef]
27. Koutouleas, A.; Sarzynski, T.; Bordeaux, M.; Bosselmann, A.S.; Campa, C.; Etienne, H.; Turreira-García, N.; Rigal, C.; Vaast, P.; Ramalho, J.C.; et al. Shaded-coffee: A nature-based strategy for coffee production under climate change? A Review. *Front. Sustain. Food Syst.* **2022**, *6*, 877476. [CrossRef]
28. Rakocevic, M.; Scholz, M.B.S.; Kitzberger, C.S.G. Berry distributions on coffee trees cultivated under high densities modulate the chemical composition of respective coffee beans during one biannual cycle. *Int. J. Fruit Sci.* **2018**, *18*, 117–137. [CrossRef]
29. Reis, A.R.; Favarin, J.L.; Gallo, L.A.; Moraes, M.F.; Tezotto, T.; Lavres, J.J. Influence of nitrogen fertilization on nickel accumulation and chemical composition of coffee plants during fruit development. *J. Plant Nutr.* **2011**, *34*, 1853–1866. [CrossRef]
30. Vinecky, F.; Davrieux, F.; Mera, A.C.; Alves, G.S.; Lavagnini, G.; Leroy, T.; Bonnot, F.; Rocha, O.C.; Bartholo, G.F.; Guerra, A.F.; et al. Controlled irrigation and nitrogen, phosphorous and potassium fertilization affect the biochemical composition and quality of Arabica coffee beans. *J. Agric. Sci.* **2017**, *155*, 902–918. [CrossRef]
31. Marcheafave, G.G.; Tormena, C.D.; Terrile, A.E.; Salamanca-Neto, C.A.; Sartori, E.R.; Rakocevic, M.; Bruns, R.E.; Scarminio, I.S.; Pauli, E.D. Ecometabolic mixture design-fingerprints from exploratory multi-block data analysis in *Coffea arabica* beans from climate changes: Elevated carbon dioxide and reduced soil water availability. *Food. Chem.* **2021**, *362*, 129716. [CrossRef] [PubMed]
32. Gokavi, N.; Mote, K.; Jayakumar, M.; Raghuramulu, Y.; Surendran, U. The effect of modified pruning and planting systems on growth, yield, labour use efficiency and economics of Arabica coffee. *Sci. Hortic.* **2021**, *276*, 109764. [CrossRef]
33. Ramalho, J.C.; Pais, I.P.; Leitão, A.E.; Guerra, M.; Reboredo, F.H.; Máguas, C.M.; Carvalho, M.L.; Scotti-Campos, P.; Ribeiro-Barros, A.I.; Lidon, F.J.; et al. Can elevated air [CO_2] conditions mitigate the predicted warming impact on the quality of coffee bean? *Front. Plant Sci.* **2018**, *9*, 287. [CrossRef] [PubMed]
34. Silva, E.B.; Nogueira, F.D.; Guimarães, P.T.G.; Chagas, S.J.R.; Costa, L. Sources and doses of potassium on the yield and quality of green coffee. *Pesq. Agropec. Bras.* **1999**, *34*, 335–345. [CrossRef]
35. Carr, M. The water relations and irrigation requirements of coffee. *Exp. Agric.* **2001**, *37*, 1–36. [CrossRef]
36. Tesfaye, S.G.; Ismail, M.R.; Kausar, H.; Marziah, M.; Ramlan, M.F. Plant water relations, crop yield and quality in coffee (*Coffea arabica* L.) as influenced by partial root zone drying and deficit irrigation. *AJCS* **2013**, *7*, 1361–1368. [CrossRef]
37. Liu, X.; Qi, Y.; Li, F.; Yang, Q.; Yu, L. Impacts of regulated deficit irrigation on yield, quality and water use efficiency of Arabica coffee under different shading levels in dry and hot regions of southwest China. *Agric. Water Manag.* **2018**, *204*, 292–300. [CrossRef]
38. Silva, E.A.; Mazzafera, P.; Brunini, O.; Sakai, E.; Arruda, F.B.; Mattoso, L.H.C.; Carvalho, C.R.L.; Pires, R.C.M. The influence of water management and environmental conditions on the chemical composition and beverage quality of coffee beans. *Braz. J. Plant Physiol.* **2005**, *17*, 229–238. [CrossRef]
39. Ahmed, S.; Brinkley, S.; Smith, E.; Sela, A.; Theisen, M.; Thibodeau, C.; Warne, T.; Anderson, E.; Van Dusen, N.; Giuliano, P.; et al. Climate change and coffee quality: Systematic review on the effects of environmental and management variation on secondary metabolites and sensory attributes of *Coffea arabica* and *Coffea canephora*. *Front. Plant Sci.* **2021**, *12*, 708013. [CrossRef] [PubMed]
40. DaMatta, F.; Ramalho, J.C. Impacts of drought and temperature stress on coffee physiology and production: A review. *Braz. J. Plant Physiol.* **2006**, *18*, 55–81. [CrossRef]
41. Anthony, F.; Quiros, O.; Topart, P.; Bertrand, B.; Lashermes, P. Detection by simple sequence repeat markers of introgression from *Coffea canephora* in *Coffea arabica* cultivars. *Plant Breed.* **2002**, *121*, 542–544. [CrossRef]
42. Pérez-Molina, J.P.; de Toledo Picoli, E.A.; Oliveira, L.A.; Silva, B.T.; de Souza, G.A.; dos Santos Rufino, J.L.; Pereira, A.A.; de Freitas Ribeiro, M.; Malvicini, G.L.; Turello, L.; et al. Treasured exceptions: Association of morphoanatomical leaf traits with cup quality of *Coffea arabica* L. cv. "Catuaí". *Food Res. Int.* **2021**, *141*, 110118. [CrossRef] [PubMed]
43. Nunes, A.L.; Cortez, G.L.; Zaro, G.C.; Zorzenoni, T.O.; Melo, T.R.; Figueiredo, A.; Aquino, G.S.; Medina, C.D.; Ralisch, R.; Caramori, P.H.; et al. Soil morphostructural characterization and coffee root distribution under agroforestry system with *Hevea Brasiliensis*. *Sci. Agric.* **2021**, *78*, e20190150. [CrossRef]
44. Meireles, E.J.; Camargo, M.B.; Pezzopane, J.R.; Thomaziello, R.A.; Fahl, J.I.; Bardin, L.; Santos, J.C.; Japiassú, L.B.; Garcia, A.W.; Miguel, A.E.; et al. *Fenologia do Cafeeiro: Condições Agrometeorológicas e Balanço Hídrico do ano Agrícola 2004–2005*; Embrapa Informação Tecnológica: Brasíla, Brazil, 2009; p. 128.
45. Cesanelli, A.; Guarracino, L. Numerical modeling of actual evapotranspiration of a coffee crop. *Sci. Agric.* **2011**, *68*, 395–399. [CrossRef]
46. Godin, C.; Caraglio, Y. A multiscale model of plant topological structures. *J. Theor. Biol.* **1998**, *191*, 1–46. [CrossRef] [PubMed]
47. Matsunaga, F.T.; Tosti, J.B.; Androcioli-Filho, A.; Brancher, J.D.; Costes, E.; Rakocevic, M. Strategies to reconstruct 3D *Coffea arabica* L. plant structure. *SpringerPlus* **2016**, *5*, 2075. [CrossRef]
48. Pradal, C.; Boudon, F.; Nouguier, C.; Chopard, J.; Godin, C. PlantGL: A Python-based geometric library for 3D plant modelling at different scales. *Graph. Models* **2009**, *71*, 1–21. [CrossRef]

49. Scholz, M.B.S.; Kitzberger, C.S.G.; Pereira, L.F.P.; Davrieux, F.; Pot, D.; Charmetant, P.; Leroy, T. Application of near infrared spectroscopy for green coffee biochemical phenotyping. *J. Near Infrared Spectrosc.* **2014**, *22*, 411–421. Available online: https://opg.optica.org/jnirs/abstract.cfm?URI=jnirs-22-6-411 (accessed on 15 December 2022). [CrossRef]
50. R Core Team. Available online: https://wwwr-projectorg/ (accessed on 18 December 2022).
51. Covre, A.M.; Oliveira, M.G.; Martins, L.D.; Bonomo, R.; Rodrigues, W.N.; Tomaz, M.A.; Duarte, H.; de Sá Paye, H.; Partelli, F.L. How is the fruit development of *Coffea canephora* trees modulated by the water supply? An analysis of growth curves for irrigated and rainfed systems. *Semin. Ciênc. Agrár. Londrina* **2022**, *43*, 2359–2374. [CrossRef]
52. Fernandes, I.; Marques, I.; Paulo, O.S.; Batista, D.; Partelli, F.L.; Lidon, F.C.; DaMatta, F.M.; Ramalho, J.C.; Ribeiro-Barros, A.I. Understanding the impact of drought in *Coffea* genotypes: Transcriptomic analysis supports a common high resilience to moderate water deficit but a genotype dependent sensitivity to severe water deficit. *Agronomy* **2021**, *11*, 2255. [CrossRef]
53. Ghosh, P.; Venkatachalapathy, N. Processing and drying of coffee—A review. *Int. J. Eng. Res. Technol.* **2014**, *3*, 784–794.
54. Pérez, V.O.; Pérez, L.G.M.; Fernandez-Alduenda, M.R.; Barreto, C.I.A.; Agudelo, C.P.G.; Restrepo, E.C.M. Chemical composition and sensory quality of coffee fruits at different stages of maturity. *Agronomy* **2023**, *13*, 341. [CrossRef]
55. Soares, A.R. Irrigação Fertirrigação Fisiologia e Produção do Cafeeiro Adulto na Região da Zona da Mata de Minas Gerais. Master's Thesis, Universidade Federal de Viçosa, Viçosa, Brazil, 2001; p. 84.
56. Baïram, E.; leMorvan, C.; Delaire, M.; Buck-Sorlin, G. Fruit and leaf response to different source–sink ratios in apple, at the scale of the fruit-bearing branch. *Front. Plant Sci.* **2019**, *10*, 1039. [CrossRef] [PubMed]
57. DaMatta, F.M.; Cunha, R.L.; Antunes, W.C.; Martins, S.C.V.; Araújo, W.L.; Fernie, A.; Moraes, G.A.B.K. In field-grown coffee trees source-sink manipulation alters photosynthetic rates, independently of carbon metabolism, via alterations in stomatal function. *New Phytol.* **2008**, *178*, 348–357. [CrossRef] [PubMed]
58. Yeager, S.E.; Batali, M.E.; Guinard, J.-X.; Ristenpart, W.D. Acids in coffee: A review of sensory measurements and meta-analysis of chemical composition. *Crit. Rev. Food Sci. Nutr.* **2021**, *23*, 1–27. [CrossRef]
59. González, A.R.; Hernández, C.Y.F.; Rios, O.G.; Quiroz, M.L.S.; Amaro, R.M.G.; Estrada, Z.J.H.; Duarte, P.R. Coffee chlorogenic acids incorporation for bioactivity enhancement of foods: A review. *Molecules* **2022**, *27*, 3400. [CrossRef]
60. Vaast, P.; Bertrand, B.; Perriot, J.-J.; Guyot, B.; Génard, M. Fruit thinning and shade improve bean characteristics and beverage quality of coffee (*Coffea arabica* L.) under optimal conditions. *J. Sci. Food Agric.* **2006**, *86*, 197–204. [CrossRef]
61. Farah, A.; Donangelo, C.M. Phenolic compounds in coffee. *Braz. J. Plant Physiol.* **2006**, *18*, 23–36. [CrossRef]
62. Silveira, S.R.; Ruas, P.M.; Ruas, C.F.; Sera, T.; Carvalho, V.P.; Coelho, A.S.G. Assessment of genetic variability within and among coffee progenies and cultivars using RAPD markers. *Genet. Mol. Biol.* **2003**, *26*, 329–336. [CrossRef]
63. Silva, F.B.; Tormena, C.D.; Pauli, E.D.; de Almeida, A.G.; Berg, A.B.; Rakocevic, M.; Bruns, R.E.; Scarminio, I.S.; Marcheafave, G.G. Time dependent berry maturation for planting density levels in *Coffea arabica* L. beans: Mixture design-fingerprinting using near-infrared transmittance spectroscopy. *J. Food Compos. Anal.* **2021**, *97*, 103795. [CrossRef]
64. Oestreich-Janzen, S. *Chemistry of Coffee: Reference Module in Chemistry, Molecular Sciences and Chemical Engineering*; Elsevier: Cambridge, MA, USA, 2013; pp. 1–28. [CrossRef]
65. Toci, A.T.; Neto, V.J.; Torres, A.G.; Farah, A. Changes in triacylglycerols and free fatty acids composition during storage of roasted coffee. *Food Sci. Technol.* **2013**, *50*, 581–590. [CrossRef]
66. Cheng, B.; Furtado, A.; Smyth, H.E.; Henry, R.J. Influence of genotype and environment on coffee quality. *Trends Food Sci. Technol.* **2016**, *57*, 20–30. [CrossRef]
67. Joët, T.; Laffargue, A.; Descroix, F.; Doulbeau, S.; Bertrand, B.; Kochko, A.; Dussert, S. Influence of environmental factors, wet processing, and their interactions on the biochemical composition of green Arabica coffee beans. *Food Chem.* **2010**, *118*, 693–701. [CrossRef]
68. Worku, M.; Meulenaer, B.; Duchateau, L.; Boeckx, P. Effect of altitude on biochemical composition and quality of green arabica coffee beans can be affected by shade and postharvest processing method. *Food Res. Int.* **2018**, *105*, 278–285. [CrossRef]
69. Santos, C.A.F.; Leitão, A.E.; Pais, I.P.; Lidon, F.C.; Ramalho, J.C. Perspectives on the potential impacts of climate changes on coffee plant and bean quality. *Emir. J. Food Agric.* **2015**, *27*, 152–163. [CrossRef]
70. Barbosa, M.S.G.; Scholz, M.B.S.; Kitzberger, C.S.G.; Benassi, M.T. Correlation between the composition of green Arabica coffee beans and the sensory quality of coffee brews. *Food Chem.* **2019**, *292*, 275–280. [CrossRef]
71. Barbosa, J.N.; Borém, F.M.; Cirillo, M.A.; Malta, M.R.; Alvarenga, A.A.; Alves, H.M.R. Coffee quality and its interactions with environmental factors in Minas Gerais. *Brazil. J. Agric. Sci.* **2012**, *4*, 181–190. [CrossRef]
72. Mazzafera, P.; Robinson, S.P. Characterization of polyphenol oxidase in coffee. *Phytochemistry* **2000**, *55*, 285–296. [CrossRef]
73. Mintesnot, A.; Dechassa, N. Effect of altitude, shade, and processing methods on the quality and biochemical composition of green coffee beans in Ethiopia. *East Afr. J. Sci.* **2018**, *12*, 87–100.
74. Upadhyay, R.; Rao, L.J.M. An outlook on chlorogenic acids-occurrence, chemistry, technology, and biological activities. *Crit. Rev. Food Sci. Nutr.* **2013**, *53*, 968–984. [CrossRef] [PubMed]
75. Marcheafave, G.G.; Pauli, E.D.; Tormena, C.D.; Ortiz, M.C.; de Almeida, A.G.; Rakocevic, M.; Bruns, R.E.; Scarminio, I.S. Factorial design fingerprint discrimination of *Coffea arabica* beans under elevated carbon dioxide and limited water conditions. *Talanta* **2020**, *209*, 120591. [CrossRef] [PubMed]

76. Semedo, J.N.; Rodrigues, A.P.; Lidon, F.C.; Pais, I.P.; Marques, I.; Gouveia, D.; Armengaud, J.; Silva, M.J.; Martins, S.; Semedo, M.C.; et al. Intrinsic non-stomatal resilience to drought of the photosynthetic apparatus in *Coffea* spp. is strengthened by elevated air [CO_2]. *Tree Physiol.* **2021**, *41*, 708–727. [CrossRef] [PubMed]
77. Geromel, C.; Ferreira, L.P.; Davrieux, F.; Guyot, B.; Ribeyre, F.; dos Santos Scholz, M.B.; Pereira, L.F.; Vaast, P.; Pot, D.; Leroy, T.; et al. Effects of shade on the development and sugar metabolism of coffee (*Coffea arabica* L.) fruits. *Plant Physiol. Biochem.* **2008**, *46*, 569–579. [CrossRef] [PubMed]

horticulturae MDPI

Article

Growth, Physicochemical, Nutritional, and Postharvest Qualities of Leaf Lettuce (*Lactuca sativa* L.) as Affected by Cultivar and Amount of Applied Nutrient Solution

Jung-Soo Lee [1], Dulal Chandra [1,2,*] and Jinkwan Son [3,*]

1 Postharvest Technology Division, National Institute of Horticultural & Herbal Science, Rural Development Administration, Iseo-myeon, Wanju-gun 55365, Jeollabuk-do, Korea; ljs808@korea.kr
2 Department of Horticulture, Faculty of Agriculture, Bangabandhu Sheikh Mujibur Rahman Agricultural University, Salna, Gazipur 1706, Bangladesh
3 Energy & Environmental Engineering Division, National Institute of Agricultural Sciences, Rural Development Administration, Iseo-myeon, Wanju-gun 55365, Jeollabuk-do, Korea
* Correspondence: dchandrajp@gmail.com (D.C.); son007005@korea.kr (J.S.); Tel.: +82-63-238-4096 (D.C.)

Abstract: The effects of different nutrient solution quantities on growth, physicochemical, nutritional, and postharvest qualities of lettuce were investigated. Two differently pigmented Korean leaf lettuce cultivars "Geockchima" and "Cheongchima" were grown in soilless perlite culture supplied with 250, 500, 1000, and 2000 $mL \cdot d^{-1} \cdot plant^{-1}$ nutrient solutions. Several growth parameters (plant height, leaf number, fresh weight, dry matter) were evaluated. The highest lettuce growth was observed when plants were supplied with 1000 $mL \cdot d^{-1} \cdot plant^{-1}$. Cultivating lettuces in the lowest nutrient solution quantity showed higher dry matter, crude fiber, osmolality, chlorophyll, and anthocyanin contents. Upon increasing the nutrient solution, the crispiness, greenness, and levels of ascorbic acid, nitrogen, and potassium, increased, while phosphorus and magnesium were unaffected, and calcium content declined. Postharvest qualities were better maintained in lettuces irrigated with the least amount of nutrient solution, extending their shelf life. We conclude that lettuce can be grown with 1000 $mL \cdot d^{-1} \cdot plant^{-1}$ for higher yield, and short-term storage and/or transportation. However, when lettuces need to be stored for a certain period, such as long-distance shipment, they should be cultivated with a limited nutrient solution, which requires further detailed investigation. The results of this study can be applied for distributing, storing, transporting, and marketing lettuce.

Keywords: hydroponics; overall visual quality; relative fresh weight; shelf life; storage

Citation: Lee, J.-S.; Chandra, D.; Son, J. Growth, Physicochemical, Nutritional, and Postharvest Qualities of Leaf Lettuce (*Lactuca sativa* L.) as Affected by Cultivar and Amount of Applied Nutrient Solution. *Horticulturae* **2022**, *8*, 436. https://doi.org/10.3390/horticulturae8050436

Academic Editors: Alessandra Durazzo and Massimo Lucarini

Received: 5 April 2022
Accepted: 9 May 2022
Published: 13 May 2022

1. Introduction

The vegetable production system has largely shifted from traditional soil culture to soilless culture in several parts of the world, especially developed countries, as it offers some clear and unique advantages, such as improved quality and safety [1–3]. The overall management and control of plant nutrition required during various growing cycle stages are also easily upheld in soilless culture. While soilless culture requires significantly higher initial capital and operational costs, it can be balanced by higher productivity and improved product quality [4]. For these reasons, greenhouse crop production is now a steadily growing agricultural sector worldwide [5]. In Korea, the total area under greenhouse cultivation has increased recently, covering about 52,444 hectares, with vegetables occupying 68.5% of the total area [6]. Lettuce production, in particular, has increased from 3387 hectares in 2016 to 3484 hectares in 2014, indicating a 4.2% increment in production areas compared to the previous year [7]. As vegetable research in Korea has traditionally prioritized the improvement of productivity and quality, commercial production of lettuce in the greenhouse system is one of the greatest challenges for maintaining those aspects, as well as the safety and postharvest behavior of the produce. Solid-medium soilless culture, one of the most popular and easy methods in greenhouse cultivation, uses artificial upper soils, such as perlite and rock wool, as plant support. Since the physical properties, such as particle size, capillary space, water retention capacity, and the presence of pathogens, of these media are better than those of soil, the rhizosphere environment can be controlled, enhancing productivity [8]. Plant nutrients or specific ingredients, which cannot be easily adjusted in soil culture, can be easily controlled in soilless media. As a result, solid-medium soilless cultures are commonly used in the cultivation of vegetables, such as tomatoes, Chinese cabbage, sweet peppers, and cucumbers [9–11], and it is currently being used in Korea for the commercial cultivation of lettuce.

In horticultural crops, water supply and/or availability plays a pivotal role in processes associated with growth, productivity, and biomass, as well as having key implications on quality aspects, especially in the case of lettuce, which comprises 95% water [12]. As the edible parts of lettuce are the photosynthetic leaves, it is crucial to maintain optimal crop growth, which can be achieved through a well-scheduled irrigation program [13]. However, a previous study reported that deficiency in irrigation water could preserve the postharvest quality and shelf life of fresh-cut lettuces [14,15]. Water stress also influences the composition and concentration of nutrient-providing and biofunctional plant compounds [16]. Components, such as chlorophyll and anthocyanin, can be accumulated when plants experience different levels of water stress or drought condition [16,17]. The application of plant nutrients, on the other hand, could play a critical role in determining the nutrient and mineral composition of the plants [18–20]. Since water is a limited natural resource, and there is an increasing pressure to optimize the water use efficiency of crop production, reduced or limited water supply to lettuce during growth could be an effective way of not only saving irrigation water, but also maintaining the postharvest qualities of the produce. One of the most significant challenges for lettuce growers, wholesalers, and retailers is preserving postharvest qualities, which depend on the type and cultivar [21,22], storage conditions [23], and postharvest treatments [24]. Therefore, improving the growth, productivity, nutritional quality, and storage potential of lettuce is a desired goal for the growers and consumers. Numerous studies have highlighted the effects of cultivation methods [2], irrigation amount and water stress condition [16,25], salinity and elevated CO_2 [26], and light quality and fertilization [27,28] on the growth and yield of lettuce; however, no study has been conducted on the effects of nutrient solution management through irrigation volume control in solid-medium soilless culture on the growth and quality aspects of lettuce at harvest, as well as their storage behavior at postharvest. Thus, the effects of various levels of applied nutrient solution on the growth, physicochemical, nutritional, and postharvest qualities of two different pigmented leaf lettuce cultivars grown in soilless perlite cultures in greenhouse conditions were investigated.

2. Materials and Methods

2.1. Plant Material, Growing Condition, and Measurement of Growth Parameters

Seeds of two Korean leaf lettuce (*Lactuca sativa* L.) cultivars, "Geockchima" (red leaf, Nongwoo Bio Co. Ltd., Gyeonggi-do, Korea) and "Cheongchima" (green leaf, Nongwoo Bio Co. Ltd., Gyeonggi-do, Korea), were sown in a plastic seedling tray containing 162 cells and filled with watered wicking material, commercially known as "*Biosangto* No. 1" (Seminis Korea, Pyeongtaek, Korea). With two months as the total duration of the trial, the seeds were sown on April 21, 2016 and on April 26. Five-day-old, seedlings were transplanted onto soilless perlite medium (particle size 5 mm, Parat No.1, Samson Co., Gyeonggi-do, Korea) in the plastic greenhouses of the National Institute of Horticultural and Herbal Science (NIHHS), Suwon-si, Republic of Korea. The average day and night air temperatures of the plastic greenhouses were maintained at 26.6 ± 2.5 °C and 17.3 ± 2.1 °C, while the relative humidity was 46.5 ± 10.3% and 74.2 ± 9.8%, respectively. The beds containing perlite medium were previously prepared for growing lettuce following the Rural Development Administration guidelines [29]. The beds were covered with black polyethylene film as mulching material, with planting holes on the film ensuring plant-to-plant distances of 20 cm. The beds contained only a single row of lettuce, and bed-to-bed distance was approximately one meter. Plants were watered, via an automated control system, with nutrient solution containing N–P–K–Ca–Mg at 14.5–3.5–9–4–2 mequiv·L^{-1}, while electrical conductivity and pH were adjusted at 1.8 and 5.5–6.0 dS·m^{-1}, respectively. The nutrient solution was prepared according to NIHHS guidelines, and adapted to the local environmental conditions [29]. The nutrient solution was supplied to the base of each plant by a drip irrigation system using a pipe during the growing period, and four different irrigation volumes, 250, 500, 1000, and 2000 mL·d^{-1}·$plant^{-1}$, were used as the irrigation treatment. The frequency of irrigation was 10 times during the day and twice at night. The experiment was arranged in a completely randomized block design with three replications consisting of 15 plants per replicate. A preliminary experiment was conducted with limited replication and plants to observe the plant growth parameters and postharvest performances. Growth parameters, including plant height, number of leaves, and fresh weight, were measured at the time of harvest. Plant height was determined by measuring the length of the plant from the shoot tip to the surface, and leaf number was determined by counting leaves with a minimum width of 1 cm. Fresh weight was measured by weighing the living biomass of each plant that was present aboveground following the harvest.

2.2. Harvest Practice, Storage Condition, and Postharvest Quality Evaluation

Lettuce was harvested by cutting the entire plant when it reached commercial maturity, i.e., 29 days after transplanting, after which plant height, number of leaves, and shoot fresh weight were measured. With regards to conducting the storage experiment, 5–12 leaves per plant from the 3rd to 14th leaf position were randomly selected and immediately transported to the laboratory (within 10 min) using a cooled chamber and maintained at 4 °C. For the postharvest quality evaluation, a single replicate under each treatment containing about 20 leaves (approx. 150–200 g) of uniform size and color, and with no physical damage, was sampled. The leaves were packaged in transparent polypropylene bags (size 32 cm × 22 cm, thickness 0.05 mm) with four holes (two holes each side) of 5 mm diameter and manually sealed with thread-like wire. Then, the packages were placed in plastic trays on a shelf in a 7 °C dark room, and stored for up to 12 d. Storage was terminated when samples became unusable due to visible deterioration symptoms, such as yellowing, wilting, decay, or spoiling. Fresh weight, color, and overall visual quality (OVQ) were measured at two-day intervals during storage. Three replicates were used for each evaluation day under each irrigation treatment. An OVQ analysis was performed by an eight-member panel, with ages between 28 and 52 years (five men and three women). Considering appearance, visual color, off-odor, and freshness, the OVQ was evaluated using a 9-point scale (9 = excellent, 7 = good, 5 = fair, 3 = poor, and 1 = unusable). A score of 5 was considered the marketability limit. The OVQ scoring and acceptance of marketability were adopted from Aguero et al. [30].

2.3. Color Measurement

Five leaves were randomly selected from the 3rd to 14th leaf position of one plant in each replicate immediately after harvest for color measurement using a chromameter under different irrigation treatments (Minolta CR-400, Minolta, Osaka, Japan). Similarly, five leaves were randomly selected from each pouch on each evaluation day for measuring color parameters. Three readings were taken from the left, middle, and right part of the upper region, and closer to the tip of the adaxial surface of the leaf; therefore, an average of 15 readings were taken for a single replication. Before measuring the color readings from leaf samples, the chromameter was calibrated using a standard white plate (Y 93.5, x 0.3155, y 0.3320) provided by the manufacturer. Color changes were quantified in the L^*, a^*, b^* color space. L^* refers to the lightness, and ranged from $L^* = 0$ (black) to $L^* = 100$ (white). Green and red indicate negative and positive values of a^*, respectively, while yellow and blue indicate positive and negative values of b^*, respectively. Total color difference (ΔE^*) was calculated using the formula of [21]:

$$\Delta E^* = \{(L_0{}^* - L^*)^2 + (a_0{}^* - a^*)^2 + (b_0{}^* - b^*)^2\}^{1/2}\ L_0{}^* \quad (1)$$

where $a_0{}^*$ and $b_0{}^*$ represent the values at harvest, and L^*, a^*, and b^* indicate reading on any evaluation day.

2.4. Measurement of Dry Matter, Texture, Crude Fiber, Osmolality, and Ascorbic Acid Content

For dry matter measurement, five intact lettuce leaves were inserted into a thin paper envelope, weighed, and dried at 70 °C for 48 h to achieve a constant weight. Three envelopes were prepared from each replicate, and the average values were expressed as a percentage of dry matter. Texture analysis of the lettuce leaves was performed in terms of force required to puncture the leaves using a texture analyzer (TA Plus, Lloyd Instruments, Model LF2303, Ametek Inc., Fareham, Hants, UK). The sample was placed on the central opening of the metal holding plate, where leaf tissues (approx. 5 cm × 5 cm) were set tightly on the stage with clips, and a 5 mm diameter flat-head stainless steel cylindrical probe was connected to the tissues for the test. The movement of the probe was adjusted to 5, 2, and 10 mm/s as the pre-test, test, and post-test speeds, respectively. The probe was run with a 20 N load cell and attached to a creep meter equipped with software (NEXYGEN™MT v 4.5, Lloyd Instruments, Ametek Inc., UK) for automatic analysis using a computer.

The crude fiber was determined following the AOAC [31] method using a crude fiber analyzer (Fibertec Systems M1020, Foss Tecator, Eden Prairie, MN, USA). The method involved the digestion of the fresh samples with boiling and via dilution in acid and alkali for the calculation for residual ash content. The osmolality of lettuce leaf sap samples was measured using a Wescor5520 vapor pressure osmometer (Wescor Inc., Logan, UT, USA) following the method of Clarkson et al. [32]. Lettuce leaves were sliced into small pieces, mixed, and 10 g was used to collect sap using a juice maker. Ascorbic acid content was determined following the method of Jagota and Dani [33]. Briefly, a 0.5 g fresh leaf sample was extracted and filtered with double-distilled water. Then, 0.2 mL homogenate was added with 0.8 mL 10% trichloroacetic acid followed by vigorous shaking, incubating in an ice bath, centrifugation, and collection of supernatant. Following this, the 2.0 mL diluted supernatant was mixed vigorously with 0.2 mL 0.2 M Folin phenol reagent. After 10 min, the absorbance of this solution was measured at 760 nm using a UV-VIS recording spectrophotometer (DU 650, Beckman Coulter™, Chaska, MN, USA). The concentration of ascorbic acid was calculated from the standard curve prepared from different concentrations of ascorbic acid following the same procedure.

2.5. Microscopic Observation of Lettuce Tissue

Samples for histological study were taken from the leaf apex at the equatorial region, at a position of 90% distance of leaf length from the base. The samples were prepared following the method of Luna et al. [34], with some modifications. Briefly, samples (about 1 mm^3) were fixed with 2.5% glutaraldehyde in 0.1 M sodium phosphate buffer (pH 7.2), post-fixed with 1% osmium tetroxide, each for 2 h at 4 °C, and then held overnight in phosphate buffer. After fixation, specimens were dehydrated in a graded series of ethyl alcohol, processed through three changes of propylene oxide (for 15, 15, 30 min per change), and gradually infiltrated with embedding medium, Epon. Specimens were sectioned (1500 nm), stained with periodic acid–Schiff, and viewed under a Carl Zeiss Axioskop 2 light microscope (Carl Zeiss, Jena, Germany).

2.6. Determination of Total Chlorophyll and Anthocyanin

Total chlorophyll content was determined following the method of Lin et al. [28], with some modifications. First, 5 g fresh leaf tissues were homogenized in 20 mL of 80% acetone using an Ultra-Turrax tissue homogenizer (T 25 B, Ika Works Sdn. Bhd, Rawang, Selangor, Malaysia) at a moderate speed for about one minute. The homogenate was filtered through four layers of cotton cloth and centrifuged at 4 °C, 15,000× g for 20 min, after which the supernatant was transferred to a volumetric flask, and the volume was determined using 80% acetone. The absorbance of the supernatant was measured at 663 and 645 nm, and total chlorophyll (4) content was calculated using the following formulae:

$$\text{Chlorophyll a (chl a)} = 12.72 \times \text{OD663} - 2.59 \times \text{OD645} \quad (2)$$

$$\text{Chlorophyll b (chl b)} = 22.88 \times \text{OD645} - 4.67 \times \text{OD663} \quad (3)$$

$$\text{Total chlorophyll} = (\text{chl a} + \text{chl b}) \quad (4)$$

Similarly, total anthocyanin was estimated using a spectrophotometric pH differential method following the procedure of Ferrante and Maggiore [35], with minor modifications. Five grams of fresh leaf tissue was homogenized in 20 mL methanol containing 1% HCl, incubated overnight at 4 °C in darkness, centrifuged at 4 °C 12,000× g for 20 min, the supernatant collected in a volumetric flask, and made up to a known volume with the same acidified methanol. An aliquot of 2 mL was diluted with 18 mL potassium chloride buffer (pH 1.0), and another 2 mL aliquot was diluted separately with 18 mL sodium acetate buffer (pH 4.5). Absorbances of these solutions were measured at 510 and 700 nm, and anthocyanin contents were calculated using the formula of [36]:

$$\text{Monomeric anthocyanin pigment } (\text{mg}\cdot\text{L}^{-1}) = (\text{A} \times \text{MW} \times \text{DF} \times 1000)/(\varepsilon \times 1) \quad (5)$$

where A = (A510–A700) pH1.0–(A510–A700) pH4.5 with molar absorptivity (ε) of 26,900 ($\text{L}\cdot\text{cm}^{-1}\cdot\text{mol}^{-1}$); MW = molecular weight of cyanidin-3-glucoside (449.2); DF = dilution factor of this experiment.

Anthocyanin content ($\text{mg}\cdot\text{L}^{-1}$) was then converted to $\text{mg}\cdot 100\ \text{g}^{-1}$ fresh sample. Unless otherwise stated, all chemicals used for biochemical assay were purchased from Sigma Aldrich Co. Ltd. St. Louis, MO, USA.

2.7. Mineral Analysis

For the analysis of mineral contents and C/N ratio, the harvested leaves were dried in a circulated drying oven for 48 h at 80 °C, and subsequently ground to make a powder. The mineral contents were determined using RDA's elemental analysis methods [29]. To measure nitrogen (N) content, a finely ground leaf sample (0.5 g) was decomposed with 1 g catalyst (K_2SO_4:$CuSO_4$ = 9:1) and 10 mL concentrated sulfuric acid in Kjeldahl flasks at 380 °C for 3–4 h, and then analyzed using a Kjeldahl analyzer (Kjeltec2300 analyzer unit, Foss Tecator AB, Hoganas, Sweden). For phosphorus (P) determination, a 0.5 g sample was digested in acid in a fume hood until the digested sample color became transparent. P was measured at 470 nm using a UV-visible spectrophotometer after coloring with ammonium vanadate. Potassium (K), calcium (Ca), and magnesium (Mg) were measured using atomic absorption spectrophotometer (AA-6800, Shimadzu, Kyoto, Japan). Dried leaf samples (0.5 g) were decomposed with 10 mL digest solution containing HNO_3 and $HClO_4$ (3:1) at 180 °C for 12–16 h, filtered, and made up to a known volume. The C/N ratio was measured by the combustion method using CN corder (CNS2000, Leco, St. Joseph, MI, USA).

2.8. Statistical Analyses

Analysis of variance (ANOVA) of the data was performed by SAS software ver. 9.1 (SAS Institute, Cary, NC, USA). A combined ANOVA was performed using cultivar as a fixed variable [37]. Based on the level of significance calculated from the F-value of the ANOVA, Duncan's multiple range tests were applied at $p \leq 0.05$ for mean comparisons among the various treatments.

3. Results and Discussion

3.1. Growth of Lettuce as Affected by Cultivar and Amount of Applied Nutrient Solution

Lettuce growth was significantly influenced by the applied nutrient solution amount and cultivars in perlite culture (Table 1). Regarding the effect of cultivar, Cheongchima produced more leaves ($p \leq 0.05$) than Geockchima, while an opposite trend was observed for plant height. Alternatively, the amount of irrigation had a significant ($p \leq 0.01$) effect on plant height, number of leaves, and fresh shoot weight. However, the irrigation volume and cultivar interaction effect on these growth parameters was not significant (Tables 1 and 2).

Table 1. Effects of the amount of applied nutrient solution on growth performance, dry matter, and crude fiber content of two leaf lettuce cultivars grown in perlite culture.

Cultivar	Amount of Applied Nutrient Solution ($mL{\cdot}d^{-1}{\cdot}plant^{-1}$)	Plant Height (cm)	No. of Leaves [x]	Fresh Weight of Shoot ($g{\cdot}plant^{-1}$)	Dry Matter (%)	Crude Fiber (%)
Geockchima	250	31.50 ± 1.58 b [z]	24.73 ± 1.84 b	76.75 ± 9.50 c	5.60 ± 0.62 a	1.29 ± 0.06 a
	500	32.18 ± 1.87 b	27.43 ± 5.54 b	119.35 ± 47.90 b	4.50 ± 0.55 b	1.43 ± 0.07 a
	1000	36.80 ± 1.76 a	34.33 ± 3.16 a	253.43 ± 65.51 a	3.25 ± 0.19 c	0.91 ± 0.03 b
	2000	35.65 ± 1.91 a	33.75 ± 3.50 a	249.05 ± 53.51 a	3.35 ± 0.51 c	0.84 ± 0.01 b
Cheongchima	250	25.70 ± 2.18 c	31.93 ± 1.58 c	78.05 ± 11.59 d	6.20 ± 0.51 a	1.31 ± 0.06 a
	500	29.93 ± 2.57 b	35.28 ± 2.60 bc	127.25 ± 26.01 c	4.75 ± 0.62 b	1.23 ± 0.06 a
	1000	33.65 ± 0.97 a	40.58 ± 2.68 a	284.28 ± 12.62 a	3.72 ± 0.36 c	0.60 ± 0.06 b
	2000	33.90 ± 1.50 a	38.68 ± 3.91 ab	228.05 ± 43.76 b	3.65 ± 0.48 c	0.61 ± 0.03 b
Significance	Cultivar (A)	*	*	NS	NS	*
	Irrigation amount (B)	**	**	**	**	**
	Interaction (A × B)	NS	NS	NS	NS	**

[x] is the average of four plants. [z] Means with the same letter(s) within a column in each cultivar are not significantly different at $p \leq 0.05$ according to Duncan's multiple range test. NS, *, ** indicate non-significant, and significant at $p \leq 0.05$, 0.01, respectively.

Table 2. Photographs of the crops cultivated in the Geockchima and Cheongchima treatment zones.

Cultivar	Amount of Nutrient Solution Applied ($mL{\cdot}d^{-1}{\cdot}plant^{-1}$)			
	250	500	1000	2000
Geockchima				
Cheongchima				

The highest plant height, number of leaves, and fresh weight were obtained by irrigating lettuces with 1000 $mL{\cdot}d^{-1}{\cdot}plant^{-1}$ nutrient solution, regardless of the cultivar.

These growth performances declined slightly when plants were irrigated with 2000 $mL{\cdot}d^{-1}{\cdot}plant^{-1}$, but remained higher than for plants irrigated with 250 or 500 $mL{\cdot}d^{-1}{\cdot}plant^{-1}$. Compared with plants irrigated with the lowest amount, fresh weight increased by 55%, 230%, and 224% when irrigation amounts were 500, 1000, and 2000 $mL{\cdot}d^{-1}{\cdot}plant^{-1}$ in Geockchima, and 63%, 264%, and 192% in Cheongchima, respectively. These results corroborated those of Kirnak et al. [25], who discovered proportionally higher plant height and yield of lettuce with increased irrigation levels. Limited irrigation was associated with reduced growth and yield in lettuce [34,38], which supports the current study. Higher leaf lettuce growth was possibly achieved via increased tissue water content coupled with cell size enlargement [39]. On comparing leaf tissue anatomy, distinct cell size differences were observed among samples grown under different irrigation regimes (Figure 1). Cellular tissues in Figure 1C,D show wide gaps owing to high fertilizer input, whereas those in Figure 1A,B are dense. Similarly, vacuoles in Figure 1C,D are larger than those in Figure 1A,B, signifying that high fertilizer input induces plant growth.

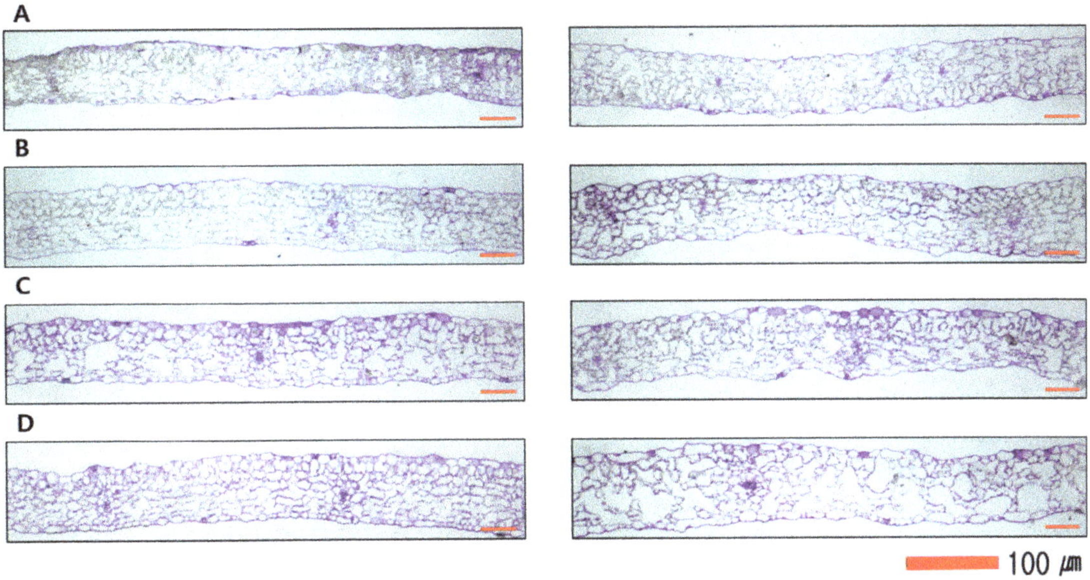

Figure 1. Leaf cross section micrographs of Geockchima (**left**) and Cheongchima (**right**) lettuce cultivars grown in the perlite culture as affected by different amount of nutrient solution. (**A**) 250 mL·d^{-1}·plant^{-1}, (**B**) 500 mL·d^{-1}·plant^{-1}, (**C**) 1000 mL·d^{-1}·plant^{-1}, and (**D**) 2000 mL·d^{-1}·plant^{-1}.

Lettuce requires sufficient water for maximum growth and yield. However, overwatering stress interrupts photosynthesis, affecting plant size and quantity [40]. In comparison, plants cultivated with the highest nutrient solution had reduced growth. Since only higher irrigation amounts did not increase growth, it is crucial to irrigate plants with the proper volume. Several studies have demonstrated how water conditions influence crop cultivation and quality [3,16,17,41,42].

3.2. Dry Matter, Crude Fiber, Texture, Osmolality, and Ascorbic Acid Content as Affected by Cultivar and Amount of Applied Nutrient Solution

The amount of irrigation significantly affected dry matter and crude fiber content in both cultivars (Table 1). With an increase in irrigation, dry matter content declined; therefore, applying 1000/2000 mL·d^{-},·plant^{-P} nutrient solution produced lower dry matter content. Eichholz et al. [16] observed significantly higher dry matter content in lettuces grown in deficient water than those produced in well-watered or water-logged conditions, supporting the results of this study. Crude fiber content was affected by cultivar ($p \leq 0.05$), irrigation amount ($p \leq 0.01$), and their interaction ($p \leq 0.01$). The crude fiber amount in both cultivars was inversely proportional to the applied nutrient solution amount. The highest crude fiber (1.31%) was observed when the least amount of nutrient solution was applied in Cheongchima. Fiber is involved in supporting growth, textural stability, and aerial plant structures [16].

Leaf texture and osmolality values showed no significant difference between cultivars, but were significantly reduced due to increased irrigation volumes ($p \leq 0.05$) (Table 3). The texture was measured by determining the puncture force, with its declining values (lower force) indicating higher crispiness of leaves, while stronger forces were required to puncture the flexible or leathery leaves. It is evident that the water content of leaf tissues plays a critical role in maintaining cellular turgor, which is a crucial component of the rigidity and firmness in plant tissue. When lettuce plants were irrigated with a higher amount of nutrient solution, the water content of tissues increased proportionally, decreasing dry matter and osmolality values (Tables 1–3).

Hence, the tissue structures, including cell size and intercellular space or vacuole volume, became larger (Figure 1). Luna et al. [34] also reported a similar observation in romaine lettuce. Ilker and Szczesniak [43] reported that textural characteristics exhibited by plant parenchymal tissues depend on cell shape and size, cytoplasm-to-vacuole ratio, intercellular space volume, cell wall thickness, osmotic pressure, and the solute types present in the cell. For lettuce, crispiness is a desirable characteristic associated with the consumer's perception and freshness of the leaves, which could be triggered by applying a higher amount of nutrient solution or irrigation water. However, maintaining the postharvest qualities of those samples would be a great challenge, as faster and more rapid water losses occurred in our study, along with other forms of quality deterioration (Figure 2).

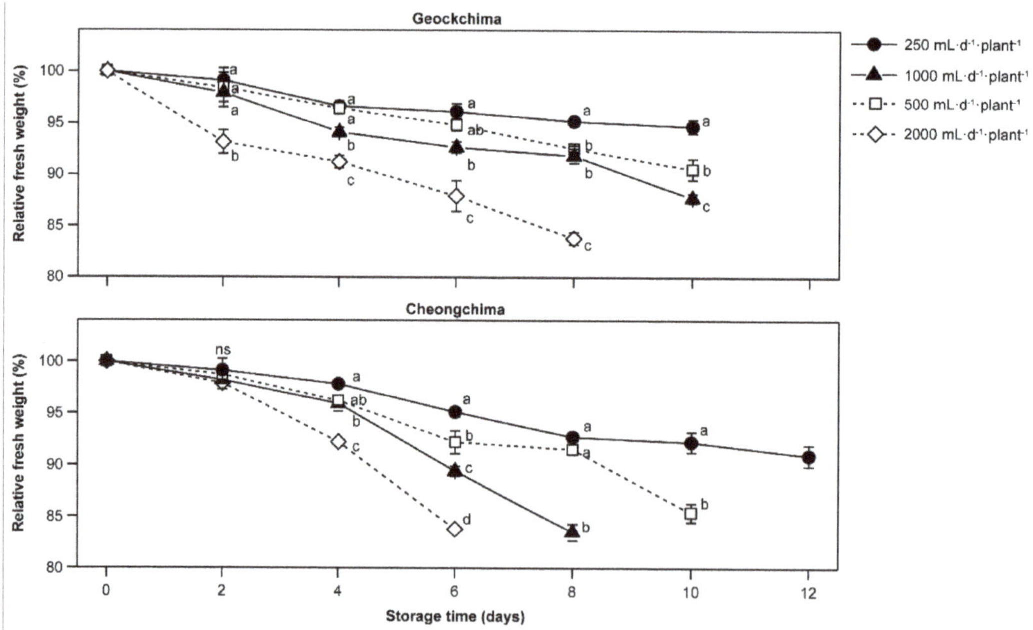

Figure 2. Effect of the amount of nutrient solution on relative fresh weight of leaf lettuce cultivars Geockchima and Cheongchima during storage at 7 °C. Mean values at each evaluation day with different lowercase letters are significantly different according to Duncan's multiple range test ($p \leq 0.05$), ns = non-significant.

The cell sap of the least-watered plants contained nearly 42% and 39% more solute than that of the highest-watered plants in Geockchima and Cheongchima, respectively. Similar results were reported in salt-stressed lettuce, which showed better organoleptic properties, postharvest processability, and shelf life [32].

Table 3. Physicochemical and nutritional quality parameters of harvested leaf lettuce as influenced by cultivar and amount of applied nutrient solution in perlite culture.

Cultivar	Amount of Applied Nutrient Solution ($mL \cdot d^{-1} \cdot plant^{-1}$)	Texture/Puncture Force ($N \cdot m^{-2}$)	Osmolality ($mmol \cdot kg^{-1}$)	Ascorbic Acid Content ($mg \cdot 100\ g^{-1}$ FW)	Color Value of Leaf			Total Chlorophyll Content ($mg \cdot g^{-1}$ FW)	Total Anthocyanin Content ($mg \cdot 100\ g^{-1}$ FW)
					L^*	a^*	b^*		
Geokchima	250	1.8 a	492.0 a	10.28 c	39.4	−0.8 a [z]	10.5 b	1.71 a	4.07 a
	500	1.7 a	442.4 b	20.01 b	41.7	−3.6 b	11.7 b	1.58 a	3.10 b
	1000	1.2 b	415.6 c	26.03 a	42.9	−7.3 c	20.1 a	0.87 b	2.10 c
	2000	1.0 b	346.7 d	24.43 a	46.2	−10.0 d	21.2 a	0.98 b	1.49 c
Cheongchima	250	1.8 a	447.3 a	16.81 b	45.4	−12.7 a	16.4 b	2.03 a	-
	500	1.5 ab	439.0 a	12.52 c	47.4	−13.9 ab	19.0 a	1.89 a	-
	1000	1.2 bc	372.3 b	21.14 a	46.5	−14.8 b	20.2 a	1.32 b	-
	2000	0.9 c	321.8 c	21.52 a	47.4	−14.1 ab	19.8 a	1.42 b	-
Significance	Cultivar (A)	NS	NS	NS	NS	**	*	*	-
	Irrigation amount (B)	**	**	**	NS	**	*	**	**
	Interaction (A × B)	NS	*	*	NS	**	NS	NS	-

[z] Means with the same letter(s) within a column in each cultivar are not significantly different at $p \leq 0.05$ according to Duncan's multiple range test. NS, *, ** indicate non-significant, and significant at $p \leq 0.05$, 0.01, respectively.

The ascorbic acid content of lettuce ranged from 10.3 to 26.0 and 12.5 to 21.5 mg 100 g^{-1} in Geockchima and Cheongchima, respectively (Table 3). A higher ascorbic acid content was measured during the highest growth stage of lettuce under optimum irrigation conditions. This may indicate appropriate ascorbic acid generation during normal crop growth. In Geockchima, about two times more ascorbic acid was determined when lettuces were grown with double the amount of nutrient solution than the initial amount (Table 3). This result corroborates the findings of Zeipina et al. [42], who discovered nearly half the amount of ascorbic acid in leaf lettuce grown in reduced moisture than for those cultivated in optimal moisture conditions. However, ascorbic acid content did not increase in such a trend for other irrigation amounts, especially in Cheongchima. The increase in ascorbic acid relative to the higher strength of nutrient solution was also observed in hydroponically grown leaf lettuce [44]. Additional to nutrient amount or fertilization, several factors, including plant genotype, growing season, and light intensity, can influence the ascorbic acid content in lettuce and other horticultural crops [45,46]. This could be a potential reason for the variability observed in ascorbic acid content in the cultivar Cheongchima.

3.3. Color Values, Chlorophyll, and Anthocyanin Contents as Affected by Cultivar and Amount of Applied Nutrient Solution

Color attributes (except L^* value) and total chlorophyll content of lettuce leaf were significantly affected by cultivar and irrigation amount (Table 3). When Geockchima was irrigated with the least amount, the a^* value (−0.8) was the highest value for red. However, a^* value (−14.8) of Cheongchima was the highest value for green when the irrigation amount was 1000 mL·d^{-1}·$plant^{-1}$. The a^* value was significantly different ($p \leq 0.05$) among the treatments in Geockchima, but not in Cheongchima. However, the b^* value increased until 1000 and 500 mL·d^{-1}·$plant^{-1}$ applied nutrient solution in Geockchima and Cheongchima, respectively. According to the amount of irrigation, the cultivar Geockchima responded more for color development than Cheongchima. Comparing the pigment contents and color values in Geockchima, the correlation coefficient (r) between chlorophyll and a^* value was 0.91, and between anthocyanin and a^* value was 0.99 (data not shown). The changing pattern of a^* and b^* observed in Geockchima were inconsistent with a previous result for leaf lettuce grown in well-watered, moderate water deficit, and severe water deficit conditions [3]. In contrast, Fallovo et al. [19] discovered no significant differences in L^*, a^*, and b^* values when leafy lettuces were grown in a floating system with different nutrient solutions, which supports our L^* values and partially supports the color data of the cultivar Cheongchima. The variation in color parameters between the cultivars could be due to the genetic constituents of the plant, as Geockchima is red-leafed and Cheongchima is green-leafed. Consequently, a comparatively lower amount of total chlorophyll contents was discovered in the former than in the latter, possibly because of the presence of anthocyanin. Varietal differences in chlorophyll contents of lettuce leaves were observed in several previous studies [16,17,26]. The lowest amount of total chlorophyll was measured in Geockchima lettuces, which indicated the highest growth Tables 1–3. Considering the total chlorophyll contents of these lettuces (irrigated with 1000 mL·d^{-1}·$plant^{-1}$), about 81% and 96% higher total chlorophyll contents were measured in plants irrigated with 500 and 250 mL d^{-1} $plant^{-1}$ nutrient solutions, respectively, in Geockchima, and 43% and 53%, respectively in Cheongchima. Our results are consistent with those of Eichholz et al. [16], who observed an increase in chlorophyll content in some varieties of lettuce grown in water deficit conditions compared to those grown in well-watered or water-logged conditions.

Similar to chlorophyll content, anthocyanin content declined ($p \leq 0.05$) gradually, corresponding to the increased levels of applied nutrient solution (Table 3). However, anthocyanin was not detected in Cheongchima, as this cultivar is characterized by its dark green color. The highest amount of anthocyanin (4.07 mg·100 g^{-1}) was produced when lettuces were irrigated with the least amount of nutrient solution, and the declines in anthocyanin contents were about 23%, 48%, and 63%, with each respective increase in level of supplied nutrient solution. Similarly, Baslam and Goicoechea [17] observed more anthocyanin in two lettuce cultivars when the plants were grown in water deficit or cyclic drought conditions than the control plants. The accumulated anthocyanin in water-stressed plants may prevent desiccation through osmotic effects, though the mechanism remains unclear [47]. Anthocyanins are the most important group of water-soluble pigments in plants, and may be induced by several environmental factors, including water stress [48], as a mechanism for adapting to the surrounding environment.

3.4. Mineral Contents as Affected by Cultivars and Amount of Applied Nutrient Solution

The mineral contents (except P and Mg) and C/N ratio of the lettuces were significantly affected by irrigation amount, whereas only N and Ca contents were influenced by cultivar (Table 4). The N content in leaves likely increases with irrigation amount because of the higher uptake of N from the supplied nutrient solution. Cheongchima had a comparatively higher N content than Geockchima. Consequently, the C/N ratio declined gradually in Cheongchima, but no obvious trend was found in Geockchima (Table 4).

Table 4. The mineral content of leaves of two lettuce cultivars grown in perlite culture as influenced by the amount of applied nutrient solution.

Cultivar	Amount of Applied Nutrient Solution ($mL{\cdot}d^{-1}{\cdot}plant^{-1}$)	Mineral Content (%, DW)					C/N Ratio
		N	P	K	Ca	Mg	
Geockchima	250	4.68 ab [z]	0.22	3.31 a	1.16 a	0.24	7.94 a
	500	4.40 b	0.20	3.24 a	1.04 ab	0.20	8.13 a
	1000	4.91 a	0.25	3.51 a	1.01 ab	0.22	6.93 b
	2000	4.87 a	0.19	2.60 b	0.91 b	0.14	7.22 b
Cheongchima	250	4.55 b	0.27	1.90 c	0.83 a	0.18	9.09 a
	500	4.88 b	0.24	2.22 c	0.74 ab	0.17	8.53 a
	1000	5.40 a	0.31	3.98 b	0.62 b	0.17	6.63 b
	2000	5.48 a	0.27	5.80 a	0.61 b	0.16	6.49 b
Significance	Cultivar (A)	**	NS	NS	**	NS	NS
	Irrigation amount (B)	**	NS	**	*	NS	**
	Interaction (A × B)	**	NS	*	NS	NS	**

[z] Means with the same letter(s) within a column in each cultivar are not significantly different at $p \leq 0.05$ according to Duncan's multiple range test. NS, *, ** indicate non-significant, and significant at $p \leq 0.05$, 0.01, respectively.

However, both cultivars showed two statistically distinct groups of C/N ratios, irrigated with a lower and higher amount of nutrient solution. These results corroborate the results of Stefanelli et al. [18], who found higher yield and N content of red oak lettuce, along with lower C:N, when nitrogen application was increased, although the trend was limited, only occurring up to certain levels of N application. Significant increases in N, P, K, and Mg contents were also reported when the concentration of nutrient solution increased in a floating culture of lettuce [20], which partly supports the current investigation. This study was based on the Carbon/Nutrient Balance (CNB) Theory's [49] N content and C:N data, and the CNB theory was validated for leaf lettuce.

P content was unaffected by all factors, while K content increased progressively as irrigation amounts increased; therefore, the highest K content (5.80%) was discovered in plants grown with the highest level of nutrient solution, and in the Cheongchima cultivar (Table 3). A similar observation was also reported in lettuce grown in floating raft culture [20]. However, the data were erratic in Geockchima, although the highest K content (3.51%) was determined in the highest growth-yielding plants, demonstrating similarity with the results of Fallovo et al. [20]. Unlike K content, the highest level of Ca was found in plants grown with the least amount of nutrient solution. Geockchima yielded more Ca ($p \leq 0.01$) than Cheongchima in all treatment conditions. Mg content did not significantly differ in any measured factor, but an insignificant decline was noticed with the increased volume of nutrient solution. The decline in Ca or Mg with increasing N, Ca, and Mg application might possibly occur because of 2^+ cation absorption–competition interactions between Ca and Mg [18].

3.5. Postharvest Qualities as Affected by Cultivars and Amount of Applied Nutrient Solution

During the storage of harvested lettuce, relative fresh weight loss declined gradually in all irrigation regimes, showing significant differences ($p \leq 0.05$) among the irrigation treatments (Figure 2). The greater fresh weight losses were found in Cheongchima than in Geockchima, and samples irrigated with higher amounts than those with lower amounts. As a result, some samples, especially those irrigated with higher nutrient solution, became unusable after a few days of storage. On day 6, for instance, about 4% and 5% water losses occurred in samples grown in the least irrigation amount, while these losses were 12% and 16% in samples irrigated with the highest nutrient solution in Geockchima and Cheongchima, respectively (Figure 2). In a similar study, Lee and Chang [22] discovered greater fresh weight loss in leaf lettuce grown in higher nutrient concentrations compared to controls during storage at 5 °C. Water loss not only induces important quality, marketability, and economic losses of fresh produce, it also negatively influences the visual, compositional, and eating quality of the plant, even when weight losses are subtle [50]. The authors of the present study also conclude that the weight loss of lettuce differs between cultivars and postharvest conditions, such as temperature, humidity, gas composition, and/or type of polymeric film used for packaging the sample during storage. The total color difference (ΔE) of Cheongchima was significantly ($p \leq 0.05$) influenced by irrigation amounts throughout storage, whereas minimal differences were observed in Geockchima (Figure 3). The experimental crops cultivated in the Geockchima and Cheongchima treatment zones are shown in Table 2. Relative fresh weight, total color difference, and overall visual quality score measurements were discontinued when they became worthless as products.

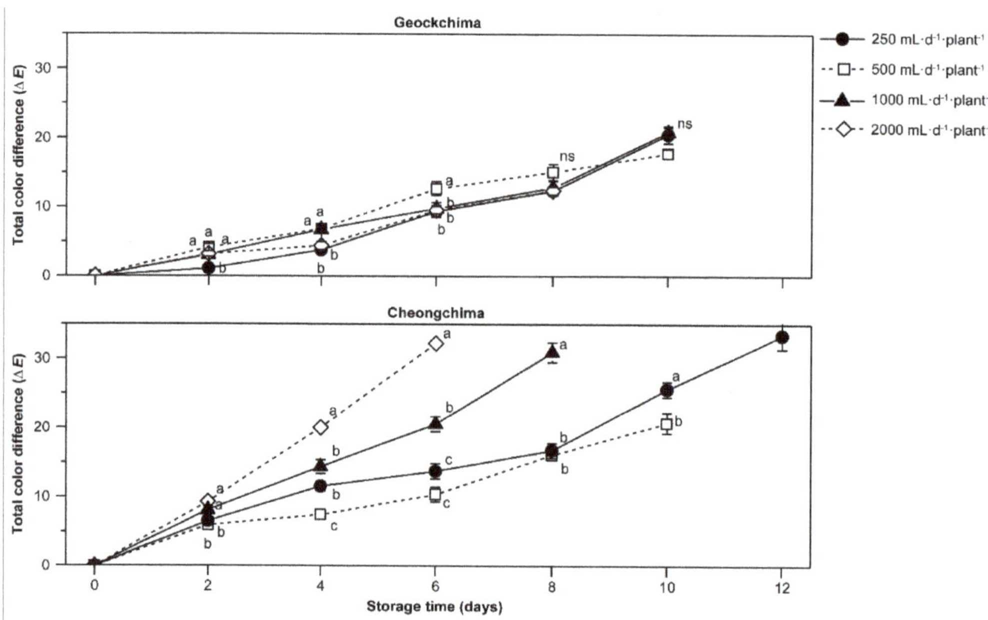

Figure 3. Changes in total color difference in two leaf lettuce cultivars (Geockchima and Cheongchima) grown in perlite culture with different amounts of the nutrient solution during storage at 7 °C. Mean values at each evaluation day with different lowercase letters are significantly different according to Duncan's multiple range test ($p \leq 0.05$), ns = non-significant.

The variable pattern of color changes found between the two varieties could be attributed to the differences in leaf color, whereby the green-leafed Cheongchima might have faster color changes compared to the red-leafed Geockchima. It seems that the green-leafed cultivar is more vulnerable to color changes possibly because the pigment (chlorophyll) responsible for the leaf color usually declines with the progress of storage (Aguero et al., 2008). Leaf color change in Geockchima was minor, probably due to the presence of anthocyanin, which is comparatively more stable than chlorophyll during storage [35]. Leaves harvested from 1000 and 2000 mL·d^{-1}·plant^{-1} treatment showed faster color changes compared to those of the other two treatments, especially in the cultivar Cheongchima. Overall, the effect of different nutrient solutions on color changes of Geockchima is small compared with Cheongchima. Since the color of the leaf is a crucial quality character of leafy vegetables such as lettuce at postharvest stages, the changing pattern of total color difference in accordance with irrigation treatment found in this study may suggest the necessity of further detailed studies on the color quality of lettuce at postharvest stages.

OVQ, measured by an expert panel, showed no variation among the irrigation treatments until the second day of storage, but significant differences ($p \leq 0.05$) were noticed after that (Figure 4). Overall, samples grown in higher nutrient amounts showed a corresponding rapid decline in OVQ during storage, to greater degrees in Cheongchima. Consequently, samples of 2000 mL·d^{-1}·plant^{-1} treatment received lower OVQ scores, which were insufficient for maintaining marketable life after four and six days of storage, in Cheongchima and Geockchima, respectively. Our results are in accordance with the results of Hoque et al. [51], who found an increase in lettuce yield with an increased rate of nitrogen application, whereas postharvest quality declined. During the storage of fresh-cut iceberg lettuce, the highest quality was discovered in samples harvested from deficit irrigation [14]. A possible reason for the lower OVQ scores of samples irrigated with the higher nutrient solution could be the higher fresh weight loss that occurred in those samples (Figures 2 and 4). The difference in color change per the storage period of the two varieties was lower in Geockchima than in Cheongchima. In Geockchima, no difference was noted based on the amount of fertilizer input; however, a large difference was confirmed in Cheongchima. This implies that for storage–distribution in Cheongchima, fertilizer input quantity must be considered.

Nunes and Emond [50] discovered a significant correlation between weight loss and visual quality in several fresh fruits and vegetables, which supports our observation. As well as water loss, total color difference data showed that samples with higher ΔE values received lower OVQ scores, suggesting the implication of color change in OVQ scoring. Lettuce leaves harvested from 250 mL·d^{-1}·plant^{-1} treatment, on the other hand, showed marketable life until 8 and 10 days after harvest in Geockchima and Cheongchima, respectively (Figure 4). The overall visual quality evaluation began with 9 points based on Table 2, and was performed by three researchers in the storage room. This result may not ensure the higher storage potentiality of the Cheongchima cultivar because the obtained OVQ value just reached the brink of minimum marketability value (a score of 5) on 10 days of storage, when a higher water loss was recorded compared to that of Geockchima (Figures 2 and 4). Overall, lettuce grown with lower irrigation volumes (250 and 500 mL·d^{-1}·plant^{-1}) in Cheongchima showed slightly higher OVQ scores compared to those cultivated with higher volumes. However, the differences in OVQ affected by irrigation volume were unclear in Geockchima, which might need further detailed investigation.

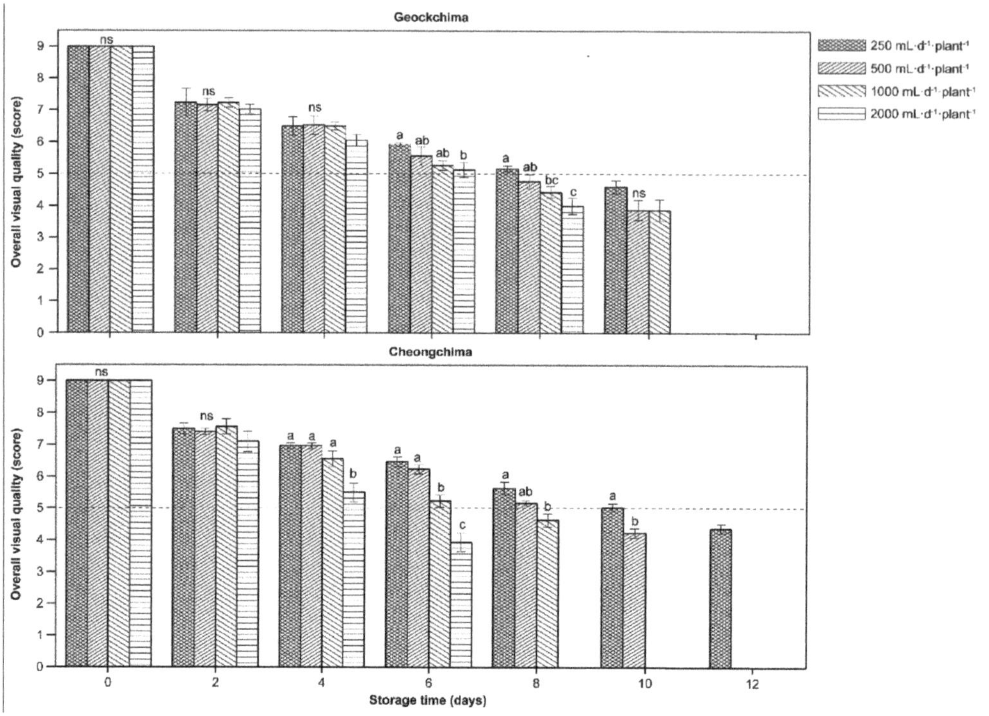

Figure 4. The overall visual quality scores of leaf lettuce cultivars, Geockchima and Cheongchima, grown in perlite culture with different amounts of the nutrient solution during storage at 7 °C (the dotted line corresponds to the limit of acceptability). Mean values in each evaluation day with different lowercase letters are significantly different according to Duncan's multiple range test ($p \leq 0.05$), ns = non-significant.

4. Conclusions

This study clearly demonstrated the amount of applied nutrient solution that influences the growth, physicochemical, nutritional, and postharvest qualities of leaf lettuce in perlite culture. Lettuces grown in nutrient-deficient solution showed significantly lower yield and contributing characters, as well as lower ascorbic acid concentration, but a higher amount of dry matter, crude fiber, total chlorophyll, and anthocyanin contents than those grown in sufficiently or excessively supplied nutrient solution. Lettuce grown with 1000 mL·d^{-1}·plant^{-1} nutrient solution had the highest growth and nutritional components. However, postharvest performances, such as relative fresh weight, color changes, and OVQ, were better maintained in lettuces grown with limited nutrient solutions. Based on these results, cultivation conditions advantageous for growth and yield do not always enhance storability after harvest. Therefore, the supply of plant nutrients with irrigation water during the growth of lettuce should be adjusted based on their uses, so that a higher yield could be targeted for domestic or immediate uses only, while a limited supply of nutrient solution could ensure a longer postharvest life when shipment or storage of lettuce for certain period is required. Considering these views and based on these findings, it can be concluded that leaf lettuce could be irrigated with 1000 or 250 mL·d^{-1}·plant^{-1} nutrient solution to increase yield or maintain postharvest qualities. As a result, further studies on the proper nutrient solution control system during cultivation in relation to postharvest quality are required. Additionally, these results can be applied to lettuce distribution, storage duration requirement, transportation conditions, and can be used for lettuce consumption and marketing.

Author Contributions: Conceptualization, J.S. and J.-S.L.; methodology, D.C.; software, D.C.; validation, J.S. and J.-S.L.; and D.C.; formal analysis, J.S.; investigation, J.S.; resources, J.S.; data curation, J.-S.L.; writing—original draft preparation, D.C.; writing—review and editing, J.S.; visualization, J.-S.L.; supervision, J.-S.L.; project administration, J.-S.L.; funding acquisition, J.-S.L. All authors have read and agreed to the published version of the manuscript.

Funding: This research was funded by NIHHS, RDA, Republic of Korea (projects PJ010194 and PJ01271801).

Institutional Review Board Statement: Not applicable.

Informed Consent Statement: Not applicable.

Data Availability Statement: The datasets generated during and/or analyzed during the current study are available from the corresponding author on reasonable request.

Acknowledgments: We are thankful to Kyung Ran Do for providing us with the laboratory facilities for measuring some parameters in this study. This study was supported by the RDA Fellowship Program of National Institute of Agricultural Sciences (Jinkwan, Son) and National Institute of Horticultural & Herbal Science (Dulal Chandra), Rural Development Administration, Republic of Korea.

Conflicts of Interest: The authors declare no conflict of interest. The funders had no role in the design of the study; in the collection, analyses, or interpretation of data; in the writing of the manuscript; or in the decision to publish the results.

References

1. Nicola, S.; Hoeberechts, J.; Fontana, E. Comparison between traditional and soilless culture systems to produce rocket (*Eruca sativa*) with low nitrate content. *Acta Hortic.* **2005**, *697*, 549–555. [CrossRef]
2. Selma, M.V.; Luna, M.C.; Martínez-Sánchez, A.; Tudela, J.A.; Beltrán, D.; Baixauli, C.; Gil, M.I. Sensory quality, bioactive constituents and microbiological quality of green and red fresh-cut lettuces (*Lactuca sativa* L.) are influenced by soil and soilless agricultural production systems. *Postharvest Biol. Technol.* **2012**, *63*, 16–24. [CrossRef]
3. Blanch, M.; Alvarez, M.D.; Sanchez-Ballesta, M.T.; Escribano, M.I.; Merodio, C. Water relations, short-chain oligosaccharides and rheological properties in lettuces subjected to limited water supply and low temperature stress. *Sci. Hortic.* **2017**, *225*, 726–735. [CrossRef]
4. Siomos, A.S.; Beis, G.; Papadopoulou, P.P.; Nasi, N.; Kaberidou, I.; Barbayiannis, N. Quality and composition of lettuce (cv. "Plenty") grown in soil and soilless culture. *Acta Hortic.* **2001**, *548*, 445–450. [CrossRef]
5. Food and Agriculture Organization of the United Nations FAO. *Good Agricultural Practices for Greenhouse Vegetable Crops*; FAO Plant Production and Protection Paper 217; FAO: Rome, Italy, 2013.
6. Ministry of Agriculture, Food and Ministry of Agricultural Food and Rural Affairs (MAFRA). 2020 Greenhouse Status and Vegetable Production Performance. Available online: https://www.mafra.go.kr/bbs/mafra/71/251301/download.do (accessed on 28 March 2022).
7. Ministry of Agricultural Food and Rural Affairs (MAFRA). 2017 Greenhouse Status and Vegetable Production Performance. Available online: https://www.mafra.go.kr/bbs/mafra/131/322442/artclView.do (accessed on 28 March 2022).
8. Xu, H.L.; Gauthier, L.; Gosselin, A. Effects of fertigation management on growth and photosynthesis of tomato plants grown in peat, rockwool and NFT. *Sci. Hortic.* **1995**, *6*, 11–20. [CrossRef]
9. Choi, E.Y.; Lee, H.J.; Lee, Y.B. (Mineral elements control of the nutrient solution in perlite culture of cucumber. *J. Korean Soc. Hortic. Sci. Technol.* **2001**, *42*, 497–500.
10. Kang, N.J.; Cho, M.W.; Kweon, J.K.; Rhee, H.C.; Choi, Y.H. Effects of deficit irrigation on the total soluble solids and fruit yields of fresh tomato. *Korean J. Bio-Environ. Control* **2006**, *15*, 335–339.
11. Yun, H.K.; Seo, T.C.; Zhang, C.H.; Huang, H.Z. Effect of selenium application on growth and quality of Chinese cabbage grown hydroponically in perlite media. *J. Korean Soc. Hortic. Sci.* **2005**, *23*, 363–366.
12. Kim, M.J.; Moon, Y.; Tou, J.C.; Mou, B.; Waterland, N.L. Nutritional value, bioactive compounds and health benefits of lettuce (*Lactuca sativa* L.). *J. Food Compos. Anal.* **2016**, *49*, 19–34. [CrossRef]
13. Ahmed, A.K.; Cresswell, G.C.; Haigh, A.M. Comparison of sub-irrigation and overhead irrigation of tomato and lettuce seedlings. *J. Hortic. Sci. Biotechnol.* **2000**, *75*, 350–354. [CrossRef]
14. Luna, M.C.; Tudela, J.A.; Martínez-Sánchez, A.; Allende, A.; Marín, A.; Gil, M.I. Long term deficit and excess of irrigation influences quality and browning related enzymes and phenolic metabolism of fresh-cut iceberg lettuce (*Lactuca sativa* L.). *Postharvest Biol. Technol.* **2012**, *73*, 37–45. [CrossRef]
15. Vickers, L.H.; Grove, I.G.; Monaghan, J.M. Irrigation affects postharvest discolouration and yield in iceberg lettuce. *Acta Hortic.* **2015**, *1091*, 253–258. [CrossRef]

16. Eichholz, I.; Forster, N.; Ulrichs, C.; Schreiner, M.; Huyskens-Keil, S. Survey of bioactive metabolites in selected cultivars and varieties of *Lactuca sativa* L. under water stress. *J. Appl. Bot. Food Qual.* **2014**, *87*, 265–273.
17. Baslam, M.; Goicoechea, N. Water deficit improved the capacity of arbuscular mycorrhizal fungi (AMF) for inducing the accumulation of antioxidant compounds in lettuce leaves. *Mycorrhiza* **2012**, *22*, 347–359. [CrossRef]
18. Stefanelli, D.; Winkler, S.; Jones, R. Reduced nitrogen availability during growth improves quality in red oak lettuce leaves by minimizing nitrate content, and increasing antioxidant capacity and leaf mineral content. *Agric. Sci.* **2011**, *2*, 477–486. [CrossRef]
19. Fallovo, C.; Rouphael, Y.; Cardarelli, M.; Rea, E.; Battistelli, A.; Colla, G. Yield and quality of leafy lettuce in response to nutrient solution composition and growing season. *J. Food Agric. Environ.* **2009**, *7*, 456–462.
20. Fallovo, C.; Rouphael, Y.; Rea, E.; Battistelli, A.; Colla, G. Nutrient solution concentration and growing season affect yield and quality of *Lactuca sativa* L. var. *acephala* in floating raft culture. *J. Sci. Food Agric.* **2009**, *89*, 1682–1689.
21. Kotsiras, A.; Vlachodimitropoulou, A.; Gerakaris, A.; Bakas, N.; Darras, A.I. Innovative harvest practices of Butterhead, Lollorosso and Bataviagreen lettuce (*Lactuca sativa* L.) types grown in floating hydroponic system to maintain the quality and improve storability. *Sci. Hortic.* **2016**, *201*, 1–9. [CrossRef]
22. Lee, J.S.; Chang, M.S. Effect of nutrient solution concentration in the second half of growing period on the growth and postharvest quality of leaf lettuce (*Lactuca sativa* L.) in a deep flow technique system. *Hortic. Sci. Technol.* **2017**, *35*, 456–464.
23. Chandra, D.; Matsui, T.; Suzuki, H.; Kosugi, Y.; Fujimura, K.; Bhowmik, P.K. Textural and compositional changes of stored iceberg lettuce in relation to harvest season and storage condition. *Int. J. Veg. Sci.* **2010**, *16*, 44–59. [CrossRef]
24. Tian, W.; Lv, Y.; Cao, J.; Jiang, W. Retention of iceberg lettuce quality by low temperature storage and postharvest application of 1-methylcyclopropene or gibberellic acid. *J. Food Sci. Technol.* **2014**, *51*, 943–949. [CrossRef] [PubMed]
25. Kirnak, H.; Tas, I.; Gokalp, Z.; Karaman, S. Effects of different irrigation levels on yield of lettuce grown in an unheated greenhouse. *Curr. Trends Nat. Sci.* **2016**, *5*, 145–151.
26. Perez-Lopez, U.; Miranda-Apodaca, J.; Lacuesta, M.; Mena-Petite, A.; Munoz-Rueda, A. Growth and nutritional quality improvement in two differently pigmented lettuce cultivars grown under elevated CO_2 and/or salinity. *Sci. Hortic.* **2015**, *195*, 56–66. [CrossRef]
27. Fu, Y.; Lia, H.Y.; Yu, J.; Liu, H.; Cao, Z.Y.; Manukovsky, N.S.; Liu, H. Interaction effects of light intensity and nitrogen concentration on growth, photosynthetic characteristics and quality of lettuce (*Lactucasativa* L. Var. *youmaicai*). *Sci. Hortic.* **2017**, *214*, 51–57. [CrossRef]
28. Lin, K.H.; Huang, M.Y.; Huang, W.D.; Hsu, M.H.; Yang, Z.W.; Yang, C.M. The effects of red, blue, and white light-emitting diodes on the growth, development, and edible quality of hydroponically grown lettuce (*Lactuca sativa* L. var. *capitata*). *Sci. Hortic.* **2013**, *150*, 86–91. [CrossRef]
29. Rural Development Administration. *Standardized Methods of Analysis for Agricultural Science and Technology*; RDA: Suwon, Korea, 2012; Available online: http://lib.rda.go.kr/newlib/upload/prbook/%EC%97%B0%EA%B5%AC%EC%A1%B0%EC%82%AC%EB%B6%84%EC%84%9D%EA%B8%B0%EC%A4%80.pdf (accessed on 28 March 2022).
30. Aguero, M.V.; Barg, M.V.; Yommi, A.; Camelo, A.; Roura, S.I. Postharvest changes in water status and chlorophyl content of lettuce (*Lactuca sativa* L.) and their relationship with overall visual quality. *J. Food Sci.* **2008**, *73*, S47–S55. [CrossRef]
31. Association of Official Agricultural Cemists. *Official Methods of Analysis*, 15th ed.; AOAC: Arlington, VA, USA, 1990.
32. Clarkson, G.J.J.; O'Byrne, E.E.; Rothwell, S.D.; Taylor, G. Identifying traits to improve postharvest processability in baby leaf salad. *Postharvest Biol. Technol.* **2003**, *30*, 287–298. [CrossRef]
33. Jagota, S.K.; Dani, H.M. A new calorimetric technique for the estimation of vitamin C using Folin phenol reagent. *Anal. Biochem.* **1982**, *127*, 178–182. [CrossRef]
34. Luna, M.C.; Tudela, J.A.; Martínez-Sánchez, A.; Allende, A.; Gil, M.I. Optimizing water management to control respiration rate and reduce browning and microbial load of fresh-cut romaine lettuce. *Postharvest Biol. Technol.* **2013**, *80*, 9–17. [CrossRef]
35. Ferrante, A.; Maggiore, T. Chlorophyl a fluorescence measurements to evaluate storage time and temperature of *Valeriana* leafy vegetables. *Postharvest Biol. Technol.* **2007**, *45*, 73–80. [CrossRef]
36. Giusti, M.M.; Wrolstad, R.E. Characterization and measurement of anthocyanins by uv-visible spectroscopy. *Curr. Prot. Food Anal. Chem.* **2001**, *1*, F1.2.1–F1.2.13. [CrossRef]
37. Gomez, K.A.; Gomez, A.A. (Eds.) Comparison between treatments means. In *Statistical Procedures for Agricultural Research*, 2nd ed.; John Wiley & Sons: New York, NY, USA, 1983; pp. 187–240.
38. Yazgan, S.; Ayas, S.; Demirtas, C.; Buyukcangaz, H.; Candogan, B.N. Deficit irrigation effects on lettuce(*Lactuca sativa* var. *Olenka*) yield in unheated greenhouse condition. *J. Food Agric. Environ.* **2008**, *6*, 357–361.
39. Reference Chunli, L.; Nicki, J.E. Comparison of growth characteristics, functional qualities, and texture of hydroponically grown and soil-grown lettuce. *LWT* **2021**, *150*, 111931. [CrossRef]
40. Seo, T.C.; Kim, Y.C.; Kim, K.Y.; Lee, J.W.; Yun, H.K.; Lee, S.G. Optimal supply amount and strength of nutrient solution for ripe-harvesting tomatoes grown under perlite culture system of semi-forcing cropping. *J. Korean Soc. Hortic. Sci.* **2003**, *21*, 79–85.
41. Rajabbeigi, E.; Eichholz, I.; Beesk, N.; Ulrichs, C.; Kroh, L.W.; Rohn, S.; Huyskens-Keil, S. Interaction of drought stress and UV-B radiation–impact on biomass production andflavonoid metabolism in lettuce (*Lactuca sativa* L.). *J. Appl. Bot. Food Qual.* **2013**, *86*, 190–197.

42. Zeipina, S.; Alsina, I.; Lepse, L.; Duma, M. The effect of irrigation on the biochemical content of leafy vegetables. In Proceedings of the Soil and Water Management, NJF 25th Congress of Nordic View to Sustainable Rural Development, Liga, Latvia, 16–18 June 2015; pp. 209–213.
43. Ilker, R.; Szczesniak, A.S. Structural and chemical bases for texture of plant food stuffs. *J. Tex. Stud.* **1990**, *21*, 1–36. [CrossRef]
44. Shinohara, Y.; Suzuki, Y. Quality improvement of hydroponically grown leaf vegetables. *Acta Hortic.* **1988**, *230*, 279–286. [CrossRef]
45. Lee, S.K.; Kader, A.A. Preharvest and postharvest factors influencing vitamin C content of horticultural crops. *Postharvest Biol. Technol.* **2000**, *20*, 207–220. [CrossRef]
46. Llorach, R.; Martinez-Sanchez, A.; Tomas-Barberan, F.A. Characterization of polyphenols and antioxidant properties of five lettuce varieties and escarole. *Food Chem.* **2008**, *A108*, 1028–1038. [CrossRef]
47. Chalker-Scott, L. Do anthocyanins function as osmoregulators in leaf tissues? *Adv. Bot. Res.* **2002**, *37*, 103–106.
48. Chalker-Scott, L. Environmental significance of anthocyanins in plant stress responses. *Photochem. Photobiol.* **1999**, *70*, 1–9. [CrossRef]
49. Bryant, J.P.; Chapin, F.S.; Klein, D.R. Carbon/nutrient balance of boreal plants in relation to vertebrate herbivory. *Oikos* **1983**, *40*, 357–368. [CrossRef]
50. Nunes, C.N.; Emond, J.P. Relationship between weight loss and visual quality of fruits and vegetables. *Proc. Fla. State Hortic. Soc.* **2007**, *120*, 235–245.
51. Hoque, M.M.; Ajwa, H.; Othman, M.; Smith, R.; Cahn, M. Yield and postharvest quality of lettuce in response to nitrogen, phosphorus, and potassium fertilizers. *Hortic. Sci.* **2010**, *45*, 1539–1544. [CrossRef]

horticulturae

MDPI

Article

Yield and Compositional Profile of Eggplant Fruits as Affected by Phosphorus Supply, Genotype and Grafting

Rosario Paolo Mauro [1,*], Silvia Rita Stazi [2], Miriam Distefano [1], Francesco Giuffrida [1], Rosita Marabottini [3], Leo Sabatino [4], Enrica Allevato [5], Claudio Cannata [1], Federico Basile [1] and Cherubino Leonardi [1]

1 Dipartimento di Agricoltura, Alimentazione e Ambiente (Di3A), University of Catania, Via Valdisavoia, 5, 95123 Catania, Italy; miriam.distefano@unict.it (M.D.); francesco.giuffrida@unict.it (F.G.); claudio.cannata@phd.unict.it (C.C.); federico.basile@phd.unict.it (F.B.); cherubino.leonardi@unict.it (C.L.)

2 Dipartimento di Scienze Chimiche, Farmaceutiche ed Agrarie (DOCPAS), University of Ferrara, Via Borsari, 46, 44121 Ferrara, Italy; silviarita.stazi@unife.it

3 Dipartimento per la Innovazione nei Sistemi Biologici, Agroalimentari e Forestali (DIBAF), University of Tuscia, Via San Camillo de Lellis, snc, 01100 Viterbo, Italy; marabottini@unitus.it

4 Dipartimento Scienze Agrarie, Alimentari e Forestali (SAAF), University of Palermo, Viale delle Scienze, ed., 90128 Palermo, Italy; leo.sabatino@unipa.it

5 Dipartimento di Scienze dell'Ambiente e della Prevenzione (DiSAP), University of Ferrara, Via Borsari, 46, 44121 Ferrara, Italy; enrica.allevato@unife.it

* Correspondence: rosario.mauro@unict.it

Abstract: The present experiment addressed the effects of two phosphorus regimes (30 and 90 kg ha^{-1}, hereafter P_{30} and P_{90}) on yield and composition of eggplant fruits in 'Birgah' and 'Dalia', whether or not these cultivars were grafted onto *Solanum torvum* 'Espina'. The P_{30} regime did not reduce yield, and promoted fruits' dry matter and total phenols content, along with their concentrations of macronutrients, mesonutrients (S and Na) and micronutrients (mostly Cu, B, Zn); however, their Fe concentrations were depressed. The rootstock 'Espina' increased fruit yield, dry matter content, epicarp chroma (in 'Birgah') and Ca content, together with their concentrations of B and Zn (especially at P_{30}), but reduced their Fe content, mostly under P_{30}. Thus, the reduced P supply and grafting proved to be effective tools to enhance fruit yield, carpometric and almost all nutritional traits in eggplant, in a framework of more sustainable crop management. However, the reduced fruit concentration of Fe suggests that the affinity of the rootstock with specific micro minerals should be taken into account, along with the option to adopt complementary practices (e.g., targeted micronutrient fertilizations) to manage the micro mineral composition of eggplants.

Keywords: *Solanum melongena* L.; rootstock; total polyphenols; mineral content

Citation: Mauro, R.P.; Stazi, S.R.; Distefano, M.; Giuffrida, F.; Marabottini, R.; Sabatino, L.; Allevato, E.; Cannata, C.; Basile, F.; Leonardi, C. Yield and Compositional Profile of Eggplant Fruits as Affected by Phosphorus Supply, Genotype and Grafting. *Horticulturae* **2022**, *8*, 304. https://doi.org/10.3390/horticulturae8040304

Academic Editor: Zhihui Cheng

Received: 2 March 2022
Accepted: 31 March 2022
Published: 3 April 2022

1. Introduction

Phosphorus (P) is present in all living cells as a component of biomolecules, such as nucleic acids, phospholipids, ATP or NADPH [1]. In plants, P participates in multiple metabolic events and, in the human body, is the second most abundant mineral, being a constituent of the skeletal system. As a result of its biological importance, P fertilization is irreplaceable to ensure adequate food provision to mankind, both from a quantitative and qualitative viewpoint [2]. Despite its biological ubiquity, phosphate rocks, i.e., the only nonrenewable sources of P fertilizers, are unevenly distributed worldwide, with Morocco, China and USA accounting for 94% of the estimated 300 billion Mt of phosphate rocks worldwide [1]. Agriculture represents the main consumer of the estimated 71 billion Mt of P reserves, which, at the actual consumption rates, are expected to be significantly depleted in the next 50–100 years, consequently posing future concerns on global food security [3]. Hence, there is a compelling need to optimize P use in agriculture, especially in horticultural systems which largely contribute to the current global annual consumption of P fertilizers (~21 Mt) and related environmental impacts [4–7].

Vegetables are pivotal sources of nutraceuticals, with antioxidants and minerals being among the most prominent ones provided to the human diet. However, these plant foods do not always fit the dietary requirements, since several factors may affect their compositional profiles, including irrational fertilization practices [8,9]. Among fruit vegetables, eggplant (*Solanum melongena* L.) is a popular solanaceous crop, whose edible, immature berries are important sources of minerals and antioxidants in the Mediterranean diet [10]. Italy is the main producer among the European countries, with 304,690 out of the 960,227 t of eggplants (31.7% of total) produced in 2020 from a 9510-hectare surface area, mainly concentrated in the southern regions [11]. In the Mediterranean Basin, the crop is often grown in open field, on soils with high limestone content, a feature that is usually correlated to strong responses of many vegetable crops to P fertilization [12]. This leads to implications for eggplant cultivation, as high P fertilization rates are often adopted under field conditions, being a reputed primary tool to maximize fruit yield under P-limiting conditions. Currently, rational P fertilizations are desirable in horticultural systems, in order to address the concerns of agriculture-derived burden on P reserves and ecosystems, food provision to future generations, and rising consumer demand for environmentally sound plant foods [13]. In this view, vegetable grafting has been proven as a useful means to modulate yield and product composition of vegetable crops in a wide array of growth conditions [14,15]. In greenhouse-grown eggplants, grafting has been able to modify the fruit yield and quality in terms of phenolic profile or mineral distribution inside the plant [16,17]. Currently, no attempts have been focused on defining the relationships among grafting and P supply in open-field eggplants, in terms of product yield and quality traits, including its nutraceutical composition. To this end, the central role of P fertilization is well known to impact the product quality and mineral profile in many crops (e.g., grain crops) [4], given the multiple and complex interactions among P and other nutrients [18]. Thus, the abovementioned information may be highly instrumental in the double perspective to obtain adequate nutrient-dense eggplants in a more sustainable way. For these reasons, the present study addressed the effects of different P fertilization regimes and grafting on yield, quality traits and mineral composition of two field-grown eggplant genotypes largely cultivated around the Mediterranean Basin.

2. Materials and Methods

2.1. Experimental Site

A field experiment was conducted during 2019 at the experimental farm of the University of Catania (37°24′27″ N, 15°03′35″ E, 5 m a.s.l.), on a xerofluvent soil, having the characteristics reported in Table 1. The soil is characterized by an alkaline pH, a very high content of exchangeable K, total Mg and Ca, and a low content of total N and available P. The local climate is semiarid-Mediterranean (*Cs* climate according to Köppen classification), with warm and rainless summers. The meteorological conditions were recorded by an integrated station near the crop (~40 m). During the experiment (from May 2 to September 1), mean monthly air temperature progressively increased from May (17.7 °C) to June (22.0 °C) and July (25.8 °C), then slightly decreased in August (25.3 °C) (Figure 1). These values were similar to those of the long-term average (2009–2018), which were equal to 17.9 (May), 21.9 (June), 25.5 (July) and 25.6 °C (August). The lowest daily temperatures were recorded at 22 days after transplanting (DAT) (19.9 and 8.1 °C for maximum and minimum mean temperature, respectively) and the highest ones at 90 DAT (35.6 and 19.7 °C, respectively) (Figure 1). Mean relative humidity showed similar values in May and June, peaking at 6, 8 and 44 DAT (82%, on average), and showing the lowest values in July, at 73, 74 and 80 DAT (47%, on average) (Figure 1). No rainfall occurred during the experimental period, such that the overall difference compared to the long-term average was equal to 34.1 mm. Before transplanting, the soil was ploughed (0.40 m depth) and disk-harrowed (0.20 m depth).

Table 1. Soil characteristics at the experimental site (0–40 cm depth).

Soil Characteristic	Unit	Value
Clay	%	30
Silt	%	13
Sand	%	57
Organic matter	%	2.8
pH	log[H$^+$]	7.5
Cation exchange capacity	cmol$_c$ kg^{-1}	19.1
Total N	mg kg^{-1}	1.4
Available P	mg kg^{-1}	1.5
Exchangeable K	mg kg^{-1}	278
Mg	mg kg^{-1}	433
Ca	mg kg^{-1}	5245
Na	mg kg^{-1}	125
Fe	mg kg^{-1}	107
Cu	mg kg^{-1}	12
Zn	mg kg^{-1}	6

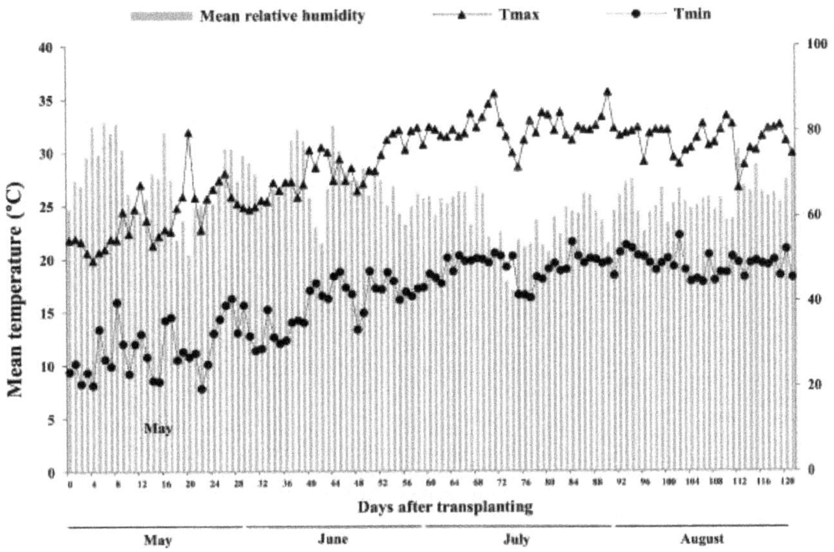

Figure 1. Meteorological conditions recorded during the trial.

2.2. Experimental Design and Crop Management

The experiment was arranged as a randomized split-split-plot design with four replications based on 3.60 × 3.50 m net experimental units (12 plants). A few days before transplanting by hand, the soil along the rows was mulched with black polyethylene sheets (40-centimeter width, 40-micrometer thickness), whereas plantlets at the stage of three true-leaves were transplanted on May 2 in a 0.50 × 1.20 m format, then trained at two stems up to September 1. Two P regimes were arranged as main plots, corresponding to 30 and 90 kg P ha^{-1} (hereafter P_{30} and P_{90}, respectively), whereas two eggplant cultivars ('Birgah' and 'Dalia'; Seminis, Milan, Italy) were imposed as sub-plots, either ungrafted or grafted onto the *Solanum torvum* Swartz rootstock 'Espina' (Esasem, Casaleone, Italy) (sub-sub-plots). Overall, we had 24 experimental units (2 P regimes × 2 eggplant cultivars × 2 graft configurations × 3 replications). The P_{90} level was chosen since it represents a high supply commonly encountered in the reference area, whereas P_{30} represents the dose calculated for a better balance with the other administered macronutrients. 'Birgah' and 'Dalia' are two cultivars largely grown in south Italy, differing for their fruit typology (globose/violet and ovoidal/black, respectively). The rootstock 'Espina' was chosen because of its extensive root system and adaptation to warm/dry conditions; thus, it is among the most widespread in the reference area. The crop was drip irrigated, restituting 100% of ET_M when accumulated daily evapotranspiration (calculated using a class A-Pan evaporimeter near the crop) reached 40 mm. Fertigation was performed once a week from transplanting up to August 22 (17 interventions), overall administering the different P levels together with 120, 140, 16 and 5 kg ha^{-1} for N, K_2O, MgO and CaO, respectively. Micronutrients were provided in the following amounts (in kg ha^{-1}): B (1), Cu (1), Fe (4), Mn (4), Mo (0.4) and Zn (2). Pests control was performed as per local custom.

2.3. Fruit Collection, Carpometric Determinations and Sample Preparation

Commercially ripe fruits were harvested by hand from June 18 when they were near the maximum size, but before the onset of the epicarp color turning [19]. At each harvest, marketable fruits were counted and weighed, in order to calculate yield and number of fruits per plant. After harvesting, 12 fruits per plot were washed with distilled water, dried with paper, after which the epicarp color was measured along the equatorial region (2 readings per fruit) using a tristimulus Chroma meter CR-400 (Minolta Corporation, Ltd., Osaka, Japan), measuring L* (lightness), a* (green-red axis) and b* (blue-yellow axis). The epicarp color was expressed in terms of chroma (C*), calculated as $(a^{*2} + b^{*2})^{1/2}$. Fruits were then weighed to determine their fresh weight (FW) and dry matter (DM) content (after oven-desiccation at 75 °C, until constant weight). Fruits samples harvested from 20 to 22 July, i.e., when the crop attained the full yield rate, were ground after freeze-drying (−50 °C) and passed through a 1-millimeter sieve, then stored at −80 °C until further analyses.

2.4. Determination of Total Phenols Content

Fruit total phenols content (TPC) was determined from lyophilized pulp powder (0.10 g), which was put into 5 mL 80% ethanol (*v*/*v*) in centrifuge tubes. The samples were extracted for 15 min using an ultrasonic bath LBS1-3 (Falc Instruments, Treviglio, Italy), maintaining water temperature below 10 °C. The tubes were then centrifuged (5000 g) at 5 °C for 20 min and the supernatant was transferred into vials. The procedure was performed 3 times. The combined extracts were diluted in measuring flasks to a 25-milliliter volume; 1.5 mL of diluted samples were filtered through a 0.45-micrometer nylon filter. Then, 200 µL of extract solution were added to 1000 µL, 1:10 diluted Folin-Ciocâlteu reagent with ultrapure water, vortexed for 1 min, after which 800 µL of 0.7 M Na_2CO_3 were added. The liquid was vortexed and left in the dark for 60 min at room temperature. TPC was determined by reading the absorbance at 765 nm with a Jenway 7315 UV-vis spectrophotometer (Cole-Parmer, Stone, UK) and expressed as mg chlorogenic acid equivalent (CAE) kg^{-1} FW.

2.5. Determination of Total N and Mineral Profile

Two hundred mg of each sample were digested with 2 mL of 30% (m/m) H_2O_2, 0.5 mL of 37% HCl and 7.5 mL of HNO_3 69% solution. The acid digestion was performed using a high-pressure laboratory microwave oven Mars plus (CEM srl, Cologno al Serio, Italy) operating at 1200 W. The temperature was linearly increased from 25 to 180 °C in 37 min, then held at 180 °C for 15 min. The digested samples were diluted to a final volume of 25 mL with ultrapure water. Blanks were prepared in each lot of samples. All determinations were performed in triplicate. For the elemental quantification, an ICP-OES (8000 DV, PerkinElmer, Shelton, CT, USA) was used, with an axially viewed configuration equipped with an ultrasonic nebulizer. All reagents used for the microwave-assisted digestions, i.e., HCl, HNO_3 and H_2O_2, were of suprapure grade (Merck, Darmstadt, Germany). High-purity water (18 MΩ cm) from a Milli-Q purification system (Millipore, Bedford, MA, USA) was used for diluting the standards.

Multi-elemental, high-purity grade was purchased from CaPurAn (CPAchem Ltd., Bogomilovo, Bulgaria). The purity of the plasma torch argon was greater than 99.99%. The external calibration solutions were prepared from standard certified elemental solutions (CaPurAn). Total N and S were determined using the combustion analysis by CHNS Elemental Analyzer (EA Flash 2000 Thermo Fisher Scientific CHNS-O determination, Rodan, Milan, Italy).

2.6. Statistical Procedures

All collected and calculated data were firstly subjected to Shapiro–Wilk and Levene's tests, in order to check for normal distribution and homoscedasticity, respectively. Then, a 'phosphorus level × cultivar × rootstock' analysis of variance (ANOVA) was applied according to the experimental layout adopted in the field. Percentage data were Bliss' transformed before the ANOVA (untransformed data are reported), whereas means comparisons were performed through Tukey's HSD test ($p \leq 0.05$).

3. Results

3.1. Yield and Carpometric Traits

Marketable yield, number of fruits plant^{-1} and fruit FW were not affected by the P supply in the studied cultivars (Table 2). These last showed contrasting yield traits, as 'Birgah' showed the highest marketable yield and fruit FW, whereas 'Dalia' had the highest number of fruits plant^{-1} (Table 2). The rootstock 'Espina' promoted marketable yield (+38%), fruit FW (+16%) and the number of fruits plant^{-1} (Table 2), with this last variable showing the strongest increase under P_{30} (+30%) (Figure 2A). Concerning fruit traits, higher firmness, chroma and TPC were noticed in P_{30} compared to P_{90} and in 'Birgah' compared to 'Dalia' (excepting TPC), whereas grafting boosted fruit firmness (+7%) and DM content (Table 2), with this last variable only in P_{30} (+1.52%) (Figure 2B). Among the tested genotypes, 'Birgah' showed a strong increase in epicarp chroma when P fertilization was reduced from P_{90} to P_{30} (+45%) (Figure 3).

Table 2. Yield performances, carpometric traits and TPC of eggplants as affected by the main factors (mean ± standard deviation). Different letters among factors' means indicate significance at Tukey's HSD test ($p \leq 0.05$).

Variable	Phosphorus Level		Cultivar		Rootstock		Mean
	P_{30}	P_{90}	'Birgah'	'Dalia'	Control	'Espina'	
Marketable yield (kg plant^{-1})	2.80 ± 0.74 a	2.80 ± 0.50 a	2.97 ± 0.68 a	2.64 ± 0.53 b	2.36 ± 0.49 b	3.25 ± 0.57 a	2.80 ± 0.62
Marketable fruits (n. plant^{-1})	7.34 ± 2.10 a	7.49 ± 1.97 a	5.71 ± 1.86 b	9.12 ± 1.70 a	6.73 ± 1.81 b	8.11 ± 2.00 a	7.42 ± 1.99
Fruit FW (g)	402 ± 125 a	399 ± 128 a	514 ± 105 a	287 ± 127 b	371 ± 109 b	430 ± 142 a	401 ± 124
Fruit DM content (%)	9.70 ± 1.06 a	9.00 ± 0.90 b	9.43 ± 0.98 a	9.27 ± 1.13 a	9.01 ± 0.90 b	9.69 ± 1.08 a	9.35 ± 1.03
Fruit firmness (N)	14.6 ± 2.7 a	13.5 ± 2.2 b	16.0 ± 1.6 a	12.0 ± 1.2 b	13.5 ± 2.4 b	14.5 ± 2.6 a	14.0 ± 2.5
Chroma (adimensional)	5.67 ± 1.06 a	4.28 ± 1.40 b	7.95 ± 2.07 a	2.00 ± 1.09 b	4.90 ± 1.70 a	5.05 ± 1.81 a	4.97 ± 2.05
TPC (mg CAE kg^{-1} FW)	2012 ± 380 a	1721 ± 373 b	1538 ± 268 b	2196 ± 377 a	1878 ± 415 a	1795 ± 398 a	1837 ± 456

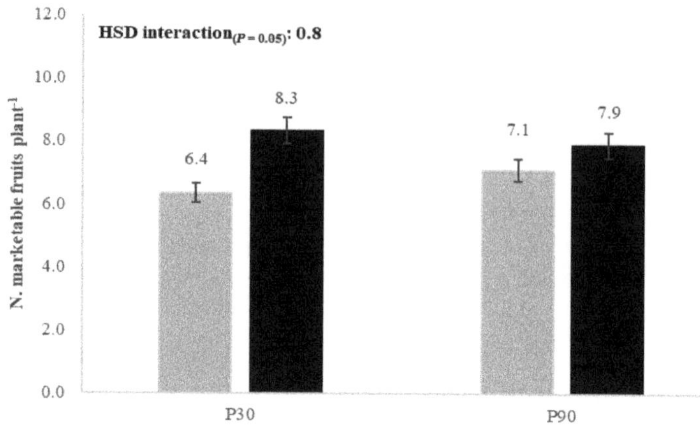

(**a**)

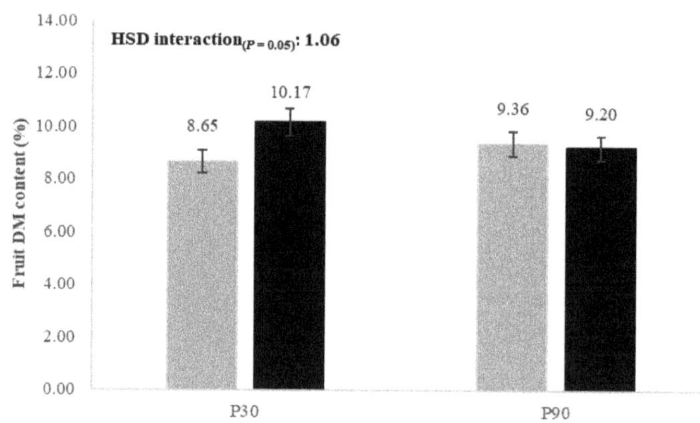

(**b**)

Figure 2. Number of marketable fruits plant^{-1} (**a**) and fruit DM content (**b**) as affected by 'phosphorus level × rootstock' interaction. Grey bars: ungrafted; black bars: grafted onto 'Espina'.

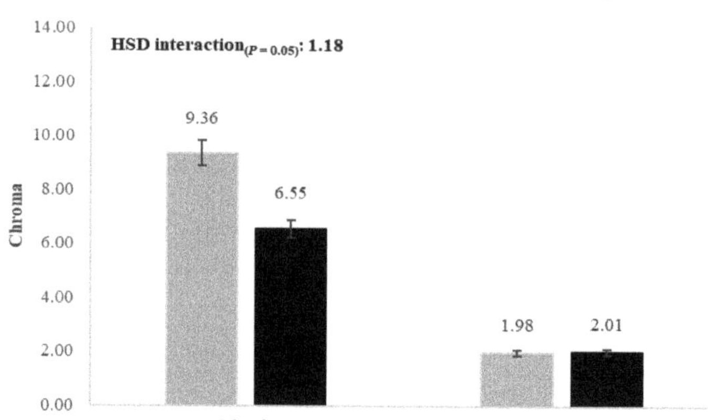

Figure 3. Fruit chroma as affected by 'phosphorus level × cultivar' interaction. Light grey bars: P_{30}; black bars: P_{90}.

3.2. Fruit N and Mineral Profile

3.2.1. Macronutrients

The P_{30} regime promoted N, P, and K accumulation in eggplants (by 18, 11 and 9%, respectively), whereas the heterograft configuration boosted K content (+7%) and depressed N (−15%) (Table 3). The genotypes differed for fruit N and K content (both higher in 'Dalia') (Table 3); however, regarding K, 'Birgah' showed a significant increase under P_{30} supply (+13%) (Table 4) whereas 'Dalia' showed the highest K rise passing from control plants to those grafted onto 'Espina' (+9%) (Table 6).

Table 3. Total N and minerals content of eggplants as affected by the main factors (mean ± standard deviation). Different letters among factors' means indicate significance at Tukey's HSD test ($p \leq 0.05$).

Variable	Phosphorus Level		Cultivar		Rootstock		Mean
	P_{30}	P_{90}	'Birgah'	'Dalia'	Control	'Espina'	
Macronutrients ($mg\ kg^{-1}$ FW)							
N	140 ± 14 a	119 ± 19 b	122 ± 22 b	137 ± 13 a	139 ± 19 a	118 ± 18 b	130 ± 19
P	165 ± 16 a	148 ± 12 b	150 ±21 a	162 ± 8 a	157 ± 16 a	155 ± 18 a	156 ± 17
K	625 ± 45 a	575 ± 75 b	561 ± 54 b	640 ± 53 a	580 ± 79 b	620 ± 53 a	600 ± 66
Mesonutrients ($mg\ kg^{-1}$ FW)							
Mg	102 ± 8 a	98 ± 9 a	97 ± 9 a	103 ± 8 a	100 ± 11 a	100 ± 6 a	100 ± 9
Ca	69 ± 20 a	78 ± 36 a	56 ± 10 b	91 ±31 a	62 ± 10 b	85 ± 37 a	73 ± 29
S	9.5 ± 1.5 a	8.4 ± 1.8 b	9.1 ± 1.9 a	8.8 ± 1.6 a	10.3 ± 0.9 a	7.5 ± 1.0 b	8.9 ± 1.7
Na	15.8 ± 8.8 a	11.5 ± 6.7 b	9.4 ± 3.1 b	17.9 ± 9.1 a	18.0 ± 9.2 a	9.2 ± 2.2 b	13.6 ± 7.9
Micronutrients ($\mu g\ kg^{-1}$ FW)							
Fe	2900 ± 857 b	3339 ± 871 a	3702 ± 757 a	2538 ± 535 b	3139 ± 930 a	3101 ± 856 a	3120 ± 874
Mn	521 ± 63 a	512 ± 76 a	512 ± 62 a	512 ± 78 a	466 ± 64 b	558 ±35 a	512 ± 69
Cu	495 ± 42 a	376 ± 105 b	423 ± 125 a	448 ± 67 a	463 ± 63 a	408 ±121 b	436 ± 99
B	1809 ± 972 a	1374 ± 661 b	1246 ± 416 b	1937 ± 1025 a	1053 ± 294 b	2131 ±874 a	1592 ± 843
Zn	3106 ±1832 a	2172 ± 1224 b	2198 ± 564 b	3080 ±2124 a	1657 ± 604 b	3621 ±1692 a	2639 ± 1597

Table 4. Potassium, magnesium, manganese and copper concentrations in eggplants (mean ± standard deviation) as affected by 'phosphorus level × cultivar' interaction.

Variable	P_{30}		P_{90}		$HSD_{interaction}$ (p = 0.05)
	'Birgah'	'Dalia'	'Birgah'	'Dalia'	
K ($mg\ kg^{-1}$ FW)	596 ± 14	636 ± 59	527 ± 38	643 ± 51	54
Mg ($mg\ kg^{-1}$ FW)	106 ± 7	106 ± 9	88 ± 7	100 ± 6	14
Mn ($\mu g\ kg^{-1}$ FW)	543 ± 31	499 ± 80	480 ± 69	525 ± 81	57
Cu ($\mu g\ kg^{-1}$ FW)	521 ± 34	469 ± 32	325 ± 101	427 ± 88	67

3.2.2. Mesonutrients

Sulphur and Na concentrations were higher in P_{30} than in P_{90} (by 13 and 42%, respectively) (Table 3), without interactive effects. Regarding Mg concentration, an analogous response was noticeable only in 'Birgah' (+20%), whereas no significant differences were recorded for 'Dalia' (Table 4). These genotypes differed in terms of fruit Ca and Na concentrations, as both minerals attained higher levels in 'Dalia' (Table 3), i.e., the cultivar that also showed strongest Ca and Na variations in response to grafting (+62 and −60%, respectively) (Table 6). On the other hand, the heterograft configuration depressed the S accumulation (−27%), irrespective of the P level and genotype (Table 3).

3.2.3. Micronutrients

Increased concentrations in Cu, B and Zn were found in response to the reduced P supply (Table 3), whereas 'Birgah' showed a stronger response to P_{30} for fruit Mn (±13%) and Cu (±62%) concentrations (Table 4). A different trend was recorded for Fe, whose fruit content was strongly decreased, passing from P_{90} to P_{30} (Table 3), with the rootstock 'Espina' depressing fruit Fe under P_{30} supply (−17%) and increasing it at P_{90} (±15%) (Table 5). Such response was opposite to that of B and Zn, whose contents were promoted by 'Espina' mostly in P_{30} (by 125 and 143%, respectively) (Table 5). Excluding Fe and Cu, grafting promoted the accumulation of micronutrients in eggplants (Table 3), although in some cases such response was genotype-dependent. Indeed, while for Mn concentration the graft-induced increase was consistent among P levels and cultivars (±20%) (Table 3), for B and Zn concentrations 'Dalia' showed the highest rise passing from control to grafted onto 'Espina' (±169 and ±320%, respectively). 'Birgah' displayed a strong reduction in Cu concentration in the heterograft configuration (−23%) (Table 6).

Table 5. Iron, boron and zinc concentrations in eggplants (mean ± standard deviation) as affected by 'phosphorus level × grafting' interaction.

Variable	P_{30}		P_{90}		$HSD_{interaction}$ (p = 0.05)
	Control	'Espina'	Control	'Espina'	
Fe (μg kg^{-1} FW)	3176 ± 1161	2625 ± 290	3102 ± 741	3577 ± 993	550
B (μg kg^{-1} FW)	1113 ± 304	2505 ± 907	993 ± 298	1756 ± 722	319
Zn (μg kg^{-1} FW)	1813 ± 600	4399 ± 1736	1501 ± 621	2843 ± 1353	441

Table 6. Potassium, calcium, sodium, copper, boron and zinc concentrations in eggplants (mean ± standard deviation) as affected by 'cultivar × grafting' interaction.

Variable	'Birgah'		'Dalia'		$HSD_{interaction}$ (p = 0.05)
	Control	'Espina'	Control	'Espina'	
K (mg kg^{-1} FW)	546 ±59	577 ± 49	612 ± 42	667 ± 53	52
Ca (mg kg^{-1} FW)	53 ± 8	59 ± 12	69 ± 3	112 ± 32	20
Na (mg kg^{-1} FW)	10.5 ± 3.3	8.3 ± 2.6	25.5 ± 6.4	10.2 ± 1.4	3.7
Cu (μg kg^{-1} FW)	477 ± 76	369 ± 147	450 ± 52	447 ± 85	67
B (μg kg^{-1} FW)	1055 ± 365	1437 ± 400	1051 ± 237	2824 ± 607	319
Zn (μg kg^{-1} FW)	2130 ± 344	2266 ± 755	1184 ± 384	4976 ± 1150	441

3.2.4. Correlation among Variables

The results of the correlation analysis are reported in Table 7. Overall, 171 correlations were analyzed, of which 67 (39% of total) showed significance, highlighting 44 positive and 23 negative relationships. In the case of yield and carpometric traits, 18 out of 21 correlations (86%) were significant, whereas they were 53 out of 150 (35%) in the case of N and mineral concentrations (Table 7). In the case of positive correlations, the highest correlation coefficients were recorded among B and Zn (0.901 ***), fruit FW and chroma (0.879 ***), fruit FW and firmness (0.868 ***), and among fruit firmness and chroma (0.864 ***). The strongest relationships in the dataset of positive correlations were recorded among marketable fruits plant^{-1} and chroma (−0.799 ***), marketable fruits plant^{-1} and fruit FW (−0.722 ***), fruit FW and TPC (−0.691 ***), and among marketable fruits plant^{-1} and Fe concentration (−0.598 **).

Table 7. Pearson's product-moment correlation coefficients (r) among variables. *, ** and *** indicate significance at $p \leq 0.05$, 0.01 and 0.001, respectively. NS: not significant.

	Marketable Yield	Marketable Fruits	Fruit FW	Fruit DM Content	Fruit Firmness	Chroma	TPC	N	P	K	Mg	Ca	S	Na	Fe	Mn	Cu	B
Marketable fruits	NS	-																
Fruit FW	0.543 **	NS	-															
Fruit DM content	0.539 **	NS	NS	-														
Fruit firmness	0.464 *	−0.535 **	0.868 ***	0.602 **	-													
Chroma	NS	NS	0.879 ***	NS	0.864 ***	-												
TPC	NS	0.574 **	NS	NS	−0.583 **	−0.550 **	-											
N	NS	NS	−0.434 *	NS	NS	NS	0.474 *	-										
P	NS	NS	NS	NS	NS	NS	0.502 *	0.658 ***	-									
K	NS	0.455 *	−0.546 **	NS	NS	NS	0.713 ***	0.450 *	0.516 **	-								
Mg	NS	NS	NS	NS	NS	NS	0.497 *	0.613 **	0.503 *	0.497 *	-							
Ca	NS	0.692 ***	−0.571 **	NS	−0.499 *	−0.597 **	NS	NS	NS	NS	NS	-						
S	−0.553 **	NS	NS	NS	NS	NS	NS	0.644 ***	NS	NS	NS	−0.462 *	-					
Na	−0.566 **	NS	−0.595 **	NS	−0.423 *	NS	0.588 **	0.551 **	NS	0.594 **	NS	NS	0.439 *	-				
Fe	NS	−0.598 **	0.669 ***	NS	0.537 **	0.535 **	−0.531 **	NS	NS	NS	NS	NS	NS	−0.497 *	-			
Mn	0.534 **	NS	NS	0.459 *	NS	NS	NS	NS	NS	NS	NS	0.437 *	−0.516 **	−0.466 *	NS	-		
Cu	NS	NS	NS	NS	NS	NS	NS	0.479 *	0.615 **	0.442 *	NS	NS	0.424 *	NS	NS		-	
B	NS	0.665 ***	NS	0.452 *	NS	NS	NS	NS	NS	NS	0.452 *	0.670 ***	−0.411 *	NS	NS	0.637 ***	NS	-
Zn	NS	0.528 **	NS	NS	NS	NS	NS	NS	NS	NS	NS	0.546 **	NS	NS	−0.460 *	0.579 **	NS	0.901 ***

4. Discussion

In the present experiment, none of the studied eggplant cultivars showed negative yield responses under reduced P fertilization, demonstrating the agronomic inefficacy to exceed the P_{30} threshold. Consequently, the P_{90} level depressed the efficiency of P fertilization; for every kg ha^{-1} of supplied P, the crop yielded 1.56 t ha^{-1} of marketable fruits in P_{30}, against 0.52 t ha^{-1} in P_{90} (data not shown). It is possible that the high fractionation of P supply we adopted contributed to these differences too, by reducing the P immobilization into the soil and subsequently yielding more efficient absorption by the plants [4]. Accordingly, the absence of any significant variation of yield components passing from P_{90} to P_{30} (i.e., number of marketable fruits plant^{-1} and fruit FW) indicates that the plants had no need for extra P supplied beyond 30 kg ha^{-1}. On the other hand, the rootstock 'Espina' promoted yield, number of marketable fruits plant^{-1} (mainly at P_{30}) and fruit FW, confirming previous results obtained by grafting eggplant onto *S. torvum* [20,21]. The highest yields of grafted vegetable crops have been often attributed to the enhanced water and nutrient uptakes, flowing in turn from a more expanded root system [22]. Both P supply and grafting are able to modify the nutritional flows between vegetative and reproductive organs [22,23]. In this view, the higher fruit yield recorded in P_{30}/grafted eggplants suggests a synergistic effect among P_{30} and 'Espina' in generating more favorable source:sink relationships, promoting fruit yield.

Fruit DM content and firmness results were clear examples of fruit quality improvement under a more rational P supply, with the former variable being maximized by combining P_{30} and grafting, likely as a consequence of an enhanced carbon flow toward the fruits. Indeed, it has been reported that the optimal tuning among macronutrient supplies through fertilization (notably among N and P) acts to improve the photosynthates translocation toward the reproductive organs [12]. Accordingly, we found that marketable yield was positively correlated to fruit FW (0.543 **) and DM content (0.539 **). This finding suggests implications for postharvest behavior of eggplants, as higher DM content and firmness values are often associated with improved vegetable shelf-life and tolerance to mechanical injuries along the distribution chain [24].

The P_{30} supply also improved the skin chroma (in 'Birgah') and TPC of eggplants, thus highlighting positive effects on both commercial (more vivid external colors) and nutraceutical traits. Indeed, the health-promoting properties of eggplants largely rely on their phenolic acids composition (mostly chlorogenic acid derivatives), whereas the typical hue of eggplant skin is primarily defined by the high accumulation of anthocyanins, such as delphinidin-3-rutinoside, delphinidin-3-rutinoside-5-glucoside or delphinidin-3-coumaroylrutinoside-5-glucoside (nasunin), which show antioxidant and antiagiogenic activities [25,26]. These outcomes could be explained by a more sustained growth rate of the fruits, allowing them an earlier achievement of perceived suitable size for harvest. In this sense, higher anthocyanin and phenol concentrations have been reported in younger eggplants, with the former variable being highly correlated to the external fruit color appearance [27,28]. For TPC, the P_{30} supply apparently acted as an eustressor, promoting their accumulation in eggplants. Accordingly, higher phenolic concentrations have been reported under reduced P supply, e.g., in lavender, strawberry or Chinese kale [29–31].

Potassium was the prevailing macronutrient in eggplants, confirming the important role of this vegetable as a K provider in the human diet. The reduced P supply promoted the fruit accumulation of K only in 'Birgah', probably as a consequence of its different K partitioning inside the plant. Differently, P_{30} boosted N and P concentrations in eggplant fruits, with differences (+17.6 and +11.5%, respectively) largely overcoming that recorded for fruit DM (+7.8%). To this end, an adequate N:P ratio into the soil has been reported to optimize root architecture and the ability of the plant to absorb both macronutrients [32]. This hypothesis seems to be confirmed by the correlation analysis, which revealed a strong relationship between both macronutrient concentrations (0.658 ***). On the other hand, fruit P and K concentrations positively responded to grafting, confirming the ability of this propagation technique to promote the accumulation of these macronutrients in fruit vegetables [10,33]. Despite the positive correlation found among these last macronutrients (0.516 **), the highest increase in K content recorded in 'Dalia' suggests a strong influence of the genotype in remodeling the fruit macromineral profile of grafted eggplants. This finding is consistent with the known importance of the "scion × rootstock" interactions reported in this crop [34]. Taken together, our results show the possibility to enhance the fruit macronutrient profile in 'Birgah' and 'Dalia' by combining a more rational P supply and grafting. Beyond its environmental importance, this outcome is nutritionally relevant mainly for P and K because of their role in the human body, regulating e.g., bone formation and cardiovascular functions (P), or blood pressure and transmission of signals in neuromuscular tissues (K) [35,36].

Where mesonutrients were concerned, S and Na showed similar responses to P fertilization and grafting, as their fruit concentrations were promoted by P_{30} and restricted by the heterograft configuration. However, no differences were found for S by comparing P_{90}/ungrafted vs. P_{30}/grafted treatments of both eggplant cultivars. Differently, the rootstock effect was largely prevalent for Na, so that the P_{30}/grafted eggplants showed the poorest Na concentration, consistent with the reduced Na uptake observed in grafted eggplant plants [16]. On the other hand, a positive response to grafting was noticed for Ca concentration (up to ±62% in 'Dalia'), confirming the positive effects of the rootstock 'Espina' on Ca accumulation in eggplants [21].

Regarding the micromineral composition, we observed strong, positive effects of P_{30} mainly on B and Zn concentrations, consistent with their improved root uptake observed under more proper P soil content [37–40]. On the other hand, beyond the differences among eggplant cultivars, grafting always promoted the accumulation of Mn, B and Zn, probably as a consequence of more expanded roots exploring deeper soil layers, where more convenient oxidative and thermal conditions foster the absorption of these minerals [41–43]. Accordingly, highly significant correlations were found between Mn and B (0.637 ***), Mn and Zn (0.579 **), and between B and Zn (0.901 ***). Consequently, in both cultivars, the P_{30}/grafted combination produced fruits with the best micronutrient compositions in terms of Zn (±193%), B (±152%), Mn (±24%) and Cu (±15%), despite grafting per se acting to reduce the Cu concentration in 'Birgah'. Many rootstock species, including *S. torvum*, are well known to restrict the translocation of some potentially toxic metals toward the shoot, a feature that, together with the peculiar scion traits, would explain the reduced fruit Cu concentration we recorded in 'Birgah' [44,45]. In any case, the overall increase in these four minerals through rational P supply and grafting is relevant from a nutritional viewpoint due to the primary role of these elements in human metabolism, including cardiovascular integrity (Cu), brain (Mn) and bones (B) development, or synthesis of nucleic acids and proteins (Zn) [8,41,42]. Differently, fruit Fe concentration was reduced under P_{30}, mostly in the grafted eggplants, thus showing an opposite trend when compared to Zn and B. The contrasting trends of Fe and Zn ($r = -0.460$ *) is one of the main outcomes we recorded, as these are among the most important trace elements for the human body. Indeed, their dietary deficiencies lead to serious metabolic disorders, including growth retardation, weakened immunological defense and anemia; their suboptimal levels in human diet are perceived as a growing concern both in developed and developing nations [8,46,47]. By comparing the P_{30}/grafted and P_{90}/ungrafted combinations, our results show a dramatic reduction of the fruit Fe:Zn ratio (from 1.26 to 0.13); this suggests an alteration of their homeostasis in fruits, despite no visual symptoms of Fe chlorosis being recorded during the experiment (data not shown). Overall, the correlation analysis supports the hypothesis that the reduced Fe concentration in fruits flows from an enhanced P_{30}-driven absorption of Zn, probably exacerbated by the rootstock, given the high capability of *S. torvum* for the uptake of Zn from the growth substrate [48]. Accordingly, it has been noted that Fe and Zn interact because of the chemical similarity of their divalent cations and basic transporter proteins, leading in many plants to reciprocal interference in their uptake and translocation, as part of the complex tripartite interaction among P, Fe and Zn homeostasis [49,50].

5. Conclusions

The present experiment provides information about the positive effects obtainable by combining a more targeted P supply and grafting in terms of fruit yield and quality in open field eggplants. Such information appears highly instrumental from the perspective of improving the sustainability and nutritional quality of this pivotal Mediterranean vegetable crop. Although the fruit constituents reported in this study (total phenols and inorganic constituents) provide no energy, their sufficient intake represents a necessary component of adequate human growth and metabolism; their suboptimal dietary concentrations can lead to metabolic disorders and increased incidence in chronic diseases. Beyond the differences among eggplant cultivars, our results show that the reduction in P fertilization did not affect fruit yield, but instead was promoted by grafting. The P_{30}/grafted eggplants showed enhanced marketable yields, along with increased concentrations in total phenols, macronutrients (mostly P and K), Ca and micronutrients (Zn, B, Mn and Cu), hence revealing an overall better nutraceutical composition. The depressed Fe concentration represented an exception having nutritional relevance, probably consequent to both complex P–Fe–Zn interactions in the soil and intrinsic rootstock characteristics. This outcome would suggest that under similar conditions, the affinity of the rootstock with specific micro minerals should be taken in due account, along with the option to adopt complementary practices (e.g., more targeted micro mineral fertilizations) to optimize the overall accumulation of these constituents in eggplant fruits.

Author Contributions: Conceptualization, R.P.M. and C.L.; methodology, R.P.M., M.D., S.R.S. and L.S.; software, F.G., M.D., C.C. and F.B.; validation, C.C., F.B., E.A. and R.M.; formal analysis, R.P.M., L.S. and M.D.; investigation, M.D., C.C., F.B., E.A. and R.M.; data curation, S.R.S., R.P.M., F.G. and E.A.; writing—original draft preparation, R.P.M., S.R.S. and L.S.; writing—review and editing, F.G., C.C., F.B. and E.A.; supervision, C.L. and F.G. All authors have read and agreed to the published version of the manuscript.

Funding: This research received no external funding.

Institutional Review Board Statement: Not applicable.

Conflicts of Interest: The authors declare no conflict of interest.

References

1. Obersteiner, M.; Peñuelas, J.; Ciais, P.; Van Der Velde, M.; Janssens, I.A. The phosphorus trilemma. *Nat. Geosci.* **2013**, *6*, 897–898. [CrossRef]
2. El Bilali, H.; Callenius, C.; Strassner, C.; Probst, L. Food and nutrition security and sustainability transitions in food systems. *Food Energy Secur.* **2019**, *8*, e00154. [CrossRef]
3. Cordell, D.; Drangert, J.O.; White, S. The story of phosphorus: Global food security and food for thought. *Glob. Environ. Chang.* **2009**, *19*, 292–305. [CrossRef]
4. Bindraban, P.S.; Dimkpa, C.O.; Pandey, R. Exploring phosphorus fertilizers and fertilization strategies for improved human and environmental health. *Biol. Fertil. Soils* **2020**, *56*, 299–317. [CrossRef]
5. Wang, R.; Min, J.; Kronzucker, H.J.; Li, Y.; Shi, W. N and P runoff losses in China's vegetable production systems: Loss characteristics, impact, and management practices. *Sci. Total Environ.* **2019**, *663*, 971–979. [CrossRef]
6. Wang, R.; Shi, W.; Li, Y. Phosphorus supply and management in vegetable production systems in China. *Front. Agric. Sci. Eng.* **2019**, *6*, 348–356. [CrossRef]
7. Kalkhajeh, Y.K.; Huang, B.; Sørensen, H.; Holm, P.E.; Hansen, H.C.B. Phosphorus accumulation and leaching risk of greenhouse vegetable soils in Southeast China. *Pedosphere* **2021**, *31*, 683–693. [CrossRef]
8. Buturi, C.V.; Mauro, R.P.; Fogliano, V.; Leonardi, C.; Giuffrida, F. Mineral biofortification of vegetables as a tool to improve human diet. *Foods* **2021**, *10*, 223. [CrossRef]
9. Ierna, A.; Mauro, R.P.; Leonardi, C.; Giuffrida, F. Shelf-life of bunched carrots as affected by nitrogen fertilization and leaf presence. *Agronomy* **2020**, *10*, 1982. [CrossRef]
10. Sabatino, L.; Iapichino, G.; Consentino, B.B.; D'Anna, F.; Rouphael, Y. Rootstock and arbuscular mycorrhiza combinatorial effects on eggplant crop performance and fruit quality under greenhouse conditions. *Agronomy* **2020**, *10*, 693. [CrossRef]
11. Food and Agriculture Organization of the United Nations. Available online: http://www.fao.org/faostat/en/#home (accessed on 26 January 2022).

12. Ierna, A.; Mauro, R.P.; Mauromicale, G. Improved yield and nutrient efficiency in two globe artichoke genotypes by balancing nitrogen and phosphorus supply. *Agron. Sustain. Dev.* **2012**, *32*, 773–780. [CrossRef]
13. Zhang, B.; Fu, Z.; Huang, J.; Wang, J.; Xu, S.; Zhang, L. Consumers' perceptions, purchase intention, and willingness to pay a premium price for safe vegetables: A case study of Beijing, China. *J. Clean. Prod.* **2018**, *197*, 1498–1507. [CrossRef]
14. Colla, G.; Rouphael, Y.; Leonardi, C.; Bie, Z. Role of grafting in vegetable crops grown under saline conditions. *Sci. Hortic.* **2010**, *127*, 147–155. [CrossRef]
15. Mauro, R.P.; Agnello, M.; Distefano, M.; Sabatino, L.; San Bautista Primo, A.; Leonardi, C.; Giuffrida, F. Chlorophyll fluorescence, photosynthesis and growth of tomato plants as affected by long-term oxygen root zone deprivation and grafting. *Agronomy* **2020**, *10*, 137. [CrossRef]
16. Semiz, G.D.; Suarez, D.L. Impact of grafting, salinity and irrigation water composition on eggplant fruit yield and ion relations. *Sci. Rep.* **2019**, *9*, 19373. [CrossRef]
17. Maršić, N.K.; Mikulić-Petkovšek, M.; Štampar, F. Grafting influences phenolic profile and carpometric traits of fruits of greenhouse-grown eggplant (*Solanum melongena* L.). *J. Agric. Food Chem.* **2014**, *62*, 10504–10514. [CrossRef]
18. Hopkins, B.G. Phosphorus. In *Handbook of Plant Nutrition*; Barker, A.V., Pilbeam, D.J., Eds.; CRC Press: Boca Raton, FL, USA, 2015; pp. 65–126.
19. Leonardi, C.; Giuffrida, F. Growth rate and carpometric characteristics during eggplant fruit growth. *Acta Hortic.* **2009**, *807*, 175–180. [CrossRef]
20. Sabatino, L.; Iapichino, G.; Maggio, A.; D'Anna, E.; Bruno, M.; D'Anna, F. Grafting affects yield and phenolic profile of *Solanum melongena* L. landraces. *J. Integr. Agric.* **2016**, *15*, 1017–1024. [CrossRef]
21. Cassaniti, C.; Giuffrida, F.; Scuderi, D.; Leonardi, C. The effect of rootstock and nutrient solution concentration on eggplant grown in a soilless system. *J. Food Agric. Environ.* **2011**, *9*, 252–256.
22. Mauro, R.P.; Agnello, M.; Onofri, A.; Leonardi, C.; Giuffrida, F. Scion and rootstock differently influence growth, yield and quality characteristics of cherry tomato. *Plants* **2020**, *9*, 1725. [CrossRef]
23. Zubillaga, M.M.; Aristi, J.P.; Lavado, R.S. Effect of phosphorus and nitrogen fertilization on sunflower (*Helianthus annus* L.) nitrogen uptake and yield. *J. Agron. Crop Sci.* **2002**, *188*, 267–274. [CrossRef]
24. Giuffrida, F.; Agnello, M.; Mauro, R.P.; Ferrante, A.; Leonardi, C. Cultivation under salt stress conditions influences postharvest quality and glucosinolates content of fresh-cut cauliflower. *Sci. Hortic.* **2018**, *236*, 166–174. [CrossRef]
25. Niño-Medina, G.; Urías-Orona, V.; Muy-Rangel, M.D.; Heredia, J.B. Structure and content of phenolics in eggplant (*Solanum melongena*)—A review. *S. Afr. J. Bot.* **2017**, *111*, 161–169. [CrossRef]
26. Komatsu, W.; Itoh, K.; Akutsu, S.; Kishi, H.; Ohhira, S. Nasunin inhibits the lipopolysaccharide-induced pro-inflammatory mediator production in RAW264 mouse macrophages by suppressing ROS-mediated activation of PI3 K/Akt/NF-κB and p38 signaling pathways. *Biosci. Biotechnol. Biochem.* **2017**, *81*, 1956–1966. [CrossRef]
27. Mennella, G.; Lo Scalzo, R.; Fibiani, M.; DAlessandro, A.; Francese, G.; Toppino, L.; Acciarri, N.; De Almeida, A.E.; Rotino, G.L. Chemical and bioactive quality traits during fruit ripening in eggplant (*S. melongena* L.) and allied species. *J. Agric. Food Chem.* **2012**, *60*, 11821–11831. [CrossRef]
28. Mauro, R.P.; Agnello, M.; Rizzo, V.; Graziani, G.; Fogliano, V.; Leonardi, C.; Giuffrida, F. Recovery of eggplant field waste as a source of phytochemicals. *Sci. Hortic.* **2020**, *261*, 109023. [CrossRef]
29. Chrysargyris, A.; Panayiotou, C.; Tzortzakis, N. Nitrogen and phosphorus levels affected plant growth, essential oil composition and antioxidant status of lavender plant (*Lavandula angustifolia* Mill.). *Ind. Crops Prod.* **2016**, *83*, 577–586. [CrossRef]
30. Valentinuzzi, F.; Mason, M.; Scampicchio, M.; Andreotti, C.; Cesco, S.; Mimmo, T. Enhancement of the bioactive compound content in strawberry fruits grown under iron and phosphorus deficiency. *J. Sci. Food Agric.* **2015**, *95*, 2088–2094. [CrossRef]
31. Chen, R.; Song, S.; Li, X.; Liu, H.; Huang, D. Phosphorus deficiency restricts plant growth but induces pigment formation in the flower stalk of Chinese kale. *Hortic. Environ. Biotechnol.* **2013**, *54*, 243–248. [CrossRef]
32. Postma, J.A.; Dathe, A.; Lynch, J.P. The optimal lateral root branching density for maize depends on nitrogen and phosphorus availability. *Plant Physiol.* **2014**, *166*, 590–602. [CrossRef]
33. Nawaz, M.A.; Imtiaz, M.; Kong, Q.; Cheng, F.; Ahmed, W.; Huang, Y.; Bie, Z. Grafting: A technique to modify ion accumulation in horticultural crops. *Front. Plant Sci.* **2016**, *7*, 1–15. [CrossRef] [PubMed]
34. Kyriacou, M.C.; Rouphael, Y.; Colla, G.; Zrenner, R.; Schwarz, D. Vegetable grafting: The implications of a growing agronomic imperative for vegetable fruit quality and nutritive value. *Front. Plant Sci.* **2017**, *8*, 1–23. [CrossRef] [PubMed]
35. Gutiérrez, O.M.; Luzuriaga-McPherson, A.; Lin, Y.; Gilbert, L.C.; Ha, S.W.; Beck, G.R. Impact of phosphorus-based food additives on bone and mineral metabolism. *J. Clin. Endocrinol. Metab.* **2015**, *100*, 4264–4271. [CrossRef] [PubMed]
36. Gumz, M.L.; Rabinowitz, L.; Wingo, C.S. Disorders of fluids and electrolytes: An integrated view of potassium homeostasis. *N. Engl. J. Med.* **2015**, *373*, 60–72. [CrossRef]
37. De Iorio, A.F.; Gorgoschide, L.; Rendina, A.; Barros, M.J. Effect of phosphorus, copper, and zinc addition on the phosphorus/copper and phosphorus/zinc interaction in lettuce. *J. Plant Nutr.* **1996**, *19*, 481–491. [CrossRef]
38. Irfan, M.; Abbas, M.; Shah, J.A.; Depar, N.; Memon, M.Y.; Sial, N.A. Interactive effect of phosphorus and boron on plant growth, nutrient accumulation and grain yield of wheat grown on calcareous soil. *Eurasian J. Soil Sci.* **2019**, *8*, 17–26. [CrossRef]
39. Awan, Z.I.; Abbasi, M.K. Interactive effect of phosphorus and copper on maize growth. *Pak. J. Agric. Res.* **2000**, *16*, 105–108.

40. Barker, A.V.; Eaton, T.E. Zinc. In *Handbook of Plant Nutrition*; Barker, A.V., Pilbeam, D.J., Eds.; CRC Press: Boca Raton, FL, USA, 2015; pp. 537–566.
41. Shaaban, M.M. Role of boron in plant nutrition and human health. *Am. J. Plant Physiol.* **2010**, *5*, 224–240. [CrossRef]
42. Alejandro, S.; Höller, S.; Meier, B.; Peiter, E. Manganese in plants: From acquisition to subcellular allocation. *Front. Plant Sci.* **2020**, *11*, 1–23. [CrossRef]
43. Natasha, N.; Shahid, M.; Bibi, I.; Iqbal, J.; Khalid, S.; Murtaza, B.; Bakhat, H.F.; Farooq, A.B.U.; Amjad, M.; Hammad, H.M.; et al. Zinc in soil-plant-human system: A data-analysis review. *Sci. Total Environ.* **2022**, *808*, 152024. [CrossRef]
44. Savvas, D.; Colla, G.; Rouphael, Y.; Schwarz, D. Amelioration of heavy metal and nutrient stress in fruit vegetables by grafting. *Sci. Hortic.* **2010**, *127*, 156–161. [CrossRef]
45. Rouphael, Y.; Cardarelli, M.; Rea, E.; Colla, G. Grafting of cucumber as a means to minimize copper toxicity. *Environ. Exp. Bot.* **2008**, *63*, 49–58. [CrossRef]
46. Olivares, M.; Walter, T.; Hertrampf, E.; Pizarro, F. Anaemia and iron deficiency disease in children. *Br. Med. Bull.* **1999**, *55*, 534–543. [CrossRef] [PubMed]
47. Beleggia, R.; Fragasso, M.; Miglietta, F.; Cattivelli, L.; Menga, V.; Nigro, F.; Pecchioni, N.; Fares, C. Mineral composition of durum wheat grain and pasta under increasing atmospheric CO_2 concentrations. *Food Chem.* **2018**, *242*, 53–61. [CrossRef]
48. Arao, T.; Takeda, H.; Nishihara, E. Reduction of cadmium translocation from roots to shoots in eggplant (*Solanum melongena*) by grafting onto *Solanum torvum* rootstock. *Soil Sci. Plant Nutr.* **2008**, *54*, 555–559. [CrossRef]
49. Saenchai, C.; Bouain, N.; Kisko, M.; Prom-u-thai, C.; Doumas, P.; Rouached, H. The involvement of OsPHO1;1 in the regulation of iron transport through integration of phosphate and Zinc deficiency signaling. *Front. Plant Sci.* **2016**, *7*, 1–9. [CrossRef]
50. Rai, S.; Singh, P.K.; Mankotia, S.; Swain, J.; Satbhai, S.B. Iron homeostasis in plants and its crosstalk with copper, zinc, and manganese. *Plant Stress* **2021**, *1*, 100008. [CrossRef]

horticulturae

MDPI

Article

Quality Characteristics of Oolong Tea Products in Different Regions and the Contribution of Thirteen Phytochemical Components to Its Taste

Zhihui Wang [1], Shuang Gan [1], Weijiang Sun [1,*] and Zhidan Chen [1,2,*]

[1] College of Horticulture, Fujian Agriculture and Forestry University, Fuzhou 350002, China; wzh1246900265@126.com (Z.W.); ganshuang1995@126.com (S.G.)
[2] Anxi College of Tea Science, Fujian Agriculture and Forestry University, Fuzhou 350002, China
* Correspondence: 000q020007@fafu.edu.cn (W.S.); xzh@fafu.edu.cn (Z.C.)

Abstract: Regionality is a term used in the tea industry to describe the particular style of tea produced by a growing region. Determining the characteristics of the tea of specific regions can help growers predict tea plant quality before harvesting and eventually production. As such, in this study, we collected representative Oolong tea samples from 15 regions in 8 countries. Quantitative description analysis (QDA) and a flavor wheel were used to analyze their sensory characteristics. Chemometrics was used to screen the phytochemical components that significantly contribute to the taste of Oolong tea. We preliminarily obtained 35 sensory characteristic descriptors and constructed a flavor wheel for Oolong tea. We found that Oolong tea in each region has unique sensory quality characteristics. The content of thirteen phytochemical components of Oolong tea in different regions widely varied, and the average coefficient of variation was 45.56%. Among of them, we found the largest difference in free amino acids. We identified the relationship between taste sub-attributes, and the thirteen phytochemical components was found through correlation analysis. Finally, we selected phytochemical components with significant effects on five taste sub-attributes that were selected from the thirteen detected phytochemical components. The construction of the Oolong tea flavor wheel can help realize the qualitative and quantitative sensory evaluation of Oolong tea from different origins and contribute to the quality identification and directional improvement of Oolong tea products.

Keywords: Oolong tea; sensory characterization; flavor wheel; phytochemical components; taste

Citation: Wang, Z.; Gan, S.; Sun, W.; Chen, Z. Quality Characteristics of Oolong Tea Products in Different Regions and the Contribution of Thirteen Phytochemical Components to Its Taste. *Horticulturae* **2022**, *8*, 278. https://doi.org/10.3390/horticulturae8040278

Academic Editors: Alessandra Durazzo and Massimo Lucarini

Received: 10 February 2022
Accepted: 21 March 2022
Published: 24 March 2022

1. Introduction

Tea (*Camellia sinensis*) plants are important horticultural plants that are widely grown in China, Japan, India, Sri Lanka, Kenya, Tanzania, and Argentina [1]. Tea is made from the leaves of tea plants and is the second most popular beverage in the world after water [2]. Oolong tea, a semi-fermented tea, was first produced in the early Song Dynasty (960–1279) and originated from the Fujian, Guangdong, and Taiwan provinces of China [2,3]. Different from green and black teas, traditional Oolong tea tastes mellow and refreshing, with an elegant fruity and floral fragrance, and enjoys wide popularity all over the world [4,5]. In the past few decades, the output and consumption of Oolong tea in the world has been increasing, and it is being increasingly produced in various countries, such as Japan, Sri Lanka, Myanmar, Thailand, and South Korea [6,7]. The special aroma and taste of tea are largely determined by its geographical and natural conditions, including growing regions, tea varieties, cultivation methods, and processing techniques [8]. Therefore, most famous teas in the world are named after regions, such as Anxi Tieguanyin, Wuyi rock tea, Xihu Longjing, Dongting Biluochun, Assam black tea, and Dajiling black tea [9]. There are great differences in the quality characteristics of Oolong tea in different regions [10]. Tea flavor experts are often asked some questions about the quality characteristics and flavor formation mechanism of Oolong tea from different regions: What are the quality differences of Oolong tea between Fujian, Guangdong, and Taiwan? What is the difference between China, Japan, and Sri Lankan Oolong tea? What words do we use to describe the taste of Oolong tea? What chemical substance is responsible for the heavy and thick or mellow flavor of Oolong tea? These answers to these questions are still unclear. As such, we must conduct an in-depth analysis of the sensory and phytochemical characteristics of Oolong tea from different regions.

Quantitative description analysis (QDA) is an effective descriptive sensory analysis method [11]. It can be used to effectively mine the flavor vocabulary of food, determine its sensory flavor characteristics, and compare the differences in sensory characteristics between foods [11]. A flavor wheel is a visual and practical tool that can be used to describe different flavor characteristics of food. Through the collection, classification, induction, and sorting of specific sensory attribute descriptors, the flavor characteristics of food are formed [12,13]. Combining QDA and the flavor wheel can not only be used to identify the flavor characteristics of food, but also to visualize the flavor. These technologies are widely used in food and tea. The Specialty Coffee Association of America (SCAA) developed the first coffee flavor wheel in 1995. Recently, the SCAA and World Coffee Research (WCR) jointly developed attribute words and mapped a new flavor wheel to further guide the production and improve the flavor of coffee [12]. The Tea Research Institute of Chinese Academy of Agricultural Sciences also constructed the primitive morpheme and flavor wheel of terms of sensory evaluation of Chinese tea [13]. The sensory descriptors and flavor wheel of alcohol [14], vinegar [15], Pu'er tea [16], Fuzhuan brick Tea [17], Huangdacha [18], and green tea [19] have also been successfully established. However, research on the flavor wheel of Oolong tea in different regions is still lacking. This is not conducive to us to distinguish and understand the quality characteristics of Oolong tea from different regions, which limits us to further explore the flavor formation mechanism of Oolong tea.

The taste of tea is the comprehensive effect of human sensory gustation from water-soluble substances in a tea infusion [20]. The main flavor components are divided into polyphenols, alkaloids, amino acids, carbohydrates, etc., which have different flavor characteristics due to their different compositions, contents and threshold values [20]. Tea polyphenols, catechins, caffeine, and free amino acids are considered to be the important chemical components affecting the taste intensity of Oolong tea [3]. Free amino acids, theanine, and total catechins are significantly correlated with the taste score of Oolong tea [21,22]. Five catechins and catechin gallates, i.e., catechin (C), epicatechin (EC), epigallocatechin (EGC), epigallocatechin gallate (ECG), and epigallocatechin gallate (EGCG), are the major phenolic compounds in Oolong tea [21]. Theanine is the main component of free amino acids [11]. These chemical components affect the taste score in the sensory evaluation of Oolong tea [21,22]. However, whether differences exist in these important chemical components in Oolong tea from different regions around the world and the contribution of these components to the taste sub-attributes of Oolong tea have not been reported. Understanding the contribution of chemical components to taste sub-attributes is very important for the quality control and improvement of Oolong tea products.

Based on this, in this study we used a combination of QDA and the flavor wheel to analyze the quality characteristics of Oolong tea from 15 regions of 8 countries for the first time. Correlation analysis, Dose-over-Threshold (DoT), and partial least square-variable import project (PLS-VIP) were used to screen compounds with important contributions to flavor properties. Our two purposes in this study were to (1) construct the flavor wheel of Oolong tea and clarify the sensory quality characteristics of Oolong tea from different regions and (2) clarify the contribution of thirteen phytochemical components, such as tea polyphenols, catechins, caffeine, and free amino acids, to the taste sub-attributes of Oolong tea. This work will help us to accurately describe and distinguish the quality characteristics of Oolong tea from different origins, and to understand the phytochemical components that have an important contribution to the taste sub-attributes of Oolong tea, so as to provide a reference for the quality improvement and directional regulation of Oolong tea products.

2. Materials and Methods

2.1. Experimental Materials

We collected 171 samples of Oolong tea from 17 regions in 8 countries. Samples from 8 regions in China including different varieties and grades were produced by different manufacturers. We purchased samples from Japan, Sri Lanka, Myanmar, Australia, Germany, Korea, and Thailand from different brands and different types sold in local markets. The sample details are shown in Table S1 in the Supplementary Materials. To prevent changes in the quality and chemical composition of the samples, they were stored in a refrigerator at −80 °C until use. Because the quality characteristics of some Oolong tea samples collected in the same region were relatively similar, we initially evaluated the appearance, infusion color, taste, and aroma of the samples by referring to GB/T 23776–2018 [23]. We excluded samples with similar quality characteristics in the same region. Representative Oolong tea samples from various regions were screened out for sensory quantitative description analysis and the detection of the thirteen phytochemical components. A total of 66 regional representative samples were finally obtained, including 7 from southern Fujian, 8 from northern Fujian, 5 from western Fujian, 8 from Guangdong, 6 from Taiwan, 1 from Hubei, 5 from Zhejiang, 3 from Sichuan, 5 from Japan, 2 from Sri Lanka, 5 from Myanmar, 2 from Australia, 4 from Germany, 2 from Korea, and 3 from Thailand (Table 1, Table S1).

Table 1. Origin and quantity of Oolong tea samples.

Region (Abbreviation)	Sample Quantity/Number	Region (Abbreviation)	Sample Quantity/Number
Southern Fujian, China (MN)	7	Japan (JP)	5
Northern Fujian, China (MB)	8	Sri Lanka (LK)	2
Western Fujian, China (MX)	5	Myanmar (MM)	5
Guangdong, China (GD)	8	Australia (AUS)	2
Taiwan, China (TW)	6	Germany (GER)	4
Hubei, China (HB)	1	Korea (KR)	2
Zhejiang, China (ZJ)	5	Thailand (THA)	3
Sichuan, China (SC)	3	Total	66

Standards of caffeine (CAF > 99%), catechin (C > 99%), epicatechin (EC > 99%), epigallocatechin (EGC > 99%), epigallocatechin gallate (ECG > 99%), epigallocatechin gallate (EGCG > 99%), theanine (> 99%), glutamic acid (> 99%), and gallic acid (GA > 99%) were purchased from Sigma-Aldrich (Saint Louis, MO, USA).

2.2. Experimental Method

2.2.1. Quantitative Description Analysis (QDA)

We first established a sensory evaluation team: according to the requirements of GB/T 16291.1–2012 (GB/T: national standards of people's republic of China) [24], five evaluators were selected from the tea science teacher team of Fujian Agriculture and Forestry University based on time, interest, and description ability. The members of the sensory evaluation team had participated in many sensory evaluation experiments for teas such as black, white, and Oolong teas, and all had more than 10 years of experience in tea sensory evaluation.

For descriptor generation and sensory evaluation, the sensory characteristics of Oolong tea were evaluated by referring to the evaluation procedures and requirements of GB/T 23776–2018 (covered-bowl method) [23]. Sensory analysis was performed in the Tea Sensory Evaluation Laboratory of the Horticulture Science and Technology Building, Fujian Agriculture and Forestry University. Samples of 100–200 g were placed in the tea evaluation tray, and the evaluators evaluated the appearance by turning the tea and changing its position. Samples of 5.0 g were weighed and placed in a covered 110 mL bowl; then we filled the bowl with boiling water and capped the bowl. After soaking the samples for 1 min, we opened the lid, and the aroma of tea was evaluated. After soaking for 2 min, we drained the tea infusion into the evaluation bowl for the color to be appraised and the infusion to be tasted. Then, we made the tea for the second brewing, covered it, soaked it for 1–2 min, opened the lid, and the aroma of tea was evaluated. After soaking for 3 min, we drained the tea soup infusion into the evaluation bowl, and the color and taste of the infusion were evaluated. We then made the tea for the third brewing, covered it, and the aroma was evaluated after 2–3 min. Then we drained the tea infusion into the evaluation bowl after 5 min for the evaluation of the color and taste of the infusion. Finally, the evaluators smelled the aroma of the leaves. The evaluators independently described the appearance of dry tea; the color, aroma, and taste characteristics of tea infusion; and we recorded all evaluator's descriptors. After that, we preliminarily screened the descriptive words and retained the descriptive words used more than 80% of the time. All evaluators discussed and screened descriptors in combination with GB/T 14487–2017 [25] and references [9,10,13,21]. Finally, the descriptors and sensory quality characteristics of Oolong tea samples from different regions were obtained. Every sample was randomly evaluated three times.

For taste sub-attributes evaluation, the method of preparing the method of tea infusion was the same as that mentioned above. The evaluators evaluated the bitterness, astringency, umami, sweet aftertaste, and "heavy and thick" of the tea infusion of all samples. A 0–5 strength scale standard was adopted, where 0 = none and 5 = very. Every sample was randomly evaluated three times.

2.2.2. Quantification of Phytochemical Components

The water content of dry tea was determined by the weight loss method according to ISO 1573: 1980 [26]. Briefly, 5.0 ± 0.001 g of the ground tea sample was put in a prepared weighing bottle and then dried at the bottle at 103 °C for 6 h, which was then cooled in a desiccator and weighed to the nearest 0.005 g. The water extract content was determined by the differential method according to ISO 9768: 1998 [27] with modifications. Briefly, 2.0 ± 0.001 g of the ground tea sample was put in a conical flask, and 300 mL of boiling water was added. After 45 min of brewing, the tea infusion was filtered under a vacuum, and tea residue was dried at 103°C for 16 h. Afterward, the tea residue was cooled in a desiccator and weighed to the nearest 0.001 g. Three biological replicates were performed of the water content and water extract content of every sample.

The tea polyphenols content was determined by Folin-Ciocalteu phenol colorimetry according to ISO 14502.1: 2005 [28]. Briefly, gallic acid was used as a standard. 0.2 ± 0.001 g of the ground tea sample was put into 10 mL centrifuge tube, and then 5 mL of 70% methanol preheated at 70 °C was added. After extraction for 30 min in a 70 °C water bath, the extracting solution was put in a centrifuge. We repeated this procedure once. We combined and fixed the extract to 10 mL. Then, 1 mL of the extract was transferred and fixed to 100 mL. Subsequently, 1 mL of the extract was drawn into a 10 mL calibration tube, 5 mL of 10% Folin-Ciocalteu was added, and after reacting for 3–8 min, 4 mL of 7.5% Na_2CO_3 was added. After waiting 60 min, the absorbance was determined at 765 nm using a TU-1810PC UV-Vis spectrophotometer (Puxi General Instrument Co., Ltd., Beijing, BJ, China) Three biological replicates were performed for every sample.

The catechins (ECG, C, EGC, EGCG, and EC) and caffeine contents were determined using a Waters 2695 high performance liquid chromatograph (HPLC) (Waters, Milford, MA, USA) according to ISO 14502.1: 2005 [29] with modifications. The method we used for extraction of the tea samples was the same as that used for tea polyphenols. The chromatographic conditions were as follows: 2998 PDA detector (Waters, Milford, MA, USA); X-select-T3 chromatographic column 4.6 mm × 250 mm × 5 μm (Waters, Milford, MA, USA); mobile phase A: 2% formic acid solution; mobile phase B: pure methanol. The gradient change of the mobile phase A was as follows: 0–9 min, 83–73%; 9–15 min, 73–58%; 15–18 min, 58–83%. The flow rate was 1 mL/min; column temperature was 35 °C; injection volume was 10 μL; and UV detection was performed at 278 nm. The ester catechin content was calculated as ECG content + EGCG content, the simple catechin content was calculated as EGC content + C content + EC content, and the total catechin content was calculated as ester catechin content + simple catechin content. Three biological replicates were performed for each sample.

The theanine content was determined by Waters 2695 HPLC (Waters, Milford, MA, USA) according to GB/T 23193-2017 [30]. Briefly, 1.0 ± 0.001 g of the ground tea sample was put in a 200 mL beaker, and then 100 mL of boiling water was added. After extracting for 30 min in 100 °C water bath, we filtered and fixed the sample to a volume in a 100 mL volumetric flask and filtered the sample through a 0.45 μm aperture polyethersulfone filter membrane (Jinteng Co., Ltd., Tianjin, TJ, China) and removed 1 mL for HPLC analysis. The chromatographic conditions were as follows: the chromatographic column was an AccQ·Tag amino acid special analytical chromatographic column 3.9 mm × 150 mm (Waters, Milford, MA, USA); mobile phase A: 100% pure water; mobile phase B: 100% acetonitrile. The gradient change of mobile phase A was 0–10 min, 100%; 10–12 min, 100–20%; 12–20 min, 20%; 20–22 min, 20–100%; 22–40 min, 100%. The injection volume was 10 μL; the column temperature was 35 °C, and the flow rate was 1 mL/min; and the detection wavelength was 210 nm. Three biological replicates were performed for every sample.

The free amino acids content was determined by the ninhydrin colorimetric method according to GB/T 8314-2013 [31]. Glutamic acid was used as a standard. Then, 1.0 ± 0.001 g of the ground tea sample was put into a 500 mL beaker, and then 300 mL of boiling water was added. After extracting for 20 min in a 100 °C water bath, the sample was filtered with 15 cm diameter quantitative filter paper (Tezhong Co., Ltd., Hangzhou, HZ, China) and fixed to volume in a 500 mL volumetric flask. After, 1.0 mL of the extracting solution was placed in a 25 mL colorimetric tube, then 0.5 mL pH 8.0 phosphate buffer and 0.5 mL 2% ninhydrin solution were added. After 15 min of being placed in a water bath in boiling water, we fixed the volume to 25 mL after cooling, and the absorbance was detected at 570 nm by a TU-1810pc UV-Vis spectrophotometer (Puxi General Instrument Co., Ltd., Beijing, BJ, China). Three biological replicates were performed for every sample.

2.2.3. Data Statistics and Analysis

The mean, standard deviation, and coefficient of variation of sensory evaluation data and phytochemical components data were calculated using Microsoft Excel 2019. Correlation analysis and heat map analysis were performed by the hiplot tool (https://hiplot.com.cn, accessed 19 December 2021). PLS-VIP analysis was performed using SIMCA 14.0. The DoT value was calculated by the phytochemical component content/phytochemical component threshold.

3. Results and Discussion

3.1. Descriptors and Flavor Wheel of Sensory Quality of Oolong Tea

We used a total of 35 descriptors to identify the sensory quality characteristics of Oolong tea from different regions, and the constructed flavor wheel was shown in Figure 1: 9 for appearance, 7 for infusion color, 7 for aroma, and 12 for taste. The frequency of descriptors of Oolong tea from different regions is shown in Table S2. The most frequently used appearance descriptors were "tight and heavy" (39.4%) and "round and tight" (28.8%). It indicated that the appearance of Oolong tea samples from different regions were mainly "tight and heavy" and "round and tight". "Orange red" (53%) and "honey yellow" (34.8%) were frequently used as infusion color descriptors, and so were the main color types of Oolong tea infusions from different regions. The most frequently used aroma descriptors were "flowery" (68.2%), "fruity" (39.4%), "clean and refreshing" (39.4%), and "roasted" (31.8%). The aromas of Oolong tea from different regions were relatively diversified, but with nearly 70% of the samples being described as flowery, this was found to be the main aroma type of Oolong tea. The proportions of "fruity", "clean and refreshing", and "roasted" were similar, being evenly distributed among the Oolong tea samples. The most frequently used taste descriptors were "astringent" (92.4%), "bitterness" (80.3%), "umami" (51.5%), "mellow and thick" (43.9%), "sweet aftertaste" (42.4%), "heavy and thick" (28.8%), "mellow" (25.8%), and "brisk and smooth" (21.1%). The descriptors obtained in this study include those reported in previous studies [9,21,32]. This finding showed that the descriptors we obtained were representative and extensive and could be used to realize the qualitative and quantitative sensory evaluation of Oolong tea from different regions.

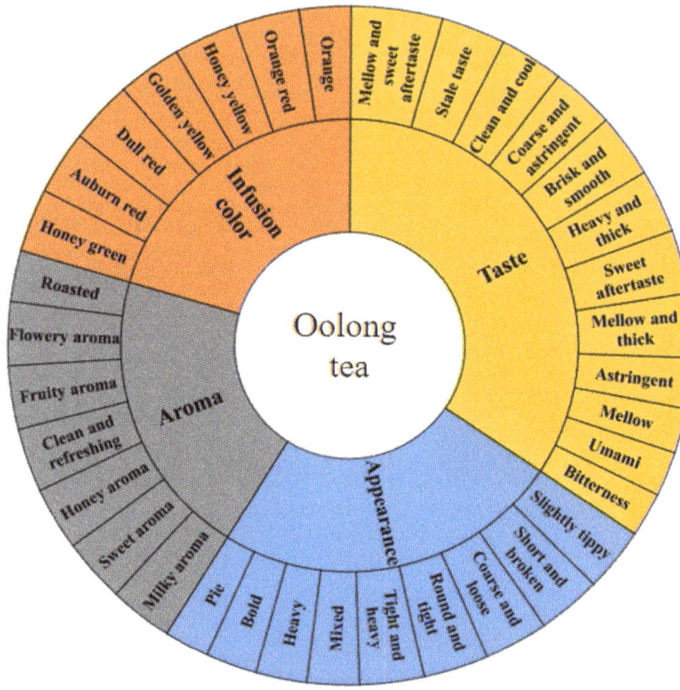

Figure 1. Flavor wheel of Oolong tea.

3.2. Analysis of Quality Characteristics of Oolong Tea from Different Regions

The quality characteristics of Oolong tea from 15 different regions, as shown in Figure 2 and Table S3, were relatively diversified, and we found that no region had completely consistent quality characteristics was discovered. Southern Fujian, northern Fujian, Guangdong, and Taiwan are the regions with the longest history of Oolong tea production and are the main Oolong tea producers in the world [33]. The Oolong tea is predominantly Tieguanyin tea, Wuyi Rock tea, Fenghuang Dancong tea, and Baozhong tea in the southern Fujian, northern Fujian, Guangdong, and Taiwan, respectively [33]. The quality characteristics of Oolong tea in these four regions were obviously different. Northern Fujian Oolong tea was "tight and heavy", and its aroma and taste were "flowery-fruity" and "roasted", and "heavy and hick" with a superior "sweet aftertaste", respectively. The appearance of Taiwan Oolong tea was mainly "round and tight", and the taste was "mellow" or "mellow and sweet aftertaste". Guangdong Oolong tea was mainly strong with a lasting flowery aroma, a taste described as "heavy and thick", superior "sweet aftertaste", and "bitter-astringency". The appearance of Oolong tea in southern Fujian was mainly "round and tight", the aroma was mainly "clean and refreshing" and "flowery", and the taste was "mellow" or "mellow and thick" and "umami and brisk". Cai et al. [9] evaluated the taste and aroma characteristics of Oolong tea from southern Fujian, northern Fujian, Guangdong, and Taiwan and found that the Oolong teas from the four regions have their own taste and aroma characteristics. These conclusions are the same as what we drew. One difference was that we further defined the appearance and infusion color characteristics of Oolong tea from these four regions (Figure 2).

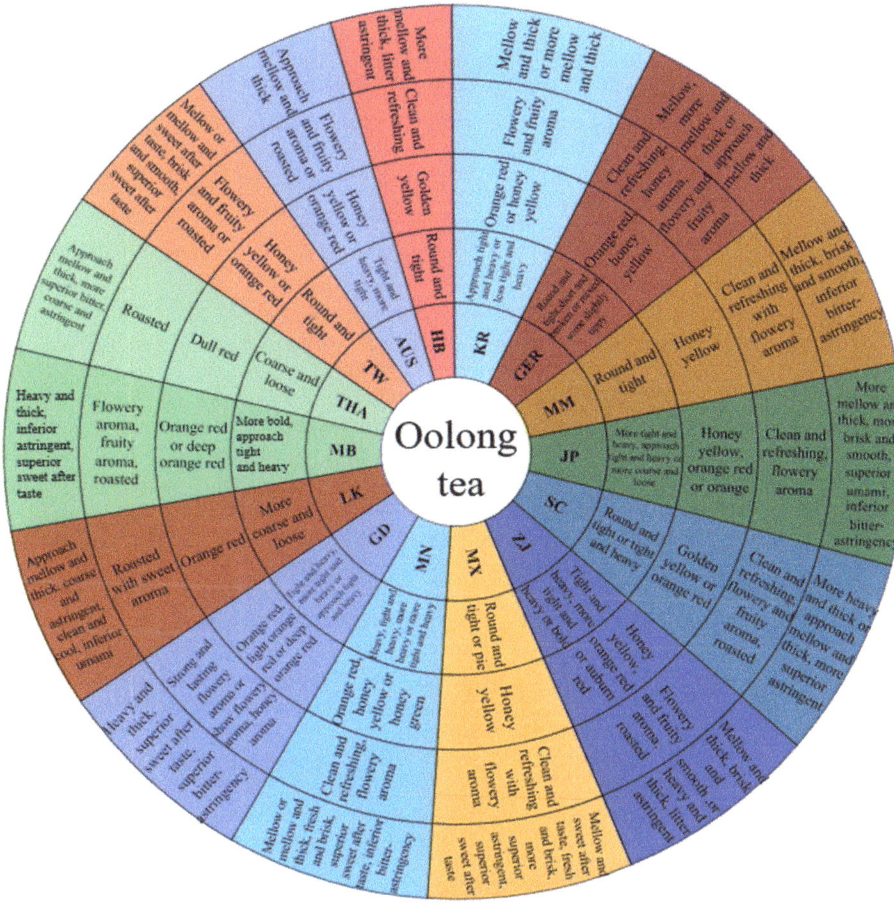

Figure 2. The quality characteristics of Oolong tea from different regions. From the inside to the outside, the flavor wheel describes the regions, appearance, infusion color, aroma, and taste of Oolong tea. MN, southern Fujian, China; MB, northern Fujian, China; MX, western Fujian, China; GD, Guangdong, China; TW, Taiwan, China; HB, Hubei, China; ZJ, Zhejiang, China; SC, Sichuan, China; JP, Japan; LK, Sri Lanka; MM, Myanmar; AUS, Australia; GER, Germany; KR, Korea; THA, Thailand.

Western Fujian, Zhejiang, Hubei, and Sichuan Oolong tea are increasing in popularity. Due to the different geographical environments and tea varieties, the teas show quality differences. For example, the appearance of some Oolong tea in western Fujian was pie, the infusion color was golden yellow, and the taste was mainly "mellow and sweet aftertaste" and "umami and brisk". The taste of Zhejiang Oolong tea was described as "mellow and thick", "smooth and brisk", or "heavy and thick"; the aroma of Hubei Oolong tea was "clean and refreshing". Sichuan Oolong tea was more diversified, having varied appearances and aromas: a "round and tight" and "tight and heavy" in appearance and a "clean and refreshing", "flowery and fruity", and "roasted" aroma.

Japan, Myanmar, Sri Lanka, Germany, Australia, Korea, and Thailand Oolong teas have appeared in recent years [34]. Oolong tea is locally produced in Japan, Myanmar, Sri Lanka, South Korea, and Thailand [34]. The Oolong tea in Germany and Australia is mainly imported from abroad, and then different types of deep processing applied are carried out to adapt to the flavor for local people [35]. Japan and Myanmar Oolong teas were similar in aroma and taste. The difference was that Japan Oolong tea had a stronger heavy and thick taste. In appearance, Myanmar Oolong tea was "round and tight", whereas Japan's was "tight and heavy" or "coarse and loose". The appearance of Oolong tea sold in the German market was "round and tight", "short and broken", or "mixed", some with tea trichome. Oolong tea sold in Australia market was "flowery and fruity" or "roasted". Sri Lanka Oolong tea was characterized by a "sweet" aroma and "clean and cool" taste. The aroma of Korean Oolong tea was "flowery and fruity", and the taste was "mellow and thick". The sensory characteristics of Thailand Oolong tea included a dull red infusion color and roasted aroma.

3.3. Profile Analysis of Phytochemical Components of Oolong Tea from Different Regions

The average coefficient of variation of the phytochemical components of Oolong tea from the different regions was 45.56% (Figure 3a), which indicated wide differences between Oolong tea samples. Their coefficients of variation from high to low were theanine > free amino acids > ECG > C > EGC > EGCG > ester catechins > simple catechins > EC > total catechins > caffeine > tea polyphenols > water extract. Free amino acids (including theanine) were the most different in the Oolong tea from different regions: their content ranged from 0.41% to 5.39%, and that of theanine was 0.04% to 3.60%. Catechins varied widely between different regions: the content of ester catechins was higher than that of simple catechins, and the content of total catechins was 0.90–12.84%. The differences in the caffeine and tea polyphenols contents in samples from different regions were medium in range. The caffeine content was 1.45–4.36%, and that of tea polyphenols was 6.78–23.18%. The difference in water extract content was the smallest, ranging from 36.59% to 52.67% (Table S4). Previously, researchers compared the differences in the chemical constituent contents of Oolong teas from Fujian, Guangdong, and Taiwan in China and found that the coefficient of variation of the contents of theanine and free amino acids were greater than 50%, and those of water extract, tea polyphenols, total catechins, and caffeine were less than 50% [21,36]. In this study, we found the same law in Oolong tea samples from eight different countries. The free amino acids content in Oolong tea from different regions varied the most. In this study, the coefficient of variation of catechins (ECG, C, EGC, and EGCG) was close to or greater than 50%, indicating large differences. The baking process of Oolong tea promotes the degradation of theanine and produces pyrazine substances, which produce a roasted aroma [37]. The difference in catechins is related to the tea plant variety and the degree of fermentation during processing. With the increase in fermentation time, the catechins content gradually decreases, especially that of ester catechins [38,39]. Therefore, the variety and processing technology may be the reasons for the wide differences in the contents of free amino acids and catechins in Oolong tea. The differences in these phytochemical components led to diversified taste characteristics of Oolong tea.

From a heat map analysis of the phytochemical components of Oolong tea from different regions, we found that samples could be divided into six categories (Figure 3b). The first category included Oolong tea from Thailand, Sri Lanka, and Germany, which had a higher content of C and the lowest contents of total catechins, ester catechins, and free amino acids. The second category included Oolong tea from Australia, Korea, Sichuan, northern Fujian, and Zhejiang, which had medium contents of free amino acids, catechins, and tea polyphenols. The third category was Oolong tea from Guangdong, which had the highest contents of total catechins, ester catechins, tea polyphenols, and caffeine. The fourth category included Oolong tea from Hubei and Myanmar, which had the highest content of free amino acids and a high content of simple catechins. The fifth category was Japanese Oolong tea, which had the highest contents of EC and C and higher contents of other substances. The sixth category was Oolong tea from Taiwan, southern Fujian, and western Fujian, which had a higher content of catechins.

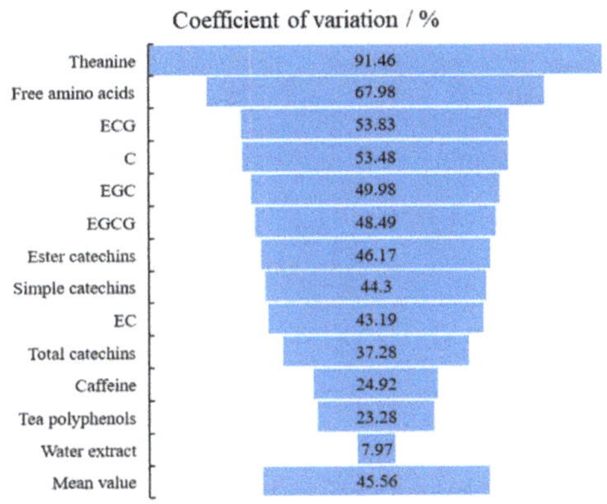

(a)

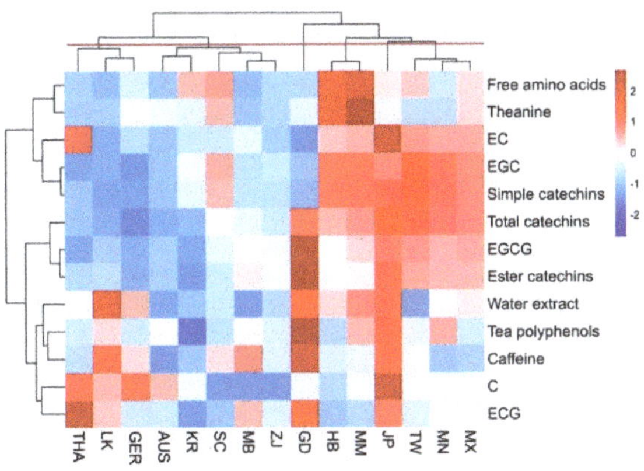

(b)

Figure 3. Coefficient of variation (**a**) and heat map analysis (**b**) of phytochemical components of Oolong tea from different regions. (**b**) Red indicates high content and green indicates low content. MN, southern Fujian, China; MB, northern Fujian, China; MX, western Fujian, China; GD, Guangdong, China; TW, Taiwan, China; HB, Hubei, China; ZJ, Zhejiang, China; SC, Sichuan, China; JP, Japan; LK, Sri Lanka; MM, Myanmar; AUS, Australia; GER, Germany; KR, Korea; THA, Thailand.

3.4. Correlation Analysis between Taste Sub-Attributes and Thirteen Phytochemical Components of Oolong Tea

The correlation analysis between sensory taste sub-attributes intensity and phytochemical components of Oolong tea from different regions (Figure 4) showed that "heavy and thick" was significantly positively correlated with "bitterness" and "sweet aftertaste" ($p < 0.05$). "Astringency" was positively correlated with "bitterness" and negatively correlated with "umami". We found a significant positive correlation between "sweet aftertaste" and "umami".

Caffeine, ECG, ester catechins, tea polyphenols, and EGCG were significantly positively correlated with "heavy and thick", while free amino acids and theanine were significantly negatively correlated with it. Free amino acids, theanine, EGC, EC, and simple catechins were significantly positively correlated with "umami". EGCG was significantly positively correlated with "sweet aftertaste". Caffeine, tea polyphenols, EGCG, ECG, ester catechins, and water extract were significantly positively correlated with "bitterness". All chemical components were correlated with "astringency", but a significant level was not reached ($p < 0.05$). However, "astringency" was significantly correlated with "bitterness" and negatively correlated with "umami". Therefore, chemical components that have an important impact on bitterness and umami affect the astringency of Oolong tea.

Correlation analysis only provided a preliminary judgment of the relationship between phytochemical components and taste sub-attributes. To understand the contribution of phytochemical components to taste sub-attributes, we performed a further comprehensive analysis by combining PLS-VIP and DoT.

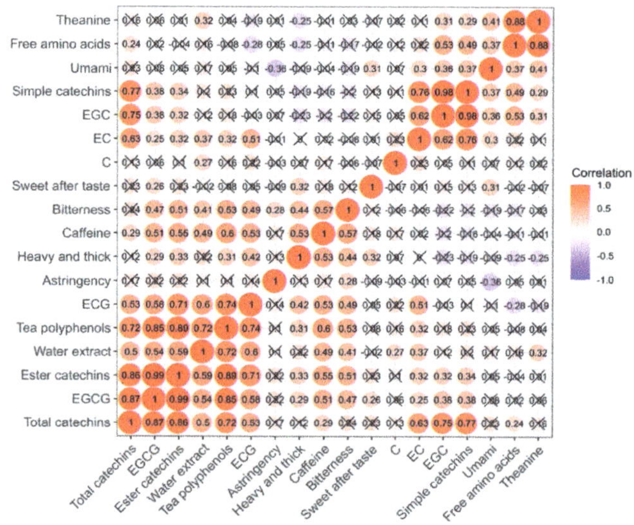

Figure 4. Correlation between taste sub-attributes and phytochemical compositions of Oolong tea. × indicates that the correlation is not significant at the level of $p < 0.05$. The numbers indicate the correlation coefficient, where positive numbers indicate positive correlation and negative numbers indicate negative correlation. Red indicates positive correlation, and blue indicates negative correlation.

3.5. Contribution of Thirteen Phytochemical Components to Taste Sub-Attributes of Oolong Tea

PLS-VIP can be used to decompose and reduce the dimensions of a huge data system and identify independent variables that have a strong impact on dependent variables among complex data variables. It can effectively avoid the interference of multicollinearity between independent variables [11]. We conducted PLS-VIP analysis with thirteen phytochemical components as the independent variable X and taste sensory sub-attributes as dependent variable Y. In this analysis, the phytochemical components with VIP > 1 were considered to provide an important contribution to each taste sub-attribute. The PLS-VIP results are shown in Figure 5. DoT comprehensively considers the contents and threshold value of taste components in tea: it is a method used to evaluate the importance of the contribution of flavor compounds to tea taste. The flavor compounds with DoT > 1 strongly contribute to the taste of tea infusions [40]. The DoT results are shown in Table S5.

3.5.1. Bitterness

Bitterness is one of the basic tastes [41]. We found that six phytochemical components contributed importantly to the bitterness of Oolong tea (Figure 5a,f): caffeine, tea polyphenols, ester catechins, ECG, EGCG, and water extract. This conclusion was exactly the same as the result of correlation analysis (Figure 4). The taste characteristic of caffeine is bitterness, which has a significant positive correlation with the bitterness intensity of tea infusion [40]. The VIP and mean DoT of caffeine in this study were 1.44 and 2.51, respectively, indicating that caffeine had an important contribution to the bitterness of Oolong tea. The taste characteristics of tea polyphenols, ester catechins, ECG, and EGCG can also be described as bitter [42]. In this study, these components substantially contributed to the bitterness of Oolong tea. The mean DoT of EGCG was greater than one in all samples, and the mean DoT of ECG was greater than one in some samples (Table S5). Both EGCG and ECG are ester catechins, which are also the main components of tea polyphenols [41]. Caffeine and tea polyphenols contribute to the bitterness of green tea [43], yellow tea [20], and black tea [11]: however, caffeine contributes the most to the bitterness of black tea, and EGCG contributes the most to the bitterness of green tea [11,43]. Our findings showed that caffeine contributed the most to the bitterness of Oolong tea. The reason for this difference might be that black tea and Oolong tea are fermented teas, whereas green tea is not fermented. In the fermentation process, ester catechins such as EGCG are degraded more, and caffeine is more stable. Therefore, some differences exist in the contribution to bitterness among different teas. In addition, flavonoids and their glycosides and anthocyanins and their glycosides can also be described as bitter [40], which may have an important impact on the bitterness of Oolong tea.

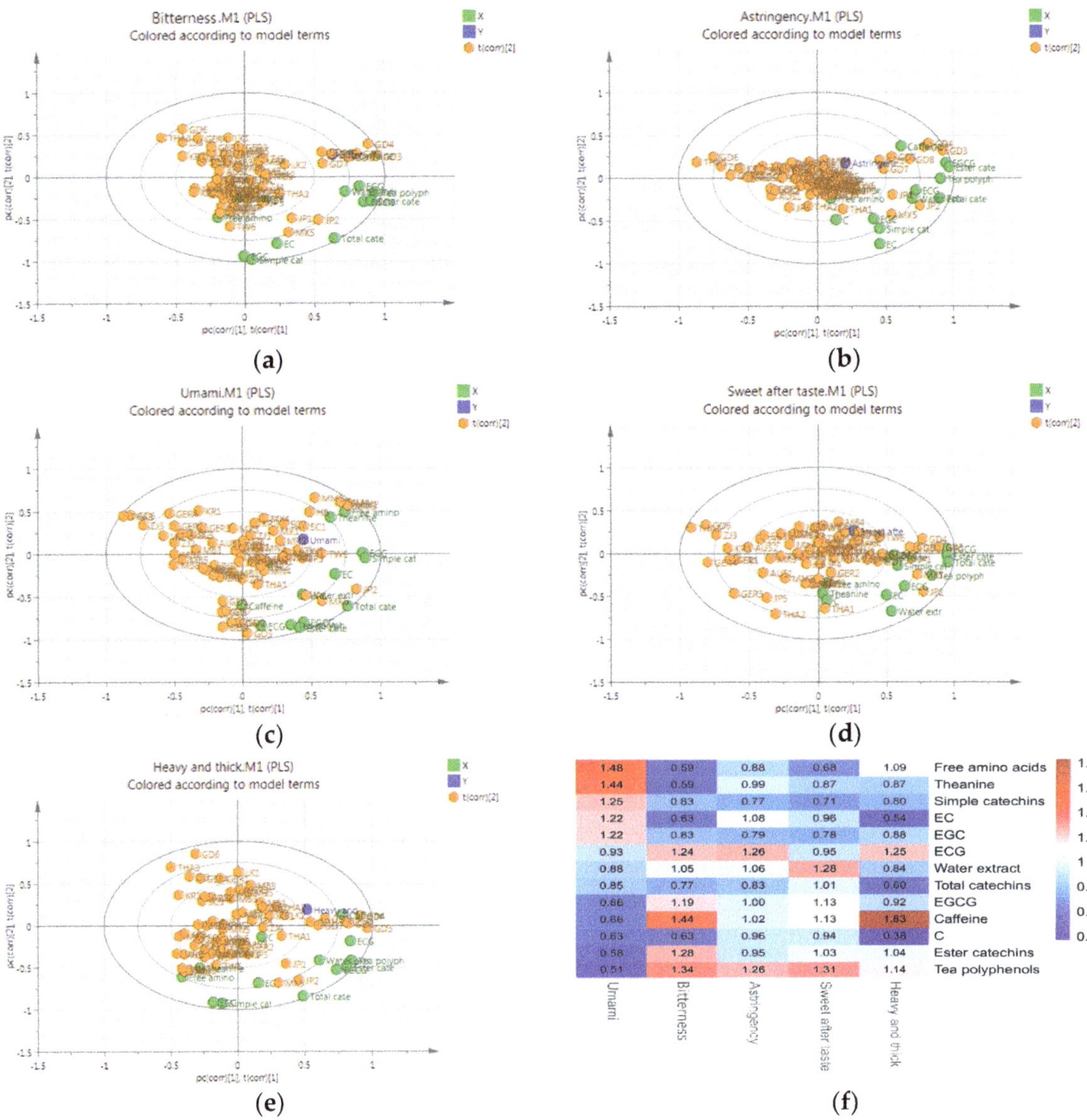

Figure 5. The contribution of phytochemical components to taste sub-attributes of Oolong tea. PLS-VIP load diagram of (**a**) bitterness; (**b**) astringency; (**c**) umami; (**d**) sweet aftertaste; and (**e**) heavy and thick. (**f**) VIP value of each phytochemical component for five taste sub-attributes. (**a**) to (**e**), X (green) indicates the phytochemical composition, Y (blue) indicates the taste powder attribute, and t(corr) [2] (yellow) indicates each Oolong tea sample. (**f**) Red and blue indicate that the VIP value of phytochemical components is large and small, respectively.

3.5.2. Astringency

Astringency is the sensation on the surface of the mouth and tongue that includes dryness, roughness, tension, and tearing [41]. Six phytochemical components significantly contributed to the astringency of Oolong tea in this study (Figure 5b,f): ECG, tea polyphenols, EC, water extract, caffeine, and EGCG. Correlation analysis showed that astringency was significantly correlated with bitterness (Figure 4), suggesting that they were promoted by similar promotion factors. Tea polyphenols, ECG, EC, and EGCG are heterogeneous compounds, producing both astringency and bitterness. The relationship between catechins, tea polyphenols, and bitterness is a quadratic curve. Bitterness and astringency also synergistically enhance each other [44]. Caffeine has no astringency, but it can significantly improve the astringency intensity of catechin, and this enhancement effect increases with caffeine concentration [45]. Theaflavins, flavonols and their glycosides, and phenolic acids can be described as astringent [42], and they may have some influence on the astringency of Oolong tea.

3.5.3. Umami

Umami is one of the basic tastes. We found five phytochemical components that significantly contributed to the umami of Oolong tea (Figure 5c,f): free amino acids, theanine, simple catechins, EC, and EGC. This conclusion is exactly the same as the result of correlation analysis (Figure 4). The overall taste of free amino acids in tea is umami and sweet, which is including theanine, glutamate, and aspartic acid. Glutamate is an important contributor to the taste of tea infusions [42]. Theanine is a monomer amino acid with a high content in tea plants, but its threshold to express umami is very high. Therefore, its DoT value is very low (DoT = 0.1 in Oolong tea, in this study) [11]. Therefore, people cannot sense umami-sweetness given the concentration of theanine in tea infusions. The contribution of theanine to umami is mainly reflected in the synergistic effect with other umami substances [46]. EC and EGC are simple catechins with a bitter taste. In this study, EC and EGC were positively correlated with the umami of Oolong tea. EC is positively correlated with the umami of green tea, whereas EGC is significantly negatively correlated with green tea umami [47]. Phenols can be complexed with caffeine and free amino acids to produce an umami and briskness [48]. Therefore, we inferred that simple catechins and umami in Oolong tea were positively correlated, which might have been caused by their complexation or interaction with other substances, but the specific reasons need to be further studied. Glutamate, aspartic acid, proline, alanine, and serine in free amino acids also impact the umami of tea [42], and they may provide an important contribution to the umami of Oolong tea.

3.5.4. Sweet Aftertaste

Sweet aftertaste is the comprehensive effect of various taste sub-attributes [42]. We identified six phytochemical components that provided important contributions to the sweet aftertaste of Oolong tea (Figure 5d,f): tea polyphenols, water extract, caffeine, EGCG, ester catechins, and total catechins. Water extract is the sum of water-soluble compounds in tea, which impacts the overall taste of tea infusions. The taste of tea polyphenols, caffeine, EGCG, ester catechins, and total catechins is bitter and astringent [20]. The sweet aftertaste of tea is caused by the activation of the sweet receptor in the taste bud cells by the promoter molecules and the transmission of information to the brain. The usual tastes of these promoter molecules are umami, sweet, bitter, and astringent [49], among which the bitterness and astringent are the key contributors to the sweet aftertaste [50]. In addition to the thirteen phytochemical components tested, other phytochemical components with a bitter and astringent taste may affect the sweet aftertaste of Oolong tea.

3.5.5. Heavy and Thick

Heavy and thick is the sensation of stickiness produced by a tea infusion in the mouth, which is related to the richness of the contents [51]. We identified five phytochemical components that importantly contributed to the heavy and thick sensation of Oolong tea (Figure 5e,f): caffeine, ECG, tea polyphenols, free amino acids, and ester catechins. The correlation analysis revealed a significant positive correlation between heavy and thick, bitterness, and umami (Figure 4), indicating that substances that contributed to bitterness and umami might also contribute to the heavy and thick sensation. Caffeine, ECG, tea polyphenols, and ester catechins are bitter and astringent; free amino acids are umami and sweet. In addition, among the tea flavor compounds, tea polyphenols, ester catechins, and caffeine are the chemical components with the highest content in Oolong tea [20]. These high-content substances play an important role in the heavy and thick strength of Oolong tea infusions. In addition to the thirteen phytochemical components tested, other phytochemical components with a bitter and astringent taste may contribute to the heavy and thick taste of Oolong tea.

In this section, we focused on the contribution of thirteen main phytochemical components of Oolong tea to its five taste sub-attributes. However, Oolong tea contains many non-volatile compounds, and other components affect the taste of Oolong tea, such as gallic acid affecting its sweetness [52], flavonoid glycosides and proanthocyanidins affecting its bitterness and astringency [52], theaflavins affecting its astringency, and soluble sugar affecting its mellowness [3]. Volatile compounds have interactive effects on the taste of tea infusions, e.g., geraniol and β-ionone can improve the sweetness of tea infusion [53]. Therefore, in future research, we will use widely targeted metabolomics to comprehensively analyze the marker compounds of Oolong tea from different origins and combine molecular sensory science to analyze the formation mechanism of the regional characteristic qualities of Oolong tea.

4. Conclusions

In this study, we established the flavor wheel of Oolong tea from different regions for the first time, systematically reported the sensory quality characteristics of Oolong tea from 15 regions of eight countries, and explored the contribution of phytochemical components to its taste sub-attributes. We preliminarily found that the sensory quality of Oolong tea can be characterized using 35 descriptors. The quality characteristics of Oolong tea in 15 regions were found to be diverse. Among the taste sub-attributes, heavy and thick was significantly positively correlated with bitterness and sweet aftertaste. Astringency was positively correlated with bitterness and negatively correlated with umami. We found a significant positive correlation between sweet aftertaste and umami. The phytochemical composition contents of Oolong tea from different regions widely varied. Among them, free amino acids contents showed the largest differences, followed by catechins, then caffeine and tea polyphenols had medium differences, and water extract showed the smallest differences. Based on the cluster analysis of phytochemical components, we clustered the Oolong tea from 15 regions into six categories. Finally, PLS-VIP was used to screen out the components that provided important contributions to the five taste sub-attributes of Oolong tea from the detected 13 phytochemical components. In the production process of Oolong tea, appropriate processes can be adopted to improve the content of phytochemical components that make an important contribution to its umami, sweet aftertaste, and heavy and thick and reduce the content of chemical components that make an important contribution to bitterness and astringency so as to produce Oolong tea products that consumers prefer more.

The flavor wheel in this study was the first flavor wheel of Oolong tea from different regions, which was preliminarily established on the basis of existing Oolong tea products. With the continuous improvement in the quality of Oolong tea products from different regions and the enrichment of categories, our findings need to be explored and supplemented in the future to ensure the richness and comprehensiveness of the flavor wheel of Oolong tea from different regions. In addition, the focus of this study was to analyze the contribution of the 13 phytochemical components to the taste sub-attributes of Oolong tea, but we did not comprehensively detect all their non-volatile compounds. Therefore, it is necessary to further include the metabolomics method to screen more chemical components that provide important contributions to the taste of Oolong tea.

Supplementary Materials: The data presented in this study are available within the manuscript and the Supplementary Materials. The following supporting information can be downloaded at: https://www.mdpi.com/article/10.3390/horticulturae8040278/s1 Table S1: Information of Oolong tea samples in different regions; Table S2: The frequency of use of descriptors in Oolong tea in different regions; Table S3: The quality characteristics of Oolong tea in different regions; Table S4: Contents of phytochemical components in Oolong tea from different regions (mean ± SD); Table S5: DoT of phytochemical components of Oolong tea in different regions.

Author Contributions: Conceptualization, W.S. and Z.C.; methodology, Z.W. and S.G.; software, Z.W.; validation, W.S. and Z.C.; formal analysis, W.S. and Z.C.; investigation, Z.W. and S.G.; resources, Z.W. and S.G.; data curation, Z.W.; writing—original draft preparation Z.W.; writing—review and editing W.S. and Z.C.; visualization, Z.W.; supervision W.S.; project administration, W.S.; funding acquisition, W.S. and Z.C. All authors have read and agreed to the published version of the manuscript.

Funding: This work was supported by the Special Project of National Key R&D Plan (2018YFF0214203), the Fujian Agriculture and Forestry University Construction Project for Technological Innovation and Service System of Tea Industry Chain (K1520005A04), and the Special Project of National Key R&D Plan (2019YFD1001601).

Institutional Review Board Statement: Not applicable.

Informed Consent Statement: Not applicable.

Data Availability Statement: The data presented in this study are available within the manuscript and the Supplementary Materials (Tables S1–S5).

Conflicts of Interest: The authors declare no conflict of interest.

References

1. Dai, Z.; Huang, H.; Zhang, Q.; Bei, J.; Chen, Z.; Liu, Q.; Gao, J.; Zhang, S.; Liu, J. Comparative multi-omics of tender shoots from a novel evergrowing tea cultivar provide insight into winter adaptation mechanism. *Plant Cell Physiol.* **2020**, *62*, 366–377.
2. Zeng, L.T.; Zhou, X.C.; Su, X.G.; Yang, Z.Y. Chinese oolong tea: An aromatic beverage produced under multiple stresses. *Trends Food Sci. Technol.* **2020**, *106*, 242–253.
3. Chen, Y.L.; Duan, J.; Jiang, Y.M.; Shi, J.; Peng, L.; Xue, S.; Kakuda, Y. Production, Quality, and Biological Effects of Oolong Tea (*Camellia sinensis*). *Food Rev. Int.* **2010**, *27*, 1–15. [CrossRef]
4. Wan, X.C. *Tea Biochemistry*, 3rd ed.; China Agricultural Press: Beijing, China, 2007; p. 248.
5. Ma, C.Y.; Li, J.X.; Chen, W.; Wang, W.W.; Qi, D.D.; Pang, S.; Mao, A.Q. Study of the aroma formation and transformation during the manufacturing process of oolong tea by solid-phase micro-extraction and gas chromatography–mass spectrometry combined with chemometrics. *Food Res. Int.* **2018**, *108*, 413–422. [CrossRef] [PubMed]
6. Sun, W.J.; Chen, Z.D.; Shang, H. Ten Years of Standardization of Oolong Tea in China. *China Tea Process.* **2020**, *3*, 91–92.
7. Zhu, S.J. Reasons and enlightenment of the rise of "Oolong tea fever" in Japan. *Tea Fujian* **2018**, *5*, 2.
8. Lin, S.-Y.; Lo, L.-C.; Chen, I.-Z.; Chen, P.-A. Effect of shaking process on correlations between catechins and volatiles in oolong tea. *J. Food Drug Anal.* **2016**, *24*, 500–507. [CrossRef]
9. Cai, L.W.; Xu, Y.Q.; Zhou, Y.H. Analysis of sensory quality and chemical composition of Oolong tea in different producing areas. *Tea Fujian* **2016**, *11*, 3.
10. Wang, C.; Lv, S.D.; Wu, Y.S.; Gao, X.M.; Li, J.B.; Zhang, W.R.; Meng, Q.X. Oolong tea made from tea plants from different locations in Yunnan and Fujian, China showed similar aroma but different taste characteristics. *SpringerPlus* **2016**, *5*, 576. [CrossRef]
11. Yue, C.N.; Yang, P.X.; Qin, D.D.; Cai, H.L.; Wang, Z.H.; Li, C.; Wu, H.L. Identification of volatile components and analysis of aroma characteristics of Jiangxi Congou black tea. *Int. J. Food Prop.* **2020**, *23*, 2160–2173. [CrossRef]
12. Spencer, M.; Sage, E.; Velez, M.; Guinard, J.-X. Using Single Free Sorting and Multivariate Exploratory Methods to Design a New Coffee Taster's Flavor Wheel. *J. Food Sci.* **2016**, *81*, S2997–S3005. [CrossRef] [PubMed]
13. Zhang, Y.B.; Liu, X.; Lu, C.Y. Study on primitive morpheme in sensory terminology and flavor wheel construction of Chinese tea. *J. Tea Sci.* **2019**, *39*, 474–483.
14. Wu, S.; Wen, L.P.; Ye, Z.H.; Cai, S.; Li, Q.H. Construction of flavor wheel of Te-flavor Baijiu and description of typical sensory characteristics. *China Brew.* **2021**, *40*, 148–153.
15. Kong, X.W.; Zhou, Z.L.; Zheng, F.P.; Liu, S.P.; Han, X.; Mao, J. Construction and application of flavor wheel of Chinese brewing vinegar. *J. Food Sci. Biotechnol.* **2020**, *39*, 74–80.

16. Deng, X.; Huang, G.; Tu, Q.; Zhou, H.; Li, Y.; Shi, H.; Wu, X.; Ren, H.; Huang, K.; He, X.; et al. Evolution analysis of flavor-active compounds during artificial fermentation of Pu-erh tea. *Food Chem.* **2021**, *357*, 129783. [CrossRef] [PubMed]
17. Li, H.-H.; Luo, L.-Y.; Wang, J.; Fu, D.-H.; Zeng, L. Lexicon development and quantitative descriptive analysis of Hunan fuzhuan brick tea infusion. *Food Res. Int.* **2019**, *120*, 275–284. [CrossRef]
18. Dai, Q.Y.; Ye, Y.J.; An, Q.; Zheng, F.L.; Xiao, M.J.; Xiao, M.X.; Wang, H.Q.; Zhang, H.W. Sensory characteristics of yellow large leaf tea by quantitative descriptive analysis and construction of flavor wheel. *J. Tea Sci.* **2021**, *41*, 535–544.
19. Lee, S.M.; Chung, S.-J.; Lee, O.-H.; Lee, H.-S.; Kim, Y.-K.; Kim, K.-O. Development of sample preparation, presentation procedure and sensory descriptive analysis of green tea. *J. Sens. Stud.* **2008**, *23*, 450–467. [CrossRef]
20. Wang, Z.-H.; Yue, C.-N.; Tong, H.-R. Analysis of taste characteristics and identification of key chemical components of fifteen Chinese yellow tea samples. *J. Food Sci. Technol.* **2021**, *58*, 1378–1388. [CrossRef]
21. Wang, K.B.; Liu, F.; Liu, Z.H.; Huang, J.A.; Xu, Z.X.; Li, Y.H.; Chen, J.H.; Gong, Y.S.; Yang, X.H. Analysis of chemical components in oolong tea in relation to perceived quality. *Int. J. Food Sci. Technol.* **2010**, *45*, 913–920.
22. Liu, H.; Zhang, Y.; Fan, Y.H.; Zhang, Q.P.; Luo, Z.G. Simultaneous identification of Oolong tea juice tastes by principal component analysis and ultra-high performance liquid chromatography. *Beverage Ind.* **2013**, *16*, 25–31.
23. *GB/T 23776-2018*; Methodology of Sensory Evaluation of Tea. Standardization Administration of the People's Republic of China: Beijing, China, 2018.
24. *GB/T 16291.1-2012*; Sensory Analysis—General Guidance for the Selection, Training and Monitoring of Assessors—Part 1: Selected Assessors. Standardization Administration of the People's Republic of China: Beijing, China, 2012.
25. *GB/T 14487-2017*; Tea Vocabulary for Sensory Evaluation. Standardization Administration of the People's Republic of China: Beijing, China, 2017.
26. *ISO 1573:1980*; Tea—Determination of Loss in Mass at 103 °C. International Organization for Standardization: Geneva, Switzerland, 1980.
27. *ISO 9768:1998*; Tea—Determination Water Extracts Content. International Organization for Standardization: Geneva, Switzerland, 1998.
28. *ISO 14502.1:2005*; Determination of Substances Characteristic of Green and Black Tea—Part 1: Content of Total Polyphenols in Tea Col-Orimetric Method Using Folin-Ciocalteu Reagent. International Organization for Standardization: Geneva, Switzerland, 2005.
29. *ISO 14502.1:2005*; Determination of Substances Characteristic of Green and Black Tea—Part 2: Content of Catechins in Green Tea—Method using High-Performance Liquid Chromatography. International Organization for Standardization: Geneva, Switzerland, 2005.
30. *GB/T 23193-2017*; Determination of Theanine in Tea—Using High-Performance Liquid Chromatography. Standardization Administration of the People's Republic of China: Beijing, China, 2017.
31. *GB/T 8314-2013*; Tea—Determination of Free Amino Acids Content. Standardization Administration of the People's Republic of China: Beijing, China, 2013.
32. Ji, W.B.; Liu, P.P.; Xu, Y.Q.; Jiang, Y.W.; Chen, J.X.; Yin, J.F. Comparative study of the aroma components of several Oolong teas. *J. Tea Sci.* **2016**, *36*, 523–530.
33. Lv, S.D.; Wu, Y.S.; Jiang, Y.F.; Meng, Q.X. Comparative analysis of aroma characteristics of Oolong tea from different geographical regions. *Food Sci.* **2014**, *35*, 146–153.
34. Liu, X.B. The first Zhuang Wanfang Tea Science Forum and International Symposium on Oolong tea innovation in 2017. *Int. Acad. Dev.* **2018**, 3, 46–47.
35. Cheng, J.B. Australia—A young market where traditional tea and mixed tea coexist. *Tea World* **2020**, *5*, 14–15.
36. Jin, Y.S.; Zheng, J.G.; Yang, J.F.; Sun, W.J. Difference analysis of biochemical quality of Tieguanyin from different producing areas. *Jiangsu Agric. Sci.* **2013**, *41*, 325–327.
37. Guo, X.Y.; Song, C.K.; Ho, C.-T.; Wan, X.C. Contribution of L-theanine to the formation of 2,5-dimethylpyrazine, a key roasted peanutty flavor in Oolong tea during manufacturing processes. *Food Chem.* **2018**, *263*, 18–28. [CrossRef]
38. Liu, P.P.; Zhong, X.Y.; Xu, Y.Q.; Chen, G.H.; Yin, J.F.; Liu, P. Study on organic acids contents in tea leaves and its extracting characteristics. *J. Tea Sci.* **2013**, *33*, 405–410.
39. Cui, J.L. *Contribution of Glycoside Aroma Precursors to Aroma Formation of Oolong Tea and Black Tea*; Southwest University: Chongqing, China, 2006.
40. Scharbert, S.; Hofmann, T. Molecular Definition of Black Tea Taste by Means of Quantitative Studies, Taste Reconstitution, and Omission Experiments. *J. Agric. Food Chem.* **2005**, *53*, 5377–5384. [CrossRef]
41. Tong, H.R.; Jin, X.F.; Gong, X.L. Sensory characteristics of tea polyphenols and it's effects on astringency of tea. *J. Tea Sci.* **2006**, *2*, 79–86.
42. Yue, C.N.; Wang, Z.H.; Mao, S.H.; Li, X.Y.; Tong, H.R. The main taste substances in tea research progress. *Food Res. Dev.* **2017**, *38*, 219–224.
43. Zhang, Y.N.; Chen, G.S.; Liu, Y.; Xu, Y.Q.; Wang, F.; Chen, J.X.; Yin, J.F. Analysis of the bitter and astringent taste of baked green tea and their chemical contributors. *J. Tea Sci.* **2015**, *35*, 377–383.
44. Shi, Z.P.; Liu, Z.H. Discussion on mathematical model of chemical essence of bitter and astringent taste of summer tea. *J. Tea Sci.* **1987**, *2*, 7–12.
45. Xu, W.P.; Li, D.X.; Zhang, Z.Z.; Tang, Q.; Wan, X.C. The nonlinear regression of bitter and astringency of main compounds in green tea and the application in organoleptic tests. *J. Tea Sci.* **2010**, *30*, 399–406.

46. Kaneko, S.; Kumazawa, K.; Masuda, H.; Henze, A.; Hofmann, T. Molecular and Sensory Studies on the Umami Taste of Japanese Green Tea. *J. Agric. Food Chem.* **2006**, *54*, 2688–2694. [CrossRef]
47. Liu, P.P.; Deng, Y.L.; Yin, J.F.; Zhang, Y.N.; Chen, G.S.; Wang, F.; Chen, J.X.; Yuan, H.B.; Xu, Y.Q. Quantitative analysis of the taste and its correlation research of chemical constitutes of green tea. *J. Chin. Inst. Food Sci. Technol.* **2014**, *14*, 173–181.
48. Chen, M.X.; Guo, Y.L.; Guo, X.N. The research situation of tea color and flavor constituents. *J. Food Saf. Qual.* **2014**, *6*, 1818–1823.
49. Zhong, Z.Y.; Zeng, Z.Z. Molecular mechanism of tea rejuvenate. *J. Agric. For.* **2014**, *63*, 91–97.
50. Yue, C.N.; Qin, D.D.; Cai, H.L.; Wang, Z.H.; Li, C.; Li, Y.S.; Yang, P.X. Taste characteristics and key compounds analysis of Congou black tea in northern Jiangxi province. *Food Ferment. Ind.* **2021**, *47*, 260–267.
51. Mao, S.H. *Study on Quality Analysis and Control of Congfu Black Tea Based on Flavor Omics*; Southwest University: Chongqing, China, 2018.
52. Cao, Q.-Q.; Fu, Y.-Q.; Wang, J.-Q.; Zhang, L.; Wang, F.; Yin, J.-F.; Xu, Y.-Q. Sensory and chemical characteristics of Tieguanyin oolong tea after roasting. *Food Chem. X* **2021**, *12*, 100178. [CrossRef]
53. Yu, J.Y.; Liu, Y.; Zhang, S.R.; Luo, L.Y.; Zeng, L. Effect of brewing conditions on phytochemicals and sensory profiles of black tea infusions: A primary study on the effects of geraniol and β-ionone on taste perception of black tea infusions. *Food Chem.* **2021**, *354*, 129504. [PubMed]

horticulturae

Article

The Bioactive Compounds and Fatty Acid Profile of Bitter Apple Seed Oil Obtained in Hot, Arid Environments

Mukesh Kumar Berwal [1,*], Chet Ram [1], Pawan Singh Gurjar [1], Jagan Singh Gora [1], Ramesh Kumar [1], Ajay Kumar Verma [1], Dhurendra Singh [1], Boris Basile [2], Youssef Rouphael [2,*] and Pradeep Kumar [3]

1 ICAR-Central Institute for Arid Horticulture (CIAH), Bikaner 334006, India; chet.ram2@icar.gov.in (C.R.); pawan.gurjar@icar.gov.in (P.S.G.); jagan.gora@icar.gov.in (J.S.G.); ramesh.kumar11@icar.gov.in (R.K.); ajay.verma2@icar.gov.in (A.K.V.); dhurendra.singh@icar.gov.in (D.S.)

2 Department of Agricultural Sciences, University of Naples Federico II, 80055 Portici, Italy; boris.basile@unina.it

3 ICAR-Central Arid Zone Research Institute, Jodhpur 342003, India; pradeep.kumar4@icar.gov.in

* Correspondence: mukesh.kumar4@icar.gov.in (M.K.B.); youssef.rouphael@unina.it (Y.R.)

Abstract: Bitter apple or tumba (*Citrullus colocynthis* L.) is a prostrate annual herb belonging to the *Cucurbitaceae* family. It is highly tolerant against multiple abiotic stresses like drought, heat, and soil salinity and can easily grow on very marginal soil, even on sand dunes in hot, arid regions. Tumba fruit is a fleshy berry 5–10 cm in diameter and of a pale yellow color at ripening. The tumba fruit used in this research was harvested from the ICAR-CIAH, Bikaner research farm. The seeds were separated, and their oil was extracted to analyze its physical characteristics and composition (phytochemical compounds, fatty acid profile, etc.). The seeds of the tumba fruit contained 23–25% golden-yellow-colored oil with a specific gravity of 0.92 g/mL. The extracted oil contained appreciable amounts of phytochemical (bioactive) compounds like phenolics (5.39 mg GAE/100 g), flavonoids (938 mg catechin eq./100 g), carotenoids (79.5 mg/kg), oryzanol (0.066%), and lignans (0.012%), along with 70–122 mg AAE/100 g total antioxidant activity (depending on the determination method). The results of fatty acid profiling carried out by GC-MS/MS demonstrated that tumba seed oil contained about 70% unsaturated fatty acids with more than 51% polyunsaturated fatty acids. It mainly contained linoleic acid (C18:2n6; 50.3%), followed by oleic acid (C18:1n9; 18.0%), stearic acid (C18:0; 15.2%), and palmitic acid (C16:0; 12.4%). Therefore, this oil can be considered as a very good source of essential fatty acids like omega-6 fatty acid (linoleic acid), whereas it contains a lower concentration of omega-3 fatty acids (α-linolenic acid) and hydroxy polyunsaturated fatty acids. In addition, it also contains some odd chain fatty acids like pentadecanoic and heptadecanoic acid (C15:0 and C17:0, respectively), which have recently been demonstrated to be bioactive compounds in reducing the risk of cardiometabolic diseases. The results of this study suggest that tumba seed oil contains several health-promoting bioactive compounds with nutraceutical properties; hence, it can be an excellent dietary source.

Keywords: tumba; physico-chemicals; fatty acid profile; medicinal uses

Citation: Berwal, M.K.; Ram, C.; Gurjar, P.S.; Gora, J.S.; Kumar, R.; Verma, A.K.; Singh, D.; Basile, B.; Rouphael, Y.; Kumar, P. The Bioactive Compounds and Fatty Acid Profile of Bitter Apple Seed Oil Obtained in Hot, Arid Environments. *Horticulturae* **2022**, *8*, 259. https://doi.org/10.3390/horticulturae8030259

Academic Editor: Charalampos Proestos

Received: 14 February 2022
Accepted: 14 March 2022
Published: 17 March 2022

1. Introduction

In recent years, there has been an increasing interest in vegetable oils with functional properties. Vegetable oils are in great demand because they have diverse applications for ensuring safe food, for nutraceuticals and medicines, and in industry. India has a long history of cultivation and use of medicinal plants. However, speedy industrialization, urbanization, and overgrazing have caused the loss of medicinally and industrially important floras [1]. These floras may be an important source of functional food and nutraceutical compounds for the pharmaceutical industry.

The introduction and domestication of some economically important plants will not only help increase the vegetation cover in India's Thar Desert, but these actions may also improve the socio-economic status of the people living in these areas. Plants from the *Cucurbitaceae* family, endemic to this region, are well adapted to the xeric conditions of the desert. Some of the naturally and commonly occurring plants of *Cucurbitaceae* family are tumba (*C. colocynthis)*, mateera or local watermelon (*Citrullus lanatus*), kachra or snap melon (*Cucumis melo* var. *utilissimus*), kachri (*Cucumis melo* var. *agrestis*), etc. [2]. Tumba is a trailing annual scabrid herb belonging to the *Cucurbitaceae* family and is known by different names in different regions, such as Hadla in Jordan, bitter cucumber or bitter apple in English-speaking countries, Abujahl watermelon or Kadu Hanzal in Persia [3]. In India, it is also known with different names such as Ghudmba in Punjab, Indark in Gujarat, Makal in Bengal and Kartama in southern India.

C. colocynthis (tumba or bitter apple) is a closely related species of cultivated watermelon (*Citrullus lanatus*) [4] grown and geographically distributed in various parts of the world, such as the deserts of the Middle East and in southern Europe and Africa [5–7]. This plant is highly tolerant to multiple abiotic stresses such as drought, heat, and soil salinity and can easily grow on very marginal soils, even on sand dunes in hot, arid regions. It grows profusely by producing multiple branches and, thanks to tendrils located at each node, spreads over sandy undulated plains and sand dunes and plays an important role in controlling soil erosion in desert areas. It is one of the most important biomass producers along with a naturally grown *Calligonum polygonoides* (phog, a shrub) under resource-limited environmental conditions in its habitat in India's Thar Desert [8]. The tumba plant has fast growing habits and starts flowering and fruiting at just 30 and 60 days, respectively, after sowing. Its creeping nature and the soil-binding properties of its roots are very helpful for preventing desertification through controlling and stabilizing sand dunes. All the organs of this plant, including the stem, leaves, fruits, seeds, and root are used as dried or fresh, either aqueous or oil extracts, and are reported to have anti-diabetic, anti-leprosy, anti-inflammatory, analgesic, vermifuge, hyperlipidemic laxative, hair-growth-promoting, antimicrobial, and antioxidant properties [4,9–14]. In spite of the several medicinal uses of tumba, some complications have also been reported from its direct use including diarrhea, colic, vomiting, nephrosis, hematochezia, and liver dysfunctions [15–17].

The fruit of *C. colocynthis* is a large fleshy berry, globular in shape and smooth, like a gourd. It is 5–10 cm in diameter, pale yellow in color at ripening, and available during October and November. Tumba fruit has very high medicinal value and is used in indigenous medicine as a purgative agent. Tumba seed oil is used in the soap and candle industry in Rajasthan, India. It can also be used for oilseed feedstock and thus can replace lubricant or biodiesel to some extent [18]. The nutritional composition of tumba seeds of different regions of the world was reported to consist of 20–30% carbohydrate, 14–24% fat, 13–26% protein, and 2–4% ash [19,20]. The fatty acid composition of tumba seed oil differs widely according to its spatial distribution in various regions of the world; it varies between 55–74% for linoleic acid, 9–17% for oleic acid, 5.36–9.84% for stearic acid, and 8.35–11.70% for palmitic acid [19,21–27]. The present study aims to harness the tumba seed oil cultivated in the hot and arid region of Rajasthan, India, as a potential vegetable oil for culinary as well as nutraceutical applications through characterizing its bioactive compounds and fatty acid composition.

2. Materials and Methods

2.1. Experimental Site and Plant Material

Tumba fruit was collected during the Fall cropping season in 2019 from the research station of the ICAR-Central Institute for Arid Horticulture, Bikaner, India, located in the Thar Desert (28°06′0.21″ N; 73°21′22.17″ E; 224 m a.s.l.). The fruits were weighed and cut for seed extraction. The seeds were dried at room temperature and weighed.

2.2. Lipid Extraction

The recovered seeds were finely ground with an electric blender, and weighted samples were subjected to oil extraction through a Soxhlet apparatus (Lab C, The Laboratory

Glassware Co., Ambala Cantt, India) using petroleum ether as solvent at 50 °C [28]. After 3 h, the recovered hexane was evaporated and the extracted seed oil (pale yellow in color) was weighed to calculate the oil content.

2.3. Fatty Acid Profiling of Tumba Seed Oil Using GC-MS/MS

The extracted tumba seed oil was subjected to preparation of fatty acid methyl esters (FAMEs) using Boron-Trifluoride (BF_3) as per the AOAC Official Method [28]. The FAMEs were subjected to GC-MS/MS analysis using a Gas Chromatograph, consisting of an AOC-20i and interfaced to a QP 2010 Plus Mass Spectrometer (GC-2010 system, Shimadzu Corporation, Japan), equipped with a polar fused silica column, COL-ELITE-2560 (highly polar phase; biscyanopropylphenylpolysiloxane, 100 mt Length × 0.25 mm ID × 0.2 μmdf). The initial temperature of the oven was 100 °C (hold for 4 min); it was then increased to 240 °C (held for 15 min) with an injection temperature of 225 °C. The helium gas flow was 1.0 mL/min to a total of 65 min program.

The identification of FAMEs was accomplished by GC-MS mass spectrum explication and a comparison of retention times with mass spectra to those of commercial-standard fatty acid methyl esters mix, C8-C-24 (Supelco 37 Component FAME Mix, Sigma Aldrich). The amount of individual fatty acid in the tumba seed oil was determined with the help of standard curves adopted in the two methods, and the value was expressed as percentage of weight.

2.4. Nutraceutical Composition and Antioxidant Activity Determination of Tumba Seed Oil

2.4.1. Methanolic Extractives of Tumba Seed Oil

One gram of oil sample was extracted in triplicate with 10 mL of 80% aqueous methanol containing 0.1% HCl by shaking continuously for 1 h at room temperature. The extractives were stored under deep freezing conditions at −20 °C until further use.

2.4.2. Determination of Total Phenolic Content (TPC) and Total Flavonoid Content (TFC)

The TPC of the methanolic extractives from the tumba seed oil was assessed according to the Folin–Ciocalteu reagent method described by Berwal et al. [29], and the values were expressed in gallic acid equivalents (GAE)/100 g of oil.

The TFC of the tumba seed oil was determined by the aluminum chloride–based colorimetric assay previously described [30]. One mL extractive was mixed with 0.3 mL each of sodium nitrite, aluminum chloride, and NaOH solution. After 15 min incubation at room temperature, absorbance was read with an UV-VIS Spectrophotometer (Shimadzu UV-2550, Shimadzu Corporation, Japan) at 510 nm against the blank. TFC was expressed as catechol equivalents/100 g of oil.

2.4.3. Estimation of Lignan, Oryzanol, and Carotenoid Content

The content of lignans was estimated using the method described by Bhatnagar et al. [31]. Briefly, 0.01 g oil samples were dissolved in a hexane + chloroform mixture (7:3, *v/v*) to reach a final volume of 10 mL. Then absorbance was read with a UV-VIS Spectrophotometer (Shimadzu UV-2550, Shimadzu Corporation, Tokyo, Japan) at 288 nm. The lignans content was intended by using a specific extinction coefficient ($E^{1\%}$/1 cm) for sesamoline, 231.1.

Total oryzanol content was estimated using the protocol described by Gopla Krishna et al. [32]. Briefly, 1 g of tumba seed oil was mixed with hexane up to a 10 mL final volume. The mixture absorbance was read at 314 nm against hexane (blank) with a UV-VIS Spectrophotometer (Shimadzu UV-2550, Shimadzu Corporation, Japan), and oryzanol content was calculated using a specific extinction coefficient ($E^{1\%}$/1 cm) for oryzanol, i.e., 358.9.

The carotenoid content of tumba seed oil was estimated using the method described by Kumar et al. [33]. One gram of oil sample was made up to 10 mL using hexane. This mixture was 10-times diluted before its absorbance was read with an UV-VIS Spectrophotometer

(Shimadzu UV-2550, Shimadzu Corporation, Tokyo, Japan) at 446 nm. The carotenoid content was calculated with the molar extinction coefficient of carotenoids, i.e., 383.

2.4.4. Total Antioxidant Activity

The total antioxidant assay by the phosphomolybdenum method was carried out as described by Prieto et al. [34]. A 0.3 mL aliquot of methanolic extractive was mixed with 28-mM sodium phosphate and 4-mM ammonium molybdate. The resultant mix was incubated for 90 min in a water bath set at 95 °C. Then the absorbance was read at 695 nm with an UV-VIS Spectrophotometer (Shimadzu UV-2550, Shimadzu Corporation, Tokyo, Japan). The total antioxidant activity was calculated with standard curves of ascorbic acid and expressed as mg ascorbic acid equivalents/100 g of oil.

2.5. Statistical Analysis

The physico-chemical parameters measured, including fatty acid analysis in tumba seed oil, were carried out in three replications. The mean values and standard deviations (SD) were calculated using MS-Office Excel.

3. Results and Discussion

3.1. The Fatty Acid Composition of Tumba Seed Oil

Seeds accounted for 2.75 ± 0.25% of total tumba fruit dry mass and contained about 24.75 ± 1.25% golden-yellow-colored oil with a 0.92 ± 0.01 g/mL specific gravity. The fatty acid composition of tumba seed oil from the hot, arid region of Rajasthan is reported in Table 1 and Figure 1. It is extremely rich in omega-6-polyunsaturated fatty acid (n6-PUFA), where linoleic acid (C18:2n6) was the most important n6-PUFA that accounted for 50.31% of the total fatty acids (Figure 1). The saturated fatty acids in the tumba seed oil contributed 30.38% of the total fatty acids and were mainly composed of palmitic acid (C16:0; 12.41%), stearic acid (C18:0; 15.15%), and arachidic acid (C20:0; 1.08%). The monounsaturated fatty acids (MUFAs) contributed about 18.83% of total fatty acids, with oleic acid (C18:1n9) being the major component (18.02%) followed by gadoleic acid (C20:1n9; 0.52%). Omega-3 fatty acids (n3-PUFA; i.e., α-linolenic acid) were also found at a minute level and accounted for about 0.50% in tumba seed oil.

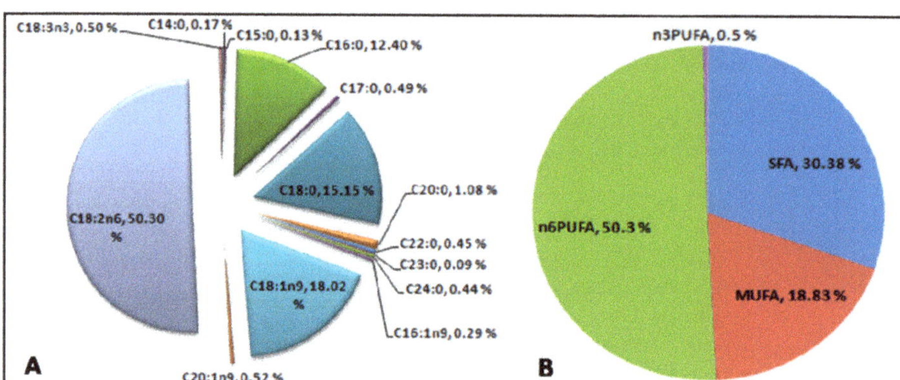

Figure 1. The fatty acid composition of tumba (*C. colocynthis*) seed oil obtained in the present study: (**A**) fatty acid profile and (**B**) fatty acid subclasses distribution in the tumba seed oil (SFA: saturated fatty acids; MUFA: monounsaturated fatty acids; n6-PUFA: omega-6 polyunsaturated fatty acids; n3-PUFA: omega-3 polyunsaturated fatty acids). The values are presented as the weight percentage for individual fatty acids.

Table 1. The fatty acid composition of tumba (*Citrullus colocynthis* L.) seed oil obtained in the present study.

	Name of the Fatty Acid	Common Name	Formula	Peak Area (%)	Type of Fatty Acid
1	Tetradecanoate	Myristic acid	C14:0	0.17 ± 0.01	Saturated
2	Pentadecanoic acid	Pentadecylic acid	C15:0	0.13 ± 0.01	Saturated
3	Hexadecanoic acid	Palmitic acid	C16:0	12.41 ± 0.02	Saturated
4	9-Hexadecenoic acid, (Z)-	Palmitelaidic acid	C16:1n9	0.29 ± 0.01	MUFA
5	Heptadecanoic acid	Margaric acid	C17:0	0.49 ± 0.03	Saturated
6	Octadecanoic acid	Stearic acid	C18:0	15.15 ± 0.46	Saturated
7	9-Octadecenoic acid (Z)-	Oleic acid	C18:1n9	18.02 ± 0.36	MUFA
8	9,12-Octadecadienoic acid (Z,Z)-	Linoleic acid	C18:2n6	50.31 ± 0.33	PUFA
9	9,12,15-Octadecatrienoic acid	α-Linolenic acid	C18:3n3	0.50 ± 0.05	PUFA
10	Eicosanoic acid	Arachidic acid	C20:0	1.08 ± 0.03	Saturated
11	11-Eicosenoic acid	Gadoleic acid	C20:1n9	0.52 ± 0.03	MUFA
12	Docosanoic acid	Behenic acid	C22:0	0.45 ± 0.02	Saturated
13	Tricosanoic acid	Tricosanoic acid	C23:0	0.09 ± 0.02	Saturated
14	Tetracosanoic acid	Lignoceric acid	C24:0	0.44 ± 0.04	Saturated

The results obtained in our study are consistent with previous reports describing the fatty acid composition of *C. colocynthis* seed oils obtained in different regions of the world, viz. India, Malaysia, Israel, Jordan, and Algeria, but the absolute values of the single fatty acids varied to some extent (Table 2) [19,21–27]. It was observed that palmitic, stearic, oleic, and linoleic acids are the major fatty acids in tumba seed oil, as they contributed more than 95 % of the total fatty acids. Among these, linoleic acid was the most prevalent fatty acid as its content ranged between 50.31 and 74.77%. The highest content of linoleic acid was previously reported in Jordan seed oil (74.77%), and the lowest content was reported in the current study (50.31%). Likewise, other major fatty acids also varied: palmitic acid ranged between 8.35 (Jordan) [26] and 12.41% (current study), stearic acid between 5.35 (Jordan) [26] and 15.15% (current study), and oleic acid between 9.04 (Jordan) [19,21–27] and 18.02% (current study) (Table 2). The fatty acid composition and unsaturated fatty acid content found in the present study were similar to those of major vegetable oils previously studied [35–37]. Orsavova et al. [37] studied the fatty acid profile of 14 major vegetable oils—safflower, grape, milk thistle, hemp, sunflower, wheat germ, pumpkin seed, sesame, rice bran, almond, rapeseed, peanut, olive, and coconut oil—and reported that palmitic, oleic and linoleic acids are major contributing fatty acids in these oils with 4.6–20%, 6.2–71.1% and 1.6–79%, respectively. Interestingly, in the present study palmitic, stearic, oleic, and linoleic acids are the major contributors, with approximately 70% unsaturated fatty acids.

In addition to the aforementioned fatty acids, some odd chain saturated fatty acids (OCFAs) were also detected in tumba seed oil at a minute level (about 0.71% of the total fatty acids). The main contributors were pentadecylic acid (C15:0; 0.13%), margaric acid (C17:0, 0.49%), and tricosanoic acid (C23:0; 0.09%) (Table 1). These results diverged from previous studies that reported only heptadecanoic acid (C17:0; 0.075–0.08%) [20] for this group of compounds. This group of fatty acids, specifically pentadecanoic acid (C15:0) and heptadecanoic acid (C17:0), is essential for human health. The higher dietary intake of these OCFAs is linked with reduced risks of cardiovascular disease, adiposity, chronic inflammation, type-2 diabetes, metabolic syndrome, nonalcoholic steatohepatitis (NASH), chronic obstructive pulmonary disease, pancreatic cancer, and other conditions [38–42].

Table 2. Comparison of the major fatty acid composition (by weight %) of tumba (*Citrullus colocynthis* L.) seed obtained in the present study with that obtained in other countries and regions of the world.

Fatty Acids	India [a]	India [b]	India [c]	India [d]	Malaysia [e]	Israel [f]	Jordan [g]	Algeria [h]	Current Study
					wt %				
Palmitic acid (C16:0)	9.38	10.43	11.70	10.30	10.48	10.10	8.35	10.22	12.41
Stearic acid (C18:0)	7.34	9.84	9.70	8.00	9.72	6.70	5.36	8.98	15.15
Oleic acid (C18:1)	17.04	15.90	11.40	24.50	17.95	13.10	9.04	9.36	18.02
Linoleic acid (C18:2)	61.05	62.81	66.10	55.90	61.41	70.10	74.77	68.49	50.31

[a] Ashish et al. [21]; [b] Gurudeeban et al. [22]; [c] Kulkarni et al. [23]; [d] Kamalakar et al. [19]; [e] Solomon et al. [24]; [f] Zohara et al. [25]; [g] Al-Hwaiti et al. [26]; [h] Bireche et al. [27].

3.2. Nutraceutical Composition and Antioxidant Activity of Tumba Seed Oil

The nutraceutical composition and antioxidant activity (total phenolic, total flavonoids, oryzanol, lignans, carotenoids, and total antioxidant activity) of the extracted tumba seed oil are reported in Table 3. Nutraceuticals are natural bioactive compounds produced by plants as their secondary metabolites, which include phenols, flavonoids, terpenes, pigments, lignans, oryzanol, etc. [43]. All these compounds exert strong antioxidant activity.

Table 3. The bioactive compound content and antioxidant activity of the Tumba (*Citrullus colocynthis* L.) seed oil obtained in the present study.

Parameters	Content
Total phenolic content (mg/100 g of oil gallic acid Eq.)	5.39 ± 0.73
Total flavonoids content (mg/100 g of oil catechin Eq.)	938.0 ± 18.0
Oryzanol (%)	0.066 ± 0.003
Lignans (%)	0.012 ±0.002
Carotenoids (mg/kg)	79.5 ± 16.1
Total antioxidant activity by phosphomolybdate method (mg/100 g oil ascorbic acid Eq.)	70.83 ± 2.37

All the values are mean ± SD of three replicates.

Phenolics constitute one of the most widely and ubiquitously distributed plant secondary metabolite groups, with more than 8000 known phenolic compounds, which have been reported to exhibit various biological functions including antimicrobial, antioxidant, and antidiabetic ones [44]. A number of studies demonstrated that phenolic content in plants is directly linked with their antioxidant potential due to their redox properties, which make them strong reducing agents, hydrogen donors, and quenchers of singlet oxygen species [44,45]. Phenolic compounds in seed oil are a relevant oil quality index because these compounds protect lipids from peroxidation through scavenging free radicals. TPC in tumba seed oil was found to be 5.39 mg gallic acid eq./100 g of oil (Table 3), which is comparable to the TPC content of other vegetable oils like coconut, ground nut, rice bran, and sunflower oil, which were reported to contain 3.09, 1.8, 0.89, and 0.49 mg per 100 g gallic acid eq., respectively [46]. Other vegetable oils that have even higher content of TPC are *Basella rubra* (32.99 mg), white mustard (150 mg), coriander seeds (20 mg), and caraway seeds (78 mg) [34,38–41,47]. The differences in TPC composition of these oils may be linked to the crop species as well as to the extraction, processing, and refining conditions. Flavonoids are an abundant sub-group of plant phenolics that include more than 4000 natural compounds [48]. The TFC of tumba seed oil (Table 3) was higher (938 mg of catechin equivalent/100 g) than that of other vegetable oils. Kumar et al. [33] also reported very high TFC content (557.88 mg/100 g) in *B. rubra* seed oils. Xuan et al. [49] studied the TFC content of 14 different vegetable oils and reported contents ranging between 3 mg/100 g (safflower oil) and 34 mg/100 g (Inca Inchi oil). TFC is reported to exhibit a high antioxidant potential and to improve the oil's shelf life. The higher level of TFC in tumba seed oil makes it a good source of dietary antioxidants and improves its shelf life.

Oryzanol is one of the most important phytochemicals in rice bran oil and exhibits important biological activities. Tumba seed oil also contains oryzanol (0.066%), nearly similar to that of *B. rubra* seed oil (0.01% oryzanol content) [33]. The highest oryzanol content is in rice bran oil, which varied from 26.7 to 61.6 mg/100 g (i.e., 0.027 to 0.06%) in different rice varieties [32,50]. It has also been demonstrated that rice bran extract–enriched (1000 ppm) soybean oil is less prone to oxidative degradation during frying than oil enriched with millet and barley bran extracts [51]. This study gives a clue about the effectiveness of oryzanol against the oxidative degradation of oil during frying. Rice bran oil is the richest source of oryzanol, and the value of oryzanol content obtained in tumba seed oil in this study is similar to that of rice bran oil. Therefore, tumba seed oil, due to its high oryzanol content, can be considered as less prone to oxidative degradation during frying and can be used for deep frying.

Lignans are a group of natural compounds derived from the oxidative coupling of β-hydroxyphenylpropane, which includes sesamin, sesamolin, sesaminol, and sesamolinol. These compounds have distinctive bioactivity and physiological and nutritional properties [31,52]. Sesamin has a typical lignan structure of β-β′ (8–8′) linked to the product of two coniferyl alcohol radicals, while sesamolin has a unique structure made up of one acetal oxygen bridge in a sesamin-type structure. Both sesamin and sesamolin are characteristic lignans of sesame seeds [31]. The lignans content of tumba seed oil is reported in Table 3. The lignans content of tumba seed oil (0.012 %) was similar to that of *B. rubra* seed oils (0.02%) [33], and among vegetable oils sesame is the richest source of lignans, containing 0.26 to 1.16 % lignans (1.08%) [52,53].

Carotenoids are a group of more than 750 types of yellow, orange, and red pigments synthesized by plants, algae, and photosynthetic bacteria. These fat-soluble phytochemicals have also gained substantial popularity for food because of their bioactivity as provitamin A and antioxidant potential [54]. The most important dietary carotenoids are the following: α-carotene, β-carotene, zeaxanthin, β-cryptoxanthin, lutein, and lycopene [55]. In our study, we found that the total carotenoid content in tumba seed oil was about 79.5 mg/kg total carotenoids (Table 3) and that they play an appreciable part in the overall antioxidant capacity of the samples. These pigments were also well represented in palm oil (53.5 mg/100 g) and *B. rubra* seed oils (30.5 mg/100 g) [33,44,56,57].

The richness of the natural antioxidants in vegetable oils delays lipid peroxidation and contributes to consumer acceptance of the food products made from them by improving their shelf life [33]. The antioxidants of oils play a key role in preserving their nutritive value and quality by scavenging the free radicals and thus protecting them from lipid peroxidation [57]. The tumba seed oil contained an appreciable amount of antioxidant compounds (about 70.83 mg ascorbic acid eq./100 g; Table 3). High antioxidant activities (evaluated with the phosphomolybdenum assay as in the present study) were previously observed also for *B. rubra* seed oils [33]. Thanks to its appreciable antioxidant activity, the use of tumba seed oil helps increase the quality and shelf life of food products.

4. Conclusions

Plant-based products, containing phytochemical-rich oils, can be useful for the preparation of food and for other uses, suggesting new topics for research and development. In recent times, plant-based food products are gaining popularity among the urban population due to their important functional and nutraceutical properties and the safety of their consumption. Tumba seed oil obtained from the hot region of Rajasthan (India) is mainly constituted of unsaturated fatty acids (about 70%) similarly to other major vegetable oils. Its major part is represented by polyunsaturated fatty acids along with some important odd chain fatty acids which have recently been suggested as essential for human health. It also contained an appreciable amount of natural bioactive compounds with strong antioxidant potential (phenolics, flavonoids, oryzanols, lignana, carotenoids, etc.), which make it more stable against oxidative degradation during frying as well as a very good source of antioxidants with nutraceutical and pharmaceutical applications. Consequently, tumba seed oil, with its good fatty acid profile and good antioxidant activity, has the potential to be a choice for vegetable oil for culinary purposes with a large number of health-promoting properties.

However, before tumba seed oil can be suggested as a vegetable oil for culinary purposes, more research is required to better evaluate its anti-nutritional factors and subsequent effects on human health.

Author Contributions: Conceptualization, M.K.B.; methodology, M.K.B. and P.S.G.; validation, C.R. and P.K.; formal analysis and investigation, M.K.B.; R.K. and A.K.V.; writing— original draft preparation, M.K.B. and J.S.G.; writing—review and editing, C.R., P.K., Y.R., B.B.; supervision, M.K.B. and D.S. All authors have read and agreed to the published version of the manuscript.

Funding: There was no external funding involved in the study.

Data Availability Statement: Data are contained in this article itself.

Conflicts of Interest: The authors declare no conflict of interest.

References

1. Berwal, M.K.; Haldhar, S.M.; Ram, C.; Shil, S.; Kumar, R.; Gora, J.S.; Singh, D.; Samadia, D.K.; Kumar, M.; Mekhemar, M. *Calligonum polygonoides* L. as Novel Source of Bioactive Compounds in Hot Arid Regions: Evaluation of Phytochemical Composition and Antioxidant Activity. *Plants* **2021**, *10*, 1156. [CrossRef] [PubMed]
2. Kumar, P.; Khapte, P.S.; Meghwal, P.R. Genetic Diversity of Vegetables in Arid Region. In *Horticulture Based Integrated Farming System*; Shukla, A.K., Gupta, D.K., Jangid, B.L., Keerthika, A., Noor Mohammad, M.B., Mehta, R.S., Eds.; New India Publishing Agency: New Delhi, India, 2021; pp. 35–41.
3. Javadzadeh, H.R.; Davoudi, A.; Davoudi, F.; Valizadegan, G.; Goodarzi, H.; Mahmoodi, S.; Ghane, M.R.; Faraji, M. *Citrullus colocynthis* as the Cause of Acute Rectorrhagia. *Case Rep. Emerg. Med.* **2013**, *2013*, 652192. [CrossRef] [PubMed]
4. Bikdeloo, M.; Colla, G.; Rouphael, Y.; Hassandokht, M.R.; Soltani, F.; Salehi, R.; Kumar, P.; Cardarelli, M. Morphological and Physio-Biochemical Responses of Watermelon Grafted onto Rootstocks of Wild Watermelon [*Citrullus colocynthis* (L.) Schrad] and Commercial Interspecific Cucurbita Hybrid to Drought Stress. *Horticulturae* **2021**, *7*, 359. [CrossRef]
5. Dane, F.; Liu, J.; Zhang, C. Phylogeography of the bitter apple, *Citrullus colocynthis*. *Genet. Resour. Crop Evol.* **2006**, *54*, 327–336. [CrossRef]
6. Hassanane, M.S.; EL-Fiky, S.; Abd EL-Bbaset, S.A. A genotoxic study of the *Citrullus colocynthis* extract. *Bull. Natl. Res. Cent.* **2001**, *26*, 223–235.
7. Rahimi, R.; Amin, G.; Ardekani, M.R. A review on *Citrullus colocynthis* Schrad.: From traditional Iranian medicine to modern phytotherapy. *J. Altern. Complement. Med.* **2012**, *18*, 551–554. [CrossRef]
8. Berwal, M.K.; Haldhar, S.M.; Ram, C.; Saroj, P.L. Determination of total phenolic & flavonoids and antioxidant activity in *Calligonum polygonoides* L. from Thar Desert. *J. Environ. Biol.* **2021**, *42*, 1347–1354. [CrossRef]
9. Rani, R.; Dushyant, S.; Monika, C.; Jaya, P.Y. Antibacterial activity of twenty different endophytic fungi isolated from *Calotropis procera* and time kill assay. *Clin. Microbiol.* **2017**, *6*, 3. [CrossRef]
10. Abdel-Hassan, I.A.; Mohammeda, S.T. The hypoglycaemic and antihyperglycaemic effect of *Citrullus colocynthis* fruit aqueous extract in normal and alloxan diabetic rabbits. *J. Ethnopharmacol.* **2000**, *71*, 325–330. [CrossRef]
11. Ardekani, M.R.S.; Rahimi, R.; Javadi, B.; Abdi, L.; Khanavi, M. Relationship between temperaments of medicinal plants and their major chemical compounds. *J. Tradit. Chin. Med.* **2011**, *31*, 27–31. [CrossRef]
12. Huseini, H.F.; Darvishzadeh, F.; Heshmat, R.; Jafariazar, Z.; Raza, M.; Larijani, B. The clinical investigation of *Citrullus colocynthis* (L.) schrad fruit in treatment of type II diabetic patients: A randomized, double blind, placebo-controlled clinical trial. *Phytother. Res.* **2009**, *23*, 1186–1189. [CrossRef]
13. Rahbar, A.R.; Nabipour, I. The hypolipidemic effect of *Citrullus colocynthis* on patients with hyperlipidemia. *Pak. J. Biol. Sci.* **2010**, *13*, 1202–1207. [CrossRef] [PubMed]
14. Marzouk, B.; Marzouk, Z.; Haloui, E.; Fenina, N.; Bouraoui, A.; Aouni, M. Screening of analgesic and anti-inflammatory activities of *Citrullus colocynthis* from southern Tunisia. *J. Ethnopharmacol.* **2010**, *128*, 15–19. [CrossRef]
15. Tannin-Spitz, T.; Bergman, M.; Grossman, S. Cucurbitacin glucosides: Antioxidant and free-radical scavenging activities. *Biochem. Biophys. Res. Commun.* **2007**, *364*, 181–186. [CrossRef] [PubMed]
16. Nmila, R.; Gross, R.; Rchid, H.; Roye, M.; Manteghetti, M.; Petit, P.; Tijane, M.; Ribes, G.; Sauvaire, Y. Insulinotropic effect of *Citrullus colocynthis* fruit extracts. *Planta Med.* **2000**, *66*, 418–423. [CrossRef] [PubMed]
17. Dehghani, F.; Panjehshahin, M.R. The toxic effect of alcoholic extract of *Citrullus colocynthis* on rat liver. *Iran. J. Pharm. Ther.* **2006**, *5*, 117–119.
18. Jouad, H.; Haloui, M.; Rhiouani, H.; El Hilaly, J.; Eddouks, M. Ethnobotanical survey of medicinal plants used for the treatment of diabetes, cardiac and renal diseases in the North centre region of Morocco (Fez-Boulemane). *J. Ethnopharmacol.* **2001**, *77*, 175–182. [CrossRef]
19. Kamalakar, K.; Sai-Manoj, G.N.V.T.; Prasad, R.B.N.; Karuna, M.S.L. Thumba (*Citrullus colocynthis* L.) seed oil: A potential bio-lubricant base-stock. *Grasas Aceites* **2015**, *66*, e055. [CrossRef]

20. Mariod, A.A.; Mirghani, M.E.S.; Hussein, I. Citrullus colocynthis Colocynth, Bitter Apple, Bitter Gourd. In *Unconventional Oilseeds and Oil Sources*; Mariod, A.A., Mirghani, M.E.S., Hussein, I., Eds.; Academic Press: Cambridge, MA, USA, 2017; pp. 99–105. [CrossRef]
21. Ashish, K.; Naveen, K.; Hasan, M.M.; Rajeev, C.; Arshad, N.S.; Zahid, A.K. Production of biodiesel from thumba oil: Optimization of process parameters. *Iran. J. Energy Environ.* **2010**, *1*, 352–358.
22. Gurudeeban, S.; Satyavani, K.; Ramanathan, T. Bitter apple (*Citrullus colocynthis*): An overview of chemical composition and biomedical potentials. *Asian J. Plant Sci.* **2010**, *9*, 394–401. [CrossRef]
23. Kulkarni, A.S.; Khotpal, R.R.; Karadbhajane, V.Y.; More, V.I. Physico-chemical Composition and lipid classes of *Aegle marmelos* (Bael) and *Citrullus colocynthis* (Tumba) Seed Oils. *J. Chem. Pharm. Res.* **2012**, *4*, 1486–1488.
24. Solomon, G.; Luqman, C.A.; Nor, M.A. Investigating "Egusi" (*Citrullus colocynthis* L.) Seed Oil as Potential Biodiesel Feedstock. *Energies* **2010**, *3*, 607–618. [CrossRef]
25. Yaniv, Z.; Shabelsky, E.; Schafferman, D. Perspectives on new crops and new uses. In *Colocynth: Potential Arid Land Oilseed from an Ancient Cucurbit*; Janick, J., Ed.; ASHS Press: Alexandria, VA, USA, 1999; pp. 257–261.
26. Al-Hwaiti, M.S.; Alsbou, E.M.; Abu Sheikha, G.; Bakchiche, B.; Pham, T.H.; Thomas, R.H.; Bardaweel, S.K. Evaluation of the anticancer activity and fatty acids composition of "Handal" (*Citrullus colocynthis* L.) seed oil, a desert plant from south Jordan. *Food Sci. Nutr.* **2021**, *9*, 282–289. [CrossRef]
27. Bireche, M.; Gherib, A.; Bakchiche, B.; Berrabah, M.; Maatallah, M. Positional distribution of fatty acids in the triglycerides of *Citrullus colocynthis* seed oil growing in Algeria. *J. Mater. Environ. Sci.* **2017**, *8*, 622–627.
28. AOCS. *Official Methods and Recommended Practices of the American Oil Chemist's Society*; AOCS: Champaign, IL, USA, 2003.
29. Berwal, M.K.; Haldhar, S.M.; Ram, C.; Shil, S.; Gora, J.S. Effect of extraction solvent on total phenolics, flavonoids and antioxidant capacity of flower bud and foliage of *Calligonum polygonoides* L. *Indian J. Agric. Biochem.* **2021**, *34*, 61–67. [CrossRef]
30. Medini, F.; Fellah, H.; Ksouri, R.; Abdelly, C. Total phenolic, flavonoid and tannin contents and antioxidant and antimicrobial activities of organic extracts of shoots of the plant *Limonium delicatulum*. *J. Taibah Univ. Sci.* **2014**, *8*, 216–224. [CrossRef]
31. Bhatnagar, A.S.; Hemavathy, J.; Gopala Krishna, A.G. Development of a rapid method for determination of lignans content in sesame oil. *Food Sci. Technol.* **2015**, *52*, 521–527. [CrossRef]
32. Gopala Krishna, A.G.; Hemakumar, K.H.; Khatoon, S. Study on the composition of rice bran oil and its higher free fatty acids value. *J. Am. Oil Chem. Soc.* **2006**, *83*, 117–120. [CrossRef]
33. Kumar, S.S.; Manasa, V.; Tumaney, A.W.; Bettadaiah, B.K.; Chaudari, S.R.; Giridhar, P. Chemical composition, nutraceuticals characterization, NMR confirmation of squalene and antioxidant activities of *Basella rubra* L. seed oil. *RSC Adv.* **2020**, *10*, 31863–31873. [CrossRef]
34. Prieto, P.; Pineda, M.; Aguilar, M. Spectrophotometric Quantitation of Antioxidant Capacity through the Formation of a Phosphomolybdenum Complex: Specific Application to the Determination of Vitamin, E. *Anal. Biochem.* **1999**, *269*, 337–341. [CrossRef]
35. Kostik, V.; Memeti, S.; Bauer, B. Fatty Acid Composition of Edible Oils and Fats. *J. Hyg. Eng. Design* **2013**, *4*, 112–116.
36. Zambiazi, R.C.; Przybylski, R.; Zambiazi, M.W.; Mendonca, C.B. Fatty Acid Composition of Vegetable Oils and Fats. *Bol. Cent. Pesqui. Processamento Aliment.* **2007**, *25*, 111–120.
37. Orsavova, J.; Misurcova, L.; Ambrozova, J.V.; Vicha, R.; Mlcek, J. Fatty Acids Composition of Vegetable Oils and Its Contribution to Dietary Energy Intake and Dependence of the Cardiovascular Mortality on Dietary Intake of Fatty Acids. *Int. J. Mol. Sci.* **2015**, *16*, 12871–12890. [CrossRef] [PubMed]
38. Venn-Watson, S.; Lumpkin, R.; Dennis, E.A. Efficacy of dietary odd-chain saturated fatty acid pentadecanoic acid parallels broad associated health benefits in humans: Could it be essential? *Sci. Rep.* **2020**, *10*, 8161. [CrossRef] [PubMed]
39. Jimenez-Cepeda, A.; Davila-Said, G.; Orea-Tejeda, A.; González-Islas, D.; Elizondo-Montes, M.; Pérez-Cortes, G.; Keirns-Davies, C.; Castillo-Aguilar, L.F.; Verdeja-Vendrell, L.; Peláez-Hernández, V.; et al. Dietary intake of fatty acids and its relationship with FEV_1/FVC in patients with chronic obstructive pulmonary disease. *Clin. Nutr. ESPEN* **2019**, *29*, 92–96. [CrossRef]
40. Huang, L.; Lin, J.S.; Aris, I.M.; Yang, G.; Chen, W.Q.; Li, L.J. Circulating Saturated Fatty Acids and Incident Type 2 Diabetes: A Systematic Review and Meta-Analysis. *Nutrients* **2019**, *11*, 998. [CrossRef] [PubMed]
41. Aglago, E.K.; Biessy, C.; Torres-Mejia, G.; Angeles-Llerenas, A.; Gunter, M.J.; Romieu, I.; Chajes, V. Association between serum phospholipid fatty acid levels and adiposity in Mexican women. *J. Lipid Res.* **2017**, *58*, 1462–1470. [CrossRef]
42. Kurotani, K.; Sato, M.; Yasuda, K.; Kashima, K.; Tanaka, S.; Hayashi, T.; Shirouchi, B.; Akter, S.; Kashino, I.; Hayabuchi, H.; et al. Even- and odd-chain saturated fatty acids in serum phospholipids are differentially associated with adipokines. *PLoS ONE* **2017**, *12*, e0178192. [CrossRef]
43. Bravo, L. Polyphenols: Chemistry, dietary sources, metabolism, and nutritional significance. *Nutr. Rev.* **1998**, *56*, 317–333. [CrossRef]
44. Thusoo, S.; Gupta, S.; Sudan, R.; Kour, J.; Bhagat, S.; Hussain, R.; Bhagat, M. Antioxidant Activity of Essential Oil and Extracts of Valeriana jatamansi Roots. *BioMed Res. Int.* **2014**, *2014*, 614187. [CrossRef]
45. Chang, S.T.; Wu, J.H.; Wang, S.Y.; Kang, P.L.; Yang, N.S.; Shyur, L.F. Antioxidant activity of extracts from *Acacia confusa* Bark and Heartwood. *J. Agric. Food Chem.* **2001**, *49*, 3420–3424. [CrossRef] [PubMed]
46. Janu, C.; Kumar, D.S.; Reshma, M.V.; Jayamurthy, P.; Sundaresan, A.; Nisha, P. Comparative Study on the Total Phenolic Content and Radical Scavenging Activity of Common Edible Vegetable Oils. *J. Food Biochem.* **2014**, *38*, 38–49. [CrossRef]

47. Kozlowska, M.; Gruczynska, E.; Scibisz, I.; Rudzinska, M. Fatty acids and sterols composition, and antioxidant activity of oils extracted from plant seeds. *Food Chem.* **2016**, *213*, 450–456. [CrossRef] [PubMed]
48. Harborne, J.B.; Baxter, H.; Moss, G.P. *Phytochemical Dictionary: Handbook of Bioactive Compounds from Plants*, 2nd ed.; Taylor & Francis: London, UK, 1999.
49. Xuan, T.D.; Gangqiang, G.; Minh, T.N.; Quy, T.N.; Khanh, T.D. An Overview of Chemical Profiles, Antioxidant and Antimicrobial Activities of Commercial Vegetable Edible Oils Marketed in Japan. *Foods* **2018**, *7*, 21. [CrossRef] [PubMed]
50. Kim, H.W.; Kim, J.B.; Cho, S.M.; Cho, I.K.; Li, O.X.; Jang, H.H.; Lee, S.H.; Lee, Y.M.; Hwang, K.A. Characterization and quantification of γ-oryzanol in grains of 16 Korean rice varieties. *Int. J. Food. Sci. Nutr.* **2015**, *66*, 166–174. [CrossRef] [PubMed]
51. Ajala, A.W.; Ghavami, A. Evaluation of the effectiveness of cereal bran extract for sunflower oil stability during frying. *Int. J. Food Stud.* **2020**, *9*, SI52–SI61. [CrossRef]
52. Moazzami, A.A.; Kamal-Eldin, A. Sesame seed is a rich source of dietary lignans. *J. Am. Oil Chem. Soc.* **2006**, *83*, 719–723. [CrossRef]
53. Reshma, M.V.; Balachandran, C.; Arumughan, C.; Sunderasan, A.; Sukumaran, D.; Thomas, S.; Saritha, S.S. Extraction, separation and characterisation of sesame oil lignan for nutraceutical applications. *Food Chem.* **2010**, *120*, 1041–1046. [CrossRef]
54. Ribayamercado, J.D.; Solon, S.F.; Tang, G.; Cabal-Borza, M.; Perfecto, S.C.; Russel, R.M. Bioconversion of plant carotenoids to vit-A in Filipino school-aged children varies inversely with vit-A status. *Am. J. Clin. Nutr.* **2000**, *72*, 455–465. [CrossRef]
55. Wang, X.D. Carotenoids. In *Modern Nutrition in Health and Disease*, 11th ed.; Ross, C.A., Caballero, B., Cousins, R.J., Tucker, K.L., Ziegler, T.R., Eds.; Lippincott Williams & Wilkins: Philadelphia, PA, USA, 2014; pp. 427–439.
56. Chandrasekaram, K.; Ng, M.H.; Choo, Y.M.; Chuah, C.H. Effect of storage temperature on the stability of phytonutrients in palm concentrates. *Am. J. Appl. Sci.* **2009**, *6*, 529–533. [CrossRef]
57. Hemalatha, S.; Ghafoorunissa. Sesame lignans enhance the thermal stability of edible vegetable oils. *Food Chem.* **2007**, *105*, 1076–1085. [CrossRef]

horticulturae

MDPI

Article

Effects of Yeast Product Addition and Fermentation Temperature on Lipid Composition, Taste and Mouthfeel Characteristics of Pinot Noir Wine

Quynh Phan, Aubrey DuBois, James Osborne and Elizabeth Tomasino *

Department of Food Science and Technology, Oregon State University, Corvallis, OR 95331, USA; quynh.phan@oregonstate.edu (Q.P.); aubrey.dubois@oregonstate.edu (A.D.); james.osborne@oregonstate.edu (J.O.)
* Correspondence: elizabeth.tomasino@oregonstate.edu; Tel.: +1-541-737-4866

Abstract: Lipids have important impacts on wine sensory. By targeting the lipid sources in wine, mainly from grape tissues and yeast cell walls, it was possible to alter the wine lipid profile thus potentially changing the final product quality. This research examined the changes of wine total lipids, lipid composition and sensory characteristics of Pinot noir wines in response to the winemaking factors, fermentation temperature and yeast product addition. Pinot noir grapes were fermented at 16 °C and 27 °C. After fermentation, Oenolees® yeast product was added to the wines at three levels (0 g/L, 0.5 g/L and 1.0 g/L). The six wine treatments were subjected to chemical analyses measuring total lipids and an untargeted lipidomic approach analyzing lipid composition. High temperature fermentation wines had significantly higher total lipid content. Random forest analysis distinguished the wine groups based on the 25 main lipids, including free fatty acids, acylcarnitines, diglycerides, triglycerides and phospholipids. Taste and mouthfeel characteristics of each treatment were assessed using descriptive analysis and check-all-that-apply (CATA) techniques. Multivariate analyses showed that changing fermentation temperature significantly impacted sweetness and drying perception in Pinot noir wines. Yeast product addition had nuanced effects on wine lipid profiles and sensory perception.

Keywords: lipidomic; winemaking; descriptive analysis; random forest; linear discriminant analysis

Citation: Phan, Q.; DuBois, A.; Osborne, J.; Tomasino, E. Effects of Yeast Product Addition and Fermentation Temperature on Lipid Composition, Taste and Mouthfeel Characteristics of Pinot Noir Wine. *Horticulturae* **2022**, *8*, 52. https://doi.org/10.3390/horticulturae8010052

Academic Editors: Alessandra Durazzo and Massimo Lucarini

Received: 16 November 2021
Accepted: 5 January 2022
Published: 6 January 2022

1. Introduction

Mouthfeel is an important aspect of wine quality but is not well understood. Studies of wine chemical contributions in mouthfeel perception have mainly focused on major constituents, such as phenolics, polysaccharides, proteins, ethanol, sugar, and acids [1–5]. However, there are other compositional elements that may influence perception of wine mouthfeel and taste [6,7]. Altering wine mouthfeel and taste qualities during winemaking processes may result in positive outcomes in consumers' acceptance and preference of wines [4,8].

This study was focused on winemaking practices that could potentially impact wine lipids and wine taste and mouthfeel perception. Lipids are minor constituents with a concentration reported to be less than 0.1% in commercial Pinot noir wines [9]. There are two main sources of wine lipids: grape tissues and yeast cell wall [10]. Previous work has demonstrated lipids contribute to taste and mouthfeel perception in model wine solutions [7]. Consumer perception of viscosity was shown to increase significantly when a phospholipid product was added to model wines [7]. Real wine matrices are much more complex than model wine solutions and the increased complexity may alter the effects of lipids on taste and mouthfeel. Therefore, it is important to study the roles of lipids in a real wine system to further understand how wine lipids affect wine taste and mouthfeel characteristics. This study evaluated lipids in real wine, attempting to alter the composition using processing aids. By evaluating lipids in wine, it is possible to see if the low composition of lipids can influence taste and mouthfeel, which is normally driven by more major compositional aspects of wine. It is most likely that the smaller compositional wine aspects will cause the more nuanced changes to mouthfeel that are regularly described, but are not linked to a specific compound.

Pinot noir wines are known to contain low concentrations of tannins and polyphenolic compounds compared to other red wines [11]. These compounds are highly correlated with the astringency mouthfeel attribute in red wine [12]. Pinot noir is an ideal wine for investigating non-astringent mouthfeel compounds due to its lower tannin concentration, as the low concentration of lipids may be able to alter or overcome the mouthfeel aspects attributed to phenolic compounds. In addition, Pinot noir is a flagship varietal of Oregon, dominating the premium and ultra-premium market [13]. Understanding the factors that could maintain or improve Pinot noir wine quality is important due to the many challenges that face winemakers each year.

Changes in winemaking procedures may lead to changes in the quality of the final product [14–17]. Unterkofler et al. (2020) listed fermentation as one of the most critical value-adding steps in the entire winemaking process [17]. Fermentation temperature is an important factor that determines the quality of the final wine. Environmental temperature is known to influence the lipid composition of yeast cells [18]. Yeasts have the ability to adapt to environmental stress, such as changes in osmotic pressure, pH, nutrient levels, exposure to heat or cold shock, high ethanol concentration or toxic compounds [19,20]. Lower fermentation temperatures can increase the total lipid content, total fatty acids, triacylglycerols and phospholipid content of yeast cells while diacylglycerols, free fatty acids, sterols and sterol ester concentrations remain unaffected [18]. To date, studies investigating fermentation temperature impacts on wine mouthfeel have focused only on phenolic compounds and not lipid contents of the wine [21,22]. In this study, lipid profiles, chemical components and sensory analysis of wines produced under different fermentation temperatures was performed to determine if fermentation temperature could influence the lipid content and sensory characteristics of wine.

The addition of external yeast products was also considered as yeast products contain several lipids and are commonly used in winemaking [23,24]. Many commercially-available yeast products claim to impart positive taste and mouthfeel characteristics onto the wine, as a way of altering quality post-harvest.

Yeast cell walls mainly consist of polysaccharides (mannoproteins, b-glucan, and chitin) [25]. In addition, lipids are critical components of yeast cell membranes with the concentration varying from less than 5% dry weight to higher than 15% dry weight depending on the species [26]. Lipids play important roles in the maintenance of yeast cell structure and controlling the growth rate [27]. The present study aims were to determine the impacts of adding different concentrations of a lipid source and the relationship between lipid additions and fermentation temperatures on the chemical and sensory characteristics of wine. Understanding these relationships could provide winemakers with additional options to achieve their desired wine quality.

2. Materials and Methods

2.1. Chemicals

A mixture of lipid standards (Splash® Lipidomix® Mass Spec Standard) was purchased from Avanti Polar Lipids, Inc. (Alabaster, AL, USA). Potassium chloride (KCl) ACS reagent grade was from EMD Chemicals Inc. (Gibbstown, NJ, USA). Sodium hydroxide (NaOH) was from Sigma–Aldrich (St. Louis, MO, USA). Dichloromethane (DCM) was from EMD Millipore Corporation (Burlington, MA, USA). 2-propanol (IPA) HPLC grade, methanol (MeOH) HPLC grade, and chloroform ($CHCl_3$) HPLC grade were from Fisher Chemical (Fair Lawn, NJ, USA). Milli-Q water was obtained from a Millipore Continental water system (EMD-Millipore, Billerica, MA, USA). Potassium metabisulfite was sourced from Institut Œnologique de Champange (Mardeuil, Grand Est, France).

2.2. Winemaking

A winemaking plan was designed to study the interactions and effects of temperature and concentration of yeast product additions on lipid composition of Pinot noir wines. Pinot noir wines were made from *V. vinifera* cv. Pinot noir grapes harvested from Oregon State University's Woodhall Vineyard (Monroe, OR, USA) in September 2019. Wines were processed at Oregon State University's research winery (Corvallis, OR, USA). After destemming, grapes were split equally into six jacketed tanks (AAA Metal Fabrication, Dalles, OR, USA) with temperature control systems (Watlow EZ-Zone, St. Louis, MO, USA). Approximately 60 kg of grapes were placed in each tank. Three tanks were randomly selected to be kept at 16 °C and the other three were maintained at 27 °C during alcoholic fermentation. Fermaid K™ (Lallemand Oenology, Montreal, Canada) was added at a rate of 0.4 g/L. An addition of 50 mg/L of SO_2 (as potassium metabisulfite) was added to all the tanks and mixed. After 20 min, *Saccharomyces cerevisiae* ZYMAFLORE® F15 (Laffort USA, Petaluma, CA, USA) was then added at a rate of 0.25 g/L after hydration according to the manufacturer's specifications. Alcoholic fermentation was monitored by changes in degree Brix over time using a digital densitometer (Anton Paar, Santner Foundation, Graz, Austria) (Figure S1).

At the completion of fermentation, treatments were pressed at 0.1 MPa for 5 min and the wine was dispensed into five-gallon (18.9 L) glass carboys and kept at room temperature (approximately 21 °C). *Oenococcus oeni* Lalvin VP41™ (Lallemand Oenology, Montreal, Canada) was inoculated at approximately 1×10^6 cfu/mL to induce malolactic fermentation (MLF) following manufacturer's instructions. After MLF (malic acid less than 100 mg/L as measured by enzymatic assay (Vintessential, Victoria, Australia), the yeast product Oenolees® (Laffort USA, Petaluma, CA, USA) was added to wines at two concentrations (low concentration at 0.5 g/L and high concentration at 1.0 g/L). The Oenolees® yeast product consisted of yeast cell walls and inactivated yeasts. The amount of yeast product addition was determined based on a preliminary study (data not shown). Controls with no yeast product addition were also obtained. After the addition of Oenolees®, all treatments were kept at 16 °C. Free sulfur dioxide (SO_2) was checked weekly using the aspiration method outlined by Iland et al. and 10% (*w*/*v*) potassium metabisulfite solution was added to the wines to maintain the concentration of free SO_2 at 30 mg/L [28]. All treatments were performed in triplicate. After 90 days, samples from each replicate were taken for total lipids and lipidomic analyses. Replicates were then combined, sterile filtered (0.45 µm PES cartridge filter) and bottled in 750 mL green glass bottles sparged with nitrogen and sealed with aluminum screw cap closures (Stelvin®, Amcor, Australia) for later sensory analyses. Detailed winemaking treatments are summarized in Figure 1.

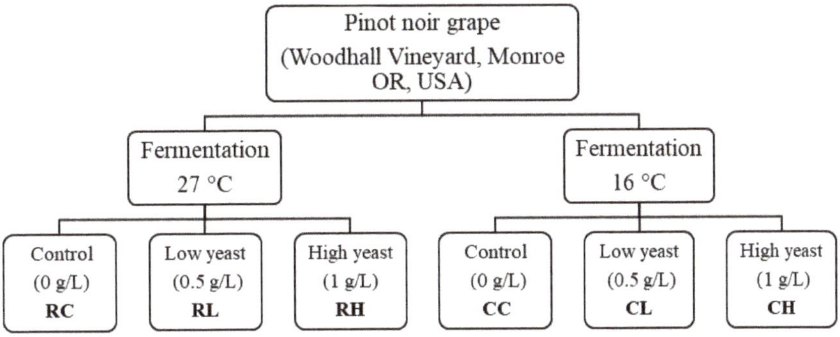

Figure 1. Pinot noir wine treatments scheme. (RC = fermentation at 27 °C, 0 g/L yeast product addition; RL = fermentation at 27 °C, 0.5 g/L yeast product addition; RH = fermentation at 27 °C, 1 g/L yeast product addition; CC = fermentation at 16 °C, 0 g/L yeast product addition; CL = fermentation at 16 °C, 0.5 g/L yeast product addition; CH = fermentation at 16 °C, 1 g/L yeast product addition).

2.3. Basic Chemistry Analyses

pH was determined by ion-selective electrode (ThermoFisher Scientific, MA, USA) and titratable acidity (TA) was measured by titration with 0.1 M NaOH [29]. Glucose/fructose, malic acid, and acetic acid were measured by enzymatic test kits (r-Biopharm, Darmstadt, Germany), while ethanol was determined using an Alcolyzer (Anton Paar, Santner Foundation, Graz, Austria). Total phenolics analysis was carried out following the Folin Ciocalteau analysis procedure [30]. Measurements were carried out in triplicate.

2.4. Lipid Analyses

Lipid extraction was carried out in triplicate for each treatment. A detailed procedure is described in Phan and Tomasino (2021). In summary, wine samples were concentrated and subjected to the liquid/liquid extraction method in a ratio of 1:1:0.9 *v*/*v* $CHCl_3$:MeOH:concentrated wine. The total lipids layer was extracted and pure lipids were collected after evaporating the organic solvent.

An untargeted lipidomic approach was used to obtain the lipidomic profiles of the wine samples. A solvent mixture of 25:10:65 *v*/*v* DCM:IPA:MeOH with 0.1% (*w*/*v*) BHT was used to dissolve the total lipid extraction. Waters Acquity UPLC system (Milford, MA, USA) coupled to a quadrupole TOF mass spectrometer AB SCIEX, TripleTOF 5600 (Sciex, Concord, ON, Canada) operated in information-dependent MS/MS acquisition mode to analyze the samples. A 10% (*v*/*v*) Splash standard in MeOH was used as the internal standard. Lipidomic data was processed using MasterView software (AB SCIEX, Framingham, MA, USA). PeakView workstation (SCIEX, Framingham, MA, USA) was used to identify lipid compounds. Compound identification was based on the exact mass, retention time, the detection of protonated molecular ion of m/z and the fragment ion peaks [9]. MultiQuant software (SCIEX, Framingham, MA, USA) was used to obtain peak areas. The peak area of each compound was used for semi-quantitation of relative abundancy of lipid species [31].

2.5. Sensory Setup

The Institutional Review Board at Oregon State University reviewed and approved this sensory study (IRB-2020-0610) at Oregon State University (Corvallis, OR, USA). The sensory panel occurred in the Arbuthnot Dairy Lab on 21 February 2021. The testing room was kept at a constant temperature of 22 °C with a stable light source consisting of both natural and artificial light. Each participant was assigned to separate white plastic booths. For all tests, 20 mL of sample were served at room temperature (21 ± 2 °C) in INAO black wineglasses (Lehmann glass, Kiyasa group, New York, NY, USA) labeled with three-digit random codes and covered with PET disposable plastic lids (Dart®, Mason, MI, USA).

The wine treatments CC, CL, CH, RC, RL and RH were evaluated in duplicate. The wine samples were served in random order. Panelists were instructed to wear nose clips (Biotronics.biz, Davie, FL, USA) during the assessment. For each wine sample, panelists swished the wine around in their mouth and expectorated into a spit cup before answering the questions described in Sections 2.6 and 2.7. After finishing each sample, they were asked to rinse their palates with a 1 g/L pectin rinsing solution in order to prevent carry-over effects from previous samples [27]. Food grade pectin was purchased from Modernist pantry (Eliot, ME, USA). Participants were required to take a one-minute break between each test.

2.6. Panelists

25 panelists (9 men, 16 women) from 21 to over 60 years old participated in this sensory study. Inclusion criteria was as follows: 21 years old or above, free of allergies to wine or wine components and were regular wine consumers (consuming on average 1 glass or more of wine per week); non-smoker; no taste deficits or other oral disorders; no piercings of the tongue, lip or cheek; and no oral lesions or canker sores. Panelists signed informed consent forms before taking the assessment.

2.7. Descriptive Analysis

Panelists were instructed to evaluate sweetness, bitterness, acidity, astringency, viscosity, warmth and drying attributes on a 100 mm visual analog scale (VAS). The definition of each term was provided during testing and are shown in Table 1. The 100 mm visual analog scales were used for rating intensity each taste and mouthfeel descriptor. The intensity ranged from none (not sweet/not bitter/not acidic/not astringent/not viscous/not warm/not dry) to extreme (very sweet/very bitter/very acidic/very astringent/very viscous/very warm/very dry).

Table 1. Definitions of the taste and mouthfeel attributes used for descriptive analysis.

Attribute	Definition
Sweetness	Being one of the five basic taste sensations that is usually pleasing to the taste and typically induced by sugar. In beverage, containing a sweet ingredient is equivalent to not dry [32].
Bitterness	Intensity of bitter taste perceived in the mouth [33].
Acidity	Intensity of the acid taste perceived in the mouth [34].
Astringency	Intensity of the drying and mouth puckering sensation in the mouth [33].
Viscosity	Perception of body, weight, or thickness of the wine in the mouth [33].
Warmth	Warming effect of the mouth surfaces primarily due to alcohol [35].
Drying	Feeling of lack of lubrication or moisture in the mouth [35].

2.8. Check-All-That-Apply (CATA)

The descriptors selected for the CATA test were: irritating, grainy, plastic, sticky, greasy, watery, harsh, gummy, velvety, unripe, thin, oily, chewy and soft [35,36]. These attributes are associated with non-astringency mouthfeel descriptors in wine. Panelists were asked to select all terms that described the mouthfeel characteristics of each wine sample.

2.9. Statistical Analyses

2.9.1. Wine Chemistry

Two-way ANOVA and Tukey's honest significant difference (HSD) post hoc were used to compare the mean differences of total lipids, peak intensities of lipid compounds and basic wine chemistry attributes (pH, tartaric acid, glucose/fructose, acetic acid, total phenolics, and malic acid) measured in wines produced at two different temperatures and three levels of yeast product addition. The analyses were performed to examine the main effects and interaction effects of fermentation temperature and concentration of yeast product on wine lipids and other wine chemical parameters. Significance was reported at $\alpha = 0.05$.

The random forest (RF) approach and linear discriminant analysis (LDA) were used to analyze the peak intensity of all lipid compounds found in each wine sample. They were used to identify and visualize the lipid compounds important to distinguishing wines produced in different conditions: CC, CL, CH, RC, RL and RH.

All statistical analyses and figures were carried out using R programming language [37]. The package ggplot2 was used to produce all figures [38]. R packages used for RF analysis were randomForest and varSelRF.

2.9.2. Wine Sensory

Two-way ANOVA was performed on the mean intensity ratings for all seven taste and mouthfeel attributes listed in Table 1 to identify the main effects and interaction effect of temperature and yeast product concentration on wine taste and mouthfeel descriptors. Correspondence analysis (CA) followed by hierarchical cluster analysis (HCA) were used to analyze CATA data [39]. FactoMineR was used for CATA analysis and factoextra for ggplot2-based elegant visualization [40,41].

3. Results

3.1. Basic Chemistry

Changes in fermentation temperature alone impacted wine chemical components. The basic chemistry attributes showed significant differences among wine treatments, including total phenolics, wine pH, TA and alcohol content (Table 2). When looking at the fermentation temperature main effect, wines fermented at 27 °C resulted in significantly higher total phenolics, lower pH, lower TA and lower alcohol content compared to wines fermented at 16 °C ($\alpha = 0.05$, Table S1).

Table 2. Mean basic chemistry measurements of the wine treatments—mean (±SE) with factor and interaction significance from Tukey HSD. Letters refer to Tukey groups.

Fermentation Temperature (°C)	**16**			**27**			***p*-Value (α = 0.05)**		
Yeast Product Concentration (g/L)	**0**	**0.5**	**1**	**0**	**0.5**	**1**	**Temp.**	**Yeast Product**	**Temp × Yeast Product**
Total phenol (mg/L)	107.76 [a] (1.91)	92.44 [b] (1.54)	101.14 [a,b] (1.71)	138.06 [c] (1.96)	135.66 [c] (2.26)	136.36 [c] (2.86)	<0.001 ***	0.001 **	0.014 **
pH	3.67 [a] (0.01)	3.70 [b] (0.00)	3.73 [c] (0.00)	3.67 [a] (0.01)	3.66 [a] (0.00)	3.67 [a] (0.01)	<0.001 ***	0.008 **	0.003 **
Titratable acidity (g/L)	2.46 [a] (0.01)	2.48 [a,b] (0.00)	2.50 [b,c] (0.00)	2.46 [a] (0.00)	2.45 [a] (0.00)	2.46 [a] (0.01)	<0.001 ***	0.008 **	0.003 **
Alcohol (% *v*/*v*)	14.47 [a] (0.01)	14.83 [b] (0.01)	14.38 [a] (0.01)	13.60 [c] (0.02)	14.10 [d] (0.02)	13.89 [e] (0.02)	<0.001 ***	<0.001 ***	<0.001 ***
Acetic acid (g/L)	0.39 [a] (0.03)	0.34 [a] (0.01)	0.39 [a] (0.01)	0.43 [a] (0.07)	0.42 [a] (0.02)	0.38 [a] (0.03)	0.252	0.686	0.404
Malic acid (g/L)	0.07 [a] (0.00)	0.06 [a,b] (0.00)	0.06 [a,b] (0.00)	0.07 [a] (0.00)	0.07 [a,b] (0.00)	0.06 [a,b] (0.00)	0.195	0.011 **	0.932
Glucose/Fructose (g/L)	0.56 [a] (0.08)	0.63 [a] (0.02)	0.58 [a] (0.03)	0.70 [a] (0.03)	0.65 [a] (0.02)	0.62 [a] (0.03)	0.055	0.558	0.24

** Significant at $\alpha = 0.01$, *** Significant at $\alpha = 0.001$.

Yeast product addition as a main effect showed significant impacts on wine measurements. Wines with low (0.5 g/L) yeast product had significantly lower total phenolics compared to the control group. Significantly higher pH, higher titratable acidity, and lower malic acid were found in wines with high (1 g/L) yeast product addition compared to the control group (no yeast product). The three groups were distinctively different only in alcohol content, in which the low (0.5 g/L) yeast product addition group had the highest (14.83% *v*/*v*), followed by the control group, and the high group (1 g/L) had the lowest ethanol (14.38% *v*/*v*).

An interaction effect between fermentation temperature and yeast product addition was observed in total phenols, pH, alcohol content and titratable acidity. Changes in both fermentation temperature and yeast product addition resulted in variation in the wine chemical composition.

3.2. Total Lipids

A significant difference ($\alpha = 0.05$) was found for the average total lipids (% *w*/*w*) in Pinot noir wines produced under two fermentation temperature conditions 16 °C and 27 °C (Table 3). Tukey's HSD showed no significant difference for the interaction of fermentation temperature × yeast product addition or no significant difference in total lipids for yeast product addition (Table 2). Wine fermented at 27 °C resulted in higher average total lipids extraction (0.083% ± 0.006%). At 16 °C, the average total lipids extracted were 0.075% ± 0.004%. There were no significant differences found among the three levels of the yeast product concentration or the fermentation temperature and yeast product concentration interaction. The total lipids extracted were at the average of 0.079% ± 0.005%, for all yeast product addition treatments including the controls. Figure 2 shows the distribution of total lipids (% *w*/*w*) in different wine treatments.

Table 3. Two-way ANOVA table for fermentation temperature and yeast product addition factorial experiment.

Source	Df	Sum of Square	Mean Square	F Value	Pr(>F)	
Fermentation Temperature	1	2.827×10^{-8}	2.827×10^{-8}	10.014	0.00815	*
Yeast Product Addition	2	1.600×10^{-10}	8.000×10^{-10}	0.028	0.97225	
Fermentation Temperature × Yeast Product Addition	2	5.650×10^{-9}	2.827×10^{-9}	1.001	0.39612	
Residuals	12	3.387×10^{-8}	2.823×10^{-9}			

* Significant at $\alpha = 0.05$.

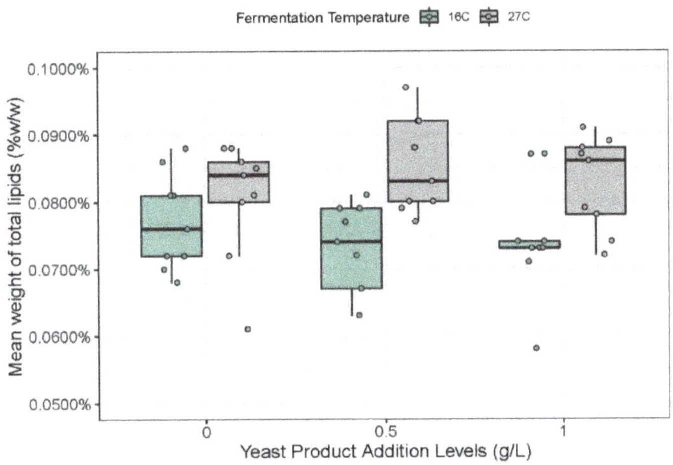

Figure 2. Boxplot of the mean of total lipids extracted from Pinot noir wines produced at two different fermentation temperatures (16 °C and 27 °C) and three yeast product concentrations (0 g/L, 0.5 g/L, and 1 g/L).

3.3. Lipid Profiles

There were 13 lipid classes detected in the wine samples. A total of 233 individual lipids were identified from the extracts of the 54 Pinot noir wine samples (6 treatments × 3 wine treatment replicates × 3 lipid extraction replicates). Of the 233 identified lipids, 142 lipid species were found to have significant differences in peak intensities among the wines that fermented at different temperatures; 127 lipids species were found to have significant differences in peak intensities among the wines that had different levels of yeast product addition. The interaction between fermentation temperature and yeast product concentration had a significant impact on 101 lipids ($\alpha = 0.05$, Table S2).

Lipid species that had the highest peak areas belong to the following three lipid classes: TG, PC and FFA (Table S3. Overall, TG with unsaturated 12, 14, 16 and/or 18 carbon chains had the greatest peak areas. The identified peaks belong to TG(14:0_18:1_18:2) combined with TG(16:0_16:1_18:2), TG(16:0_18:2_18:3), TG(16:0_16:0_18:31) and TG(12:0_14:0_18:1) combined with TG(12:0_16:0_16:1). The high peak areas for TG with carbon lengths from 12 to 18 were found among all six wine groups. In the PC class, major identified compounds were PC 38:1, PC 42:4, and PC(18:0_22:4). The mean peak areas of FFA 22:0, FFA 16:0 and FFA 18:1 were the highest. See Table 4 for lipid abbreviations and classification names.

Table 4. Abbreviations for each lipid class found in the Pinot noir wines.

Lipid Class Name	Abbreviation
Phosphatidylcholine	PC
Diglyceride	DG
Phosphatidylethanolamine	PE
Free fatty acids	FFA
Triglycerides	TG
Acylcarnitine	AC
Monoglycerides	MG
Lysophosphatidylcholine	LPC
Phosphatidic acids	PA
Lysophosphatidylethanolamine	LPE
Cholesteryl esters	CE
Phosphatidylserines	PS
Phosphatidylglycerols	PG

Figure 3 shows the total number of detected lipid compounds in each lipid class in Pinot noir wine based on a nontargeted lipidomic strategy and the number of important lipid compounds within each class selected by the random forest method. The selected lipid compounds were identified as important variables in differentiating wine treatments. DG had the highest number of important lipids, follow by FFA and TG. PC, MG, LPC, LPE, CE, PS and PG classes did not have lipid compounds important to distinguishing wines produced under the study conditions. The selected model with 25 lipid compounds showed high % variation in linear discriminant analysis (84.29%). Details of model selection and names of all 25 lipid compounds selected can be found in supplementary data Table S4.

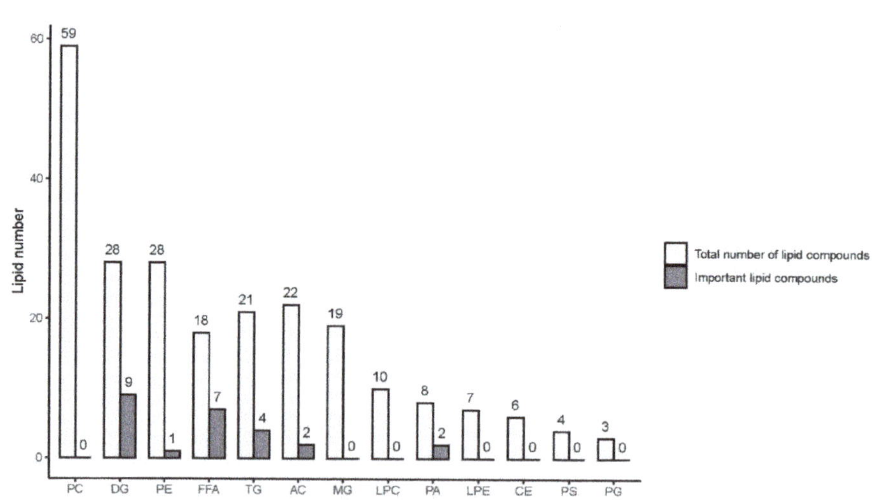

Figure 3. Total number of detected lipid compounds in each lipid class in Pinot noir wine based on a nontargeted lipidomic strategy (☐) and number of important lipid compounds within each class selected by random forest method (■). See Table 3 for lipid classification names.

Linear discriminant analysis using peak areas of 25 lipid compounds and 95% confidence intervals separated wine samples into six distinct groups based on winemaking treatments (Figure 4). The first two linear combinations explain 84% of the variance, showing a strong discrimination of wine treatments based on lipidomic data. LD1 clearly separated the wines from 16°C fermentation and 27 °C fermentation. LD2 separated the high concentration of yeast addition from the control and low yeast product concentration groups.

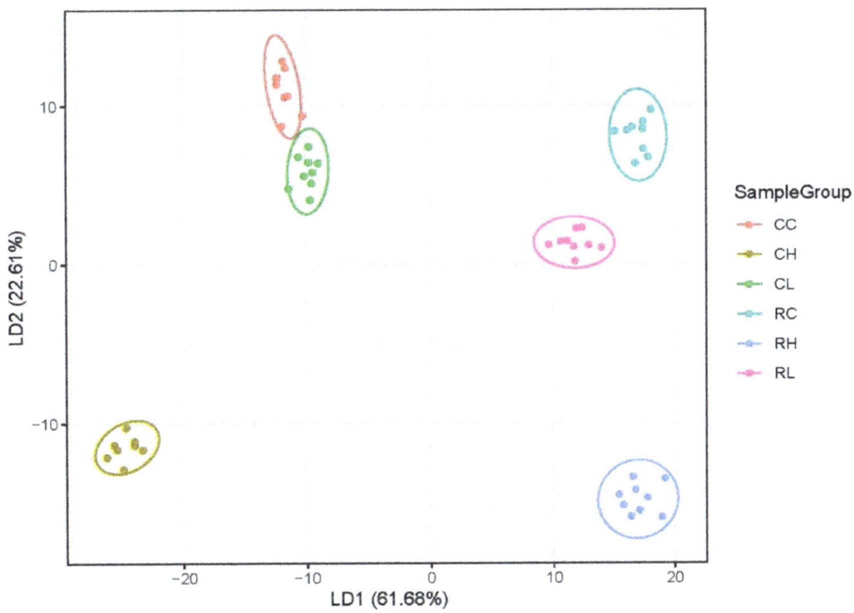

Figure 4. Linear discriminant analysis of Pinot noir wines by winemaking conditions with 25 selected important variables out of 243 lipid species as dependent variables and six winemaking conditions as independent variables. (RC: 27 °C fermentation temperature, 0 g/L yeast product addition; RL: 27 °C fermentation temperature, 0.5 g/L yeast product addition; RH: 27 °C fermentation temperature, 1 g/L yeast product addition; CC: 16 °C fermentation temperature, 0 g/L yeast product addition; CL: 16 °C fermentation temperature, 0.5 g/L yeast product addition; and CH: 16 °C fermentation temperature, 1 g/L yeast product addition).

Using random forest analysis, the most important lipidomic features were extracted, with 25 lipid compounds being the most important for explaining the differences based on fermentation temperature and yeast product addition (Figure 5). Lipids were ranked based on their importance in wine treatment separation. They showed relatively 0% OOB compared to other models (Table S4). The CH wines had relatively higher amounts of lipids compared to the wines from other treatments, mostly FFA, DG and TG. The majority of lipids in CH have relative concentrations higher than 1 (represented as the yellow boxes in Figure 5). Wines made under other treatments did not have many lipids at high relative concentrations. CC and CL wines had the lowest relative amount of lipids, while RC, RH and RL wines had higher relative amounts.

An untargeted lipidomic approach was successfully used to identify the most important lipids in differentiating the wine treatments. FFA, AC, DG and TG are the main lipid classes contributing to the differences. The effects of yeast product addition on the relative concentration of the 25 important lipids are noted in Figure 5. Adding more yeast product altered the lipidomic profile of the wines, even though it did not increase the amount of total lipids. The CH and RH wines were distinctively separated from other groups (Figure 5). In the cold fermentation groups (CC, CL and CH), the relative concentration of lipids in CH were seen to be much higher. Similar trends were also observed in the high fermentation temperature groups.

3.4. Sensory-Descriptive Analysis

Two-way ANOVA showed significant differences in sweetness and drying among wine treatments having different fermentation temperatures (p-value = 0.004 for drying and 0.005 for sweetness, Table S5). The sweetness taste attribute was higher in wines fermented at 16 °C compared to those fermented at 27 °C (Figure 6). The drying mouthfeel attribute was rated with significantly lower intensity in wines fermented at 16 °C compared to wines fermented at 27 °C (Figure 6). Descriptive analysis of wine samples showed no significant differences for acidity, astringency, bitterness, viscosity and warmth. It is worth considering that the mean intensity rating of astringency was higher in RL wines compared to other treatments, and viscosity and warmth had the highest mean intensity rating in CC.

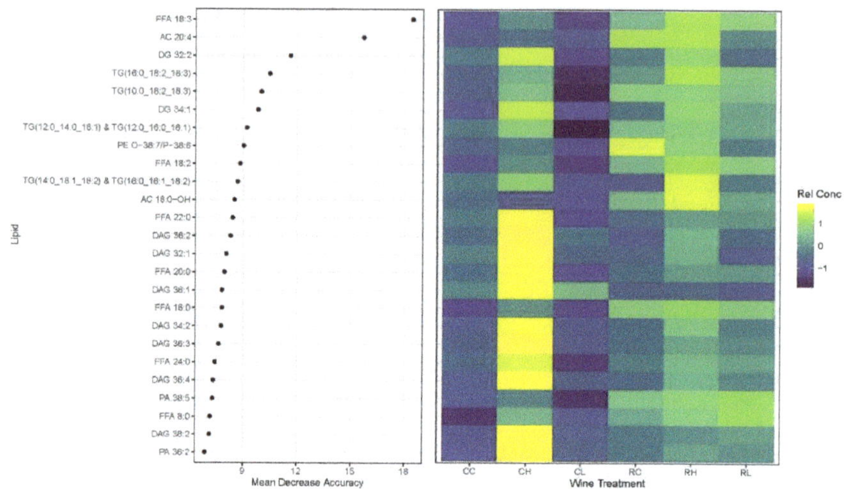

Figure 5. Variance of importance plot (VIP) for lipids that contribute the greatest variance to the classification of six wine treatments. Relative concentrations (Rel Conc) for each group are also shown (yellow for higher concentrations, dark green for lower).

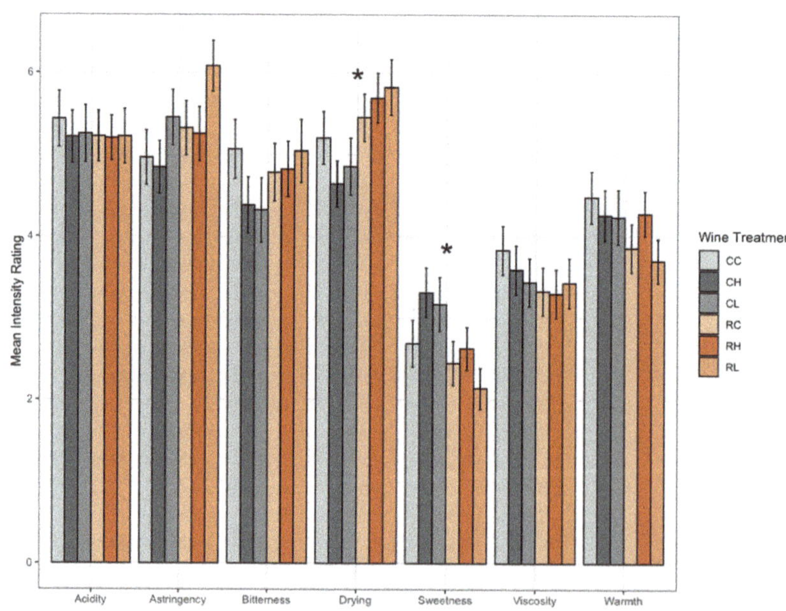

Figure 6. Mean perceived intensity rating (mean ± standard error) for taste and mouthfeel attributes in descriptive analysis test of wine treatments. * indicates significant difference (two-way ANOVA, $\alpha = 0.05$).

3.5. Sensory—CATA

CATA data for taste and mouthfeel descriptors show variations in taste and mouthfeel terms associated with wine treatments. Across the first two dimensions, 77.3% variance was found. Wines fermented at 16 °C (CC, CL, and CH) were described as oily, velvety, and soft. Wines fermented at 27 °C (RC, RL and RH) were described as grainy, gummy, harsh, unripe and sticky (Figure 7).

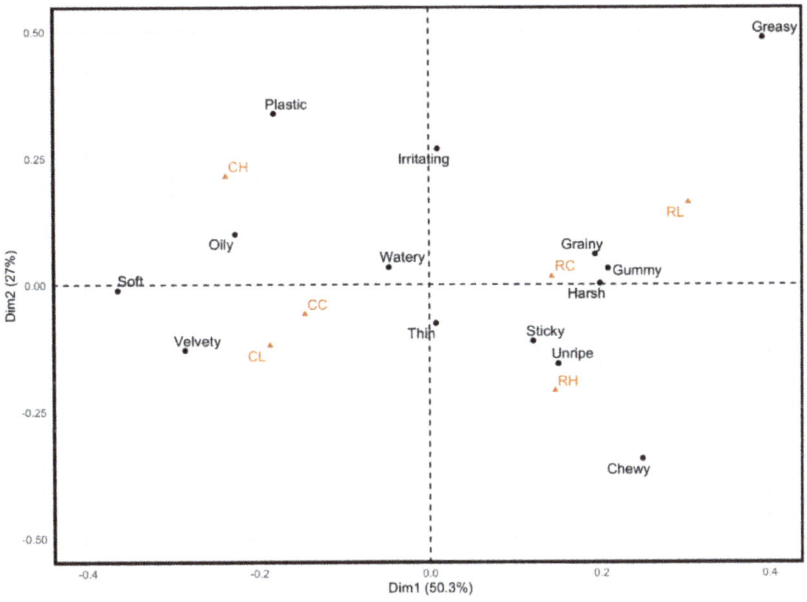

Figure 7. Corresponding analysis using CATA data. CATA terms are in black and wine treatments are in orange.

4. Discussion

4.1. Impacts of Yeast Fermentation Temperature on Wine Lipids, Taste and Mouthfeel Perception

Fermentation temperature has a significant effect on grape extraction [16]. A higher lipid concentration in a high temperature fermentation wine could be due to a greater breakdown of the grape firm tissues and the yeast cell wall during fermentation [16]. It would be of interest to determine the source of lipid differences in the present study treatments to establish if lipid extraction was from both or one specific source.

Increasing fermentation temperature produces wine with higher phenolics and color [42,43]. Our total phenolic data confirmed this finding, as the higher fermentation temperature resulted in wines with greater phenolics (Table 2). Phenolics, especially polymeric flavanols, are known to have major impacts on wine taste and mouthfeel [10,16,44]. Wines with higher phenolic content are known to have more intense bitterness, astringency, drying and sourness [44]. Among the seven attributes assessed in the descriptive analysis test, some were expected to show significant differences between wines fermented at 16 °C and 27 °C. However, drying was the only mouthfeel attribute that was significantly more intense in high fermentation temperature wines and sweetness was the only taste attribute that was significantly less intense in high fermentation temperature wines (Figure 6).

Wine fermented at 27 °C had significantly lower alcohol content and lower sweetness perception. Sweetness is the sensory attribute found to be significantly different and is not known to be impacted by lipids (Figure 5). It is known that high alcohol content can affect sweetness perception in wine [45]. The higher temperature used in the present study may have evaporated a portion of the ethanol content, resulting in a significant change of sweetness or may also be due to fermentation yield.

Fermentation temperature did not have strong impacts on bitterness, astringency, acidity, viscosity and warmth attributes in this study. Even though there was a significantly higher amount of phenolics in wines fermented at 27 °C, the impacts of phenolics on wine taste and mouthfeel may be masked by the presence of wine lipids, as lipids have been found to reduce the astringency in wine [46], although the concentration of lipids studied were much higher than those tested in our study. Research has examined the effects of phenolic compounds on lipid models in oral cell membranes and lipid droplets found in food [6]. Interactions between catechin, epicatechin and epigallocatechin gallate and lipids were examined and strong interactions were found between phenolics, especially tannins, and lipids in a model system [6]. The authors stated that astringency could be affected by potential competition between the tannin-lipid and other tannin-macromolecule interactions [47]. Similar interactions could be found in the present study treatments but further research is needed to confirm this. One of the difficulties of working in a wine system is that it is difficult to isolate the exact impact of lipids to wine mouthfeel. Working with a simpler model and adding different wine components may be a way to isolate any sensory changes specifically imparted by lipids, such as those conducted in Phan et Siebert [7].

4.2. Impacts of Yeast Product Addition on Wine Lipids, Taste and Mouthfeel Perception

Theoretically, the addition of yeast products after fermentation would be similar to having more yeast cells during fermentation and was predicted to increase the total lipids. However, yeast product addition did not have significant effects on the total lipid concentration (Figure 2). Yeast product did alter the wine lipidomic profiles but may not be a good source for enhancing total lipid content in wine. The impact of different lipids to wine sensory are discussed further in Section 4.3.

Adding yeast product had significant impacts on total phenolics. The total phenolics decreased in wines with higher yeast additions (Table 2). A similar argument can be made in high fermentation wine; since there were more organic substances present in high yeast product addition wine, there could be more chance for phenolics to interact with the components, such as lipids, polysaccharides, and proteins, to form entanglement and then precipitate out of the solution [48]. As a result, the contribution of phenolics to taste and mouthfeel perception was insignificant in wines with yeast product addition.

Astringency is a very important quality factor that influences consumer preference [49]. A recent study reported that increasing astringency decreased liking and elicited more intense negative emotions in wine consumers [4]. A combination of 27 °C fermentation and low (0.5 g/L) addition of yeast product as in RL treatment would potentially provide higher astringency mouthfeel thus an undesirable sensory outcome. It is therefore possible to craft a wine with more desirable mouthfeel by altering fermentation temperature and use of yeast product addition.

4.3. Impacts of Wine Lipids on Taste and Mouthfeel Perception

Research has shown that dietary lipids interact with wine phenolics to form wine phenolic–lipid complexes at molecular levels. Wine phenolics interact with the lipid systems, multilamellar vesicles, isotropic bicelles or lipid droplets from a phospholipid-stabilized oil-in-water emulsion [6,46]. Further investigation focusing on lipid systems present in wine would provide more information about how wine lipids interact with phenolic compounds. The interactions could potentially lead to a decrease in astringency perception in phenolic solutions [46]. The result was not observed in this study. Lipids found in Pinot noir wine did not have an effect on astringency. Astringency perception was not significantly different among wines with higher total lipids. The previous studies investigating dietary lipids found in fatty food or oil so the lipids were at higher concentrations compared to the normal lipid concentration found in wine. This could be possible because the wine lipid content was low in general and those wine treatments with high lipid content also had higher concentrations of phenolics.

The freer fatty acids in the cooler fermentation wines might be the cause of the significant increase in TA and lower pH in these wines (Table 2). Differences in wine lipidomic profiles may be driving the differences in sensory perception seen in CATA results (Figure 6). CH wine with a higher relative concentration of FFA, PA and DG was associated with mouthfeel terms, such as oily, soft and velvety. FFA lipids were one of the most important lipid classes based on treatments as determined by random forest (specifically, unsaturated FFA 18:3 and FFA 18:2 and the saturated FFA 22:0, FFA 20:0, FFA 8:0, FFA 16:0 and FFA 24:0). FFA lipids found to be different between the wine treatments have been reported as major FFAs in grapes and wine [10]. In addition, FFAs have been considered the main component related to fat taste [50]. Specifically, the longer and more saturated the fatty acids, the greater the contribution to smoother and creamier mouthfeel texture [51]. Fatty acids have also been found to modulate bitter taste in aqueous solutions [52]. Besnard et al. (2016) and Mattes (2009) both agreed that long-chain fatty acids are indeed an orosensory effective taste stimuli [53,54].

Viscosity was not affected by either fermentation temperature or yeast product addition. Sugar has been reported to have influence on viscosity perception [45]. The glucose and fructose content were not changed among wine treatments. The insignificant changes in viscosity perception is aligned with previous sensory work [7]. When low levels of lipids are present, phospholipids appear to be a factor driving viscosity perception. The wine treatments produced in this study did not utilize phospholipids, as these lipids were not found to be important in treatment differentiation (Figure 5). The lack of phospholipids additions to the wines is the most likely explanation for the lack of viscosity changes seen.

5. Conclusions

This work is one of the first to evaluate lipid composition in Pinot noir wine and its impact to taste and mouthfeel perception. Our study showed that fermentation temperature is an important factor that influences the concentrations of several wine chemical components and wine taste and mouthfeel characteristics. By increasing the fermentation temperature, the total lipids increased significantly and wine taste and mouthfeel had lower sweetness and higher drying intensities. Adding yeast product additions to a winemaking procedure can change the Pinot noir wine lipidomic profile but did not affect the total lipids or wine sensory. Interactions between the two winemaking factors only influenced the variations in wine basic components but had no impact on the total lipids nor wine sensory. Lipid compositions of wines made under different conditions show clear distinctions among wine treatments. The differences in lipid profiles of Pinot noir wine treatments suggest potential impacts of winemaking techniques on the composition of lipids and wine taste and mouthfeel perception profiles. There were distinctive groups of taste and mouthfeel terms associated with each wine treatment. Specifically, the fermentation temperature is more important than yeast product addition to alter the sweetness and drying properties of the wine. This information may be used to develop winemaking practices that produce wine with desirable taste and mouthfeel attributes, attributed to altering lipid composition.

Supplementary Materials: The following are available online at https://www.mdpi.com/article/10.3390/horticulturae8010052/s1, Table S1: Mean basic chemistry ± SE with p-values of two-way ANOVA and Tukey's HSD for the main effects, Table S2: p-values of two-way ANOVA for mean peak intensity of 233 lipid compounds. Table S3: Peak intensities of lipid compounds detected in Pinot noir wine samples from different wine treatments. Table S4: Percent out-of-bag (% OOB) ± standard deviation of lipid combination models, Table S5: p-values of the two-way ANOVA of the descriptive analysis of the wine treatments. Figure S1: Temperature and ° Brix of wine treatment tanks during alcoholic fermentation.

Author Contributions: Conceptualization, Q.P. and E.T.; methodology, Q.P, A.D., J.O. and E.T.; software, Q.P.; validation, Q.P. and E.T.; formal analysis, Q.P. and E.T.; investigation, Q.P. and E.T.;

resources, E.T.; data curation, Q.P.; writing—original draft preparation, Q.P.; writing—review and editing, E.T., A.D. and J.O.; visualization, Q.P.; supervision, E.T.; project administration, E.T.; funding acquisition, E.T. All authors have read and agreed to the published version of the manuscript.

Funding: This research was funded by E & J Gallo Winery.

Institutional Review Board Statement: The study was conducted according to the guidelines of the Declaration of Helsinki and approved by the Institutional Review Board (or Ethics Committee) of Oregon State University (Corvallis, OR, USA), protocol code IRB-2020-0610 and Date of approval: 3 June 2020.

Informed Consent Statement: Informed consent was obtained from all subjects involved in the study.

Acknowledgments: The authors thank the Mass Spectrometry Center at Oregon State University for assisting the lipidomic analysis.

Conflicts of Interest: The authors declare no conflict of interest.

References

1. Bertino, M.; Lawless, H.T. Understanding Mouthfeel Attributes: A Multidimensional Scaling Approach. *J. Sens. Stud.* **1993**, *8*, 101–114. [CrossRef]
2. Cosme, F.; Vilela, A.; Jordão, A.M.; Desk, S. Wine Phenolics: Looking for a Smooth Mouthfeel. *J. Food Sci. Technol.* **2016**, *1*, 20–28.
3. Laguna, L.; Bartolomé, B.; Moreno-Arribas, M.V. Mouthfeel Perception of Wine: Oral Physiology, Components and Instrumental Characterization. *Trends Food Sci. Technol.* **2017**, *59*, 49–59. [CrossRef]
4. Niimi, J.; Danner, L.; Li, L.; Bossan, H.; Bastian, S.E.P. Wine Consumers' Subjective Responses to Wine Mouthfeel and Understanding of Wine Body. *Food Res. Int.* **2017**, *99*, 115–122. [CrossRef]
5. Runnebaum, R.C.; Boulton, R.B.; Powell, R.L.; Heymann, H. Key Constituents Affecting Wine Body—An Exploratory Study. *J. Sens. Stud.* **2011**, *26*, 62–70. [CrossRef]
6. Furlan, A.L.; Castets, A.; Nallet, F.; Pianet, I.; Grélard, A.; Dufourc, E.J.; Géan, J. Red Wine Tannins Fluidify and Precipitate Lipid Liposomes and Bicelles. A Role for Lipids in Wine Tasting? *Langmuir* **2014**, *30*, 5518–5526. [CrossRef]
7. Phan, Q.; Hoffman, S.; Tomasino, E. Contribution of Lipids to Taste and Mouthfeel Perception in a Model Wine Solution. *ACS Food Sci. Technol.* **2021**, *1*, 1561–1566. [CrossRef]
8. Jackson, R.S. *Wine Tasting: A Professional Handbook*; Academic Press: San Diego, CA, USA, 2016; ISBN 978-0-12-801826-2.
9. Phan, Q.; Tomasino, E. Untargeted Lipidomic Approach in Studying Pinot Noir Wine Lipids and Predicting Wine Origin. *Food Chem.* **2021**, *355*, 129409. [CrossRef] [PubMed]
10. Waterhouse, A.L.; Sacks, G.L.; Jeffery, D.W. *Understanding Wine Chemistry*; John Wiley & Sons, Inc.: West Sussex, UK, 2016; ISBN 978-1-118-62780-8.
11. Harbertson, J.F.; Hodgins, R.E.; Thurston, L.N.; Schaffer, L.J.; Reid, M.S.; Landon, J.L.; Ross, C.F.; Adams, D.O. Variability of Tannin Concentration in Red Wines. *Am. J. Enol. Vitic.* **2008**, *59*, 210–214.
12. Vidal, S.; Courcoux, P.; Francis, L.; Kwiatkowski, M.; Gawel, R.; Williams, P.; Waters, E.; Cheynier, V. Use of an Experimental Design Approach for Evaluation of Key Wine Components on Mouth-Feel Perception. *Food Qual. Prefer.* **2004**, *15*, 209–217. [CrossRef]
13. Woody, R.C.; Schmidt, R. Following the Roots of Oregon Wine. *Or. Hist. Q.* **2013**, *114*, 324–339. [CrossRef]
14. Li, S.; Bindon, K.; Bastian, S.E.P.; Jiranek, V.; Wilkinson, K.L. Use of Winemaking Supplements to Modify the Composition and Sensory Properties of Shiraz Wine. *J. Agric. Food Chem.* **2017**, *65*, 1353–1364. [CrossRef] [PubMed]
15. Dequin, S.; Escudier, J.-L.; Bely, M.; Noble, J.; Albertin, W.; Masneuf-Pomarède, I.; Marullo, P.; Salmon, J.-M.; Sablayrolles, J.M. How to Adapt Winemaking Practices to Modified Grape Composition under Climate Change Conditions. *OENO One* **2017**, *51*, 205–214. [CrossRef]
16. Sacchi, K.L.; Bisson, L.F.; Adams, D.O. A Review of the Effect of Winemaking Techniques on Phenolic Extraction in Red Wines. *Am. J. Enol. Vitic.* **2005**, *56*, 197–206.
17. Unterkofler, J.; Muhlack, R.A.; Jeffery, D.W. Processes and Purposes of Extraction of Grape Components during Winemaking: Current State and Perspectives. *Appl. Microbiol. Biotechnol.* **2020**, *104*, 4737–4755. [CrossRef]
18. Hunter, K.; Rose, A.H. Lipid Composition of Saccharomyces Cerevisiae as Influenced by Growth Temperature. *Biochim. Biophys. Acta (BBA) Lipids Lipid Metab.* **1972**, *260*, 639–653. [CrossRef]
19. Coleman, M.C.; Fish, R.; Block, D.E. Temperature-Dependent Kinetic Model for Nitrogen-Limited Wine Fermentations. *Appl. Environ. Microbiol.* **2007**, *73*, 5875–5884. [CrossRef]
20. Henderson, C.M.; Block, D.E. Examining the Role of Membrane Lipid Composition in Determining the Ethanol Tolerance of Saccharomyces Cerevisiae. *Appl. Environ. Microbiol.* **2014**, *80*, 2966–2972. [CrossRef]
21. Gil-Muñoz, R.; Moreno-Pérez, A.; Vila-López, R.; Fernández-Fernández, J.I.; Martínez-Cutillas, A.; Gómez-Plaza, E. Influence of Low Temperature Prefermentative Techniques on Chromatic and Phenolic Characteristics of Syrah and Cabernet Sauvignon Wines. *Eur. Food Res. Technol.* **2009**, *228*, 777–788. [CrossRef]

22. Reynolds, A.; Cliff, M.; Girard, B.; Kopp, T.G. Influence of Fermentation Temperature on Composition and Sensory Properties of Semillon and Shiraz Wines. *Am. J. Enol. Vitic.* **2001**, *52*, 235–240.
23. Ndlovu, T.; Divol, B.; Bauer, F.F. Yeast Cell Wall Chitin Reduces Wine Haze Formation. *Appl. Environ. Microbiol.* **2018**, *84*, e00668-18. [CrossRef]
24. Ángeles Pozo-Bayón, M.; Andújar-Ortiz, I.; Moreno-Arribas, M.V. Scientific Evidences beyond the Application of Inactive Dry Yeast Preparations in Winemaking. *Food Res. Int.* **2009**, *42*, 754–761. [CrossRef]
25. Aguilar-Uscanga, B.; François, J.M. A Study of the Yeast Cell Wall Composition and Structure in Response to Growth Conditions and Mode of Cultivation. *Lett. Appl. Microbiol.* **2003**, *37*, 268–274. [CrossRef]
26. Rattray, J.B.; Schibeci, A.; Kidby, D.K. Lipids of Yeasts. *Bacteriol. Rev.* **1975**, *39*, 197–231. [CrossRef] [PubMed]
27. Sáenz-Navajas, M.-P.; Ferreira, V.; Dizy, M.; Fernández-Zurbano, P. Characterization of Taste-Active Fractions in Red Wine Combining HPLC Fractionation, Sensory Analysis and Ultra Performance Liquid Chromatography Coupled with Mass Spectrometry Detection. *Anal. Chim. Acta* **2010**, *673*, 151–159. [CrossRef] [PubMed]
28. Iland, P. *Chemical Analysis of Grapes and Wine*; Patrick Iland Wine Promotions PTYLTD: Adelaide, Australia, 2004; ISBN 978-0-9581605-1-3.
29. Gallander, J.; Briner, L.; Stetson, J.; Liu, J.-W.; Krielow, L.; Wilker, K.; Romberger, R.; Stamp, C.; Riesen, R. *Manual for Wine Analysis and Laboratory Techniques*; Ohio State University, OARDC: Wooster, OH, USA, 1991.
30. Waterhouse, A.L. Determination of Total Phenolics. *Curr. Protoc. Food Anal. Chem.* **2002**, *6*, I1.1.1–I1.1.8. [CrossRef]
31. Choi, J.; Leonard, S.W.; Kasper, K.; McDougall, M.; Stevens, J.F.; Tanguay, R.L.; Traber, M.G. Novel Function of Vitamin E in Regulation of Zebrafish (Danio Rerio) Brain Lysophospholipids Discovered Using Lipidomics. *J. Lipid Res.* **2015**, *56*, 1182–1190. [CrossRef]
32. Definition of SWEET. Available online: https://www.merriam-webster.com/dictionary/sweet (accessed on 28 July 2021).
33. Sparrow, A.M.; Holt, H.E.; Pearson, W.; Dambergs, R.G.; Close, D.C. Accentuated Cut Edges (ACE): Effects of Skin Fragmentation on the Composition and Sensory Attributes of Pinot Noir Wines. *Am. J. Enol. Vitic.* **2016**, *67*, 169–178. [CrossRef]
34. Williamson, P. *Sensory Descriptive Analysis on Commercial Wines of Varied Styles*; The Australian Wine Research Institute: Urrbrae, Australia, 2013.
35. Gawel, R.; Oberholster, A.; Francis, I.L. A 'Mouth-Feel Wheel': Terminology for Communicating the Mouth-Feel Characteristics of Red Wine. *Aust. J. Grape Wine Res.* **2000**, *6*, 203–207. [CrossRef]
36. Guinard, J.-X.; Mazzucchelli, R. The Sensory Perception of Texture and Mouthfeel. *Trends Food Sci. Technol.* **1996**, *7*, 213–219. [CrossRef]
37. R Core Team. *R: A Language and Environment for Statistical Computing*; R Foundation for Statistical Computing: Vienna, Austria, 2021.
38. Wickham, H. *Ggplot2: Elegant Graphics for Data Analysis*; Springer: New York, NY, USA, 2016; ISBN 978-3-319-24277-4.
39. Campo, E.; Ballester, J.; Langlois, J.; Dacremont, C.; Valentin, D. Comparison of Conventional Descriptive Analysis and a Citation Frequency-Based Descriptive Method for Odor Profiling: An Application to Burgundy Pinot Noir Wines. *Food Qual. Prefer.* **2010**, *21*, 44–55. [CrossRef]
40. Lê, S.; Josse, J.; Husson, F. FactoMineR: An R Package for Multivariate Analysis. *J. Stat. Softw.* **2008**, *25*, 1–18. [CrossRef]
41. Kassambara, A.; Mundt, F. Factoextra: Extract and Visualize the Results of Multivariate Data Analyses; 2020. Available online: https://cran.r-project.org/web/packages/factoextra/readme/README.html (accessed on 18 January 2021).
42. Gao, L.; Girard, B.; Mazza, G.; Reynolds, A.G. Changes in Anthocyanins and Color Characteristics of Pinot Noir Wines during Different Vinification Processes. *J. Agric. Food Chem.* **1997**, *45*, 2003–2008. [CrossRef]
43. Girard, B.; Yuksel, D.; Cliff, M.A.; Delaquis, P.; Reynolds, A.G. Vinification Effects on the Sensory, Colour and GC Profiles of Pinot Noir Wines from British Columbia. *Food Res. Int.* **2001**, *34*, 483–499. [CrossRef]
44. Frost, S.C.; Harbertson, J.F.; Heymann, H. A Full Factorial Study on the Effect of Tannins, Acidity, and Ethanol on the Temporal Perception of Taste and Mouthfeel in Red Wine. *Food Qual. Prefer.* **2017**, *62*, 1–7. [CrossRef]
45. Nurgel, C.; Pickering, G. Contribution of Glycerol, Ethanol and Sugar to the Perception of Viscosity and Density Elicited by Model White Wines. *J. Texture Stud.* **2005**, *36*, 303–323. [CrossRef]
46. Saad, A.; Bousquet, J.; Fernandez-Castro, N.; Loquet, A.; Géan, J. New Insights into Wine Taste: Impact of Dietary Lipids on Sensory Perceptions of Grape Tannins. *J. Agric. Food Chem.* **2021**, *69*, 3165–3174. [CrossRef]
47. Furlan, A.L.; Saad, A.; Dufourc, E.J.; Géan, J. Grape Tannin Catechin and Ethanol Fluidify Oral Membrane Mimics Containing Moderate Amounts of Cholesterol: Implications on Wine Tasting? *Biochimie* **2016**, *130*, 41–48. [CrossRef]
48. Scollary, G.R.; Pásti, G.; Kállay, M.; Blackman, J.; Clark, A.C. Astringency Response of Red Wines: Potential Role of Molecular Assembly. *Trends Food Sci. Technol.* **2012**, *27*, 25–36. [CrossRef]
49. Bajec, M.R.; Pickering, G.J. Astringency: Mechanisms and Perception. *Crit. Rev. Food Sci. Nutr.* **2008**, *48*, 858–875. [CrossRef]
50. Martin, C.; Passilly-Degrace, P.; Gaillard, D.; Merlin, J.; Chevrot, M.; Besnard, P. The Lipid-Sensor Candidates CD36 and GPR120 Are Differentially Regulated by Dietary Lipids in Mouse Taste Buds: Impact on Spontaneous Fat Preference. *PLoS ONE* **2011**, *6*, e24014. [CrossRef] [PubMed]
51. Zhou, X.; Shen, Y.; Parker, J.K.; Kennedy, O.B.; Methven, L. Relative Effects of Sensory Modalities and Importance of Fatty Acid Sensitivity on Fat Perception in a Real Food Model. *Chem. Percept.* **2016**, *9*, 105–119. [CrossRef] [PubMed]

52. Aglione, A.; Cassutt, K.; Dragan, S.; Gravina, S.; Kurash, Y.; Johnson, W. Modulation of Bitterness and Mouthfeel via Synergistic Mixtures of Long Chain Fatty Acids. U.S. Patent Application No. 15/106,179, 3 November 2016.
53. Besnard, P.; Passilly-Degrace, P.; Khan, N.A. Taste of Fat: A Sixth Taste Modality? *Physiol. Rev.* **2016**, *96*, 151–176. [CrossRef] [PubMed]
54. Mattes, R.D. Is There a Fatty Acid Taste? *Annu. Rev. Nutr.* **2009**, *29*, 305–327. [CrossRef]

horticulturae

Review

UVA and UVB Radiation as Innovative Tools to Biofortify Horticultural Crops with Nutraceuticals

Daniel A. Jacobo-Velázquez [1,*], Melissa Moreira-Rodríguez [2] and Jorge Benavides [2]

1 Tecnologico de Monterrey, The Institute for Obesity Research, Ave. General Ramón Corona 2514, Zapopan 45201, Jal., Mexico
2 Tecnologico de Monterrey, The Institute for Obesity Research, Ave. Eugenio Garza Sada 2501, Monterrey 64849, N.L., Mexico; melissamoreira19@hotmail.com (M.M.-R.); jorben@tec.mx (J.B.)
* Correspondence: djacobov@tec.mx

Abstract: The consumption of fruits and vegetables is related to the prevention and treatment of chronic–degenerative diseases due to the presence of secondary metabolites with pharmaceutical activity. Most of these secondary metabolites, also known as nutraceuticals, are present in low concentrations in the plant tissue. Therefore, to improve the health benefits of horticultural crops, it is necessary to increase their nutraceutical content before reaching consumers. Applying ultraviolet radiation (UVR) to fruits and vegetables has been a simple and effective technology to biofortify plant tissue with secondary metabolites. This review article describes the physiological and molecular basis of stress response in plants. Likewise, current literature on the mechanisms and effects of UVA and UVB radiation on the accumulation of different bioactive phytochemicals are reviewed. The literature shows that UVR is an effective tool to biofortify horticultural crops to enhance their nutraceutical content.

Keywords: ultraviolet radiation (UVR); UVA; UVB; nutraceuticals; functional foods; biofortification

Citation: Jacobo-Velázquez, D.A.; Moreira-Rodríguez, M.; Benavides, J. UVA and UVB Radiation as Innovative Tools to Biofortify Horticultural Crops with Nutraceuticals. *Horticulturae* **2022**, *8*, 387. https://doi.org/10.3390/horticulturae8050387

Academic Editors: Francisco Garcia-Sanchez and Luigi De Bellis

Received: 27 March 2022
Accepted: 25 April 2022
Published: 28 April 2022

1. Introduction

Being sessile, plants are constantly exposed to biotic and abiotic stresses. Their response to such stresses is complex, involving changes at the transcriptome, cellular, and physiological levels [1]. Secondary metabolites are well known to be related to the plant's defense response mechanisms, being induced in response to abiotic stresses and acting as natural phytoalexins to protect plants against these stresses [2]. Moreover, many of these secondary metabolites possess pharmacological activity that results in the prevention and/or treatment of chronic and degenerative diseases [3].

In this context, the application of abiotic stresses (i.e., wounding, modified atmospheres, exogenous phytohormones and ultraviolet radiation (UVR)) may be used as an approach to biofortify crops with specific health-promoting compounds with applications in the pharmaceutical, cosmeceutical, and nutraceutical industries [4,5]. For instance, mature crops such as broccoli [6–8], carrot [9], potato [10,11], and lettuce [12] have been used to study the effect of abiotic stresses on antioxidant biosynthesis and accumulation.

UVR comprises 7–9% of the total energy of solar radiation reaching the Earth surface and is sub-divided in three wavelength ranges: UVA (320–400 nm), which represents about 6.3% of the incoming solar radiation and is the least harmful range; UVB (280–320 nm), representing about 1.5% of the total spectrum, but causing several detrimental effects in plants; and UVC (100–280 nm), which is extremely harmful to organisms, but is completely absorbed by stratospheric ozone [13–15].

Plants are unavoidably exposed to UVR as they are sessile organisms and as they need to capture sunlight for photosynthesis. It is well known that UVR causes different responses in plants; some of them are detrimental, including damage to DNA and proteins, generation of ROS and initiation of cellular stress responses, changes on cell physiology, as well as changes in plant growth, morphology and development [16–18]. Thus, plants need to evolve different mechanisms for UV-protection and repair [13]. These mechanisms include the biosynthesis of nutraceuticals such as phenolic compounds, glucosinolates, carotenoids, chlorophylls, ascorbic acid, and betalains.

In this review paper, the physiological and molecular basis of stress response in plants is described. Likewise, current literature on the mechanisms and effects of UVA and UVB radiation on the accumulation of bioactive phytochemicals is reviewed.

2. Physiology of Stress Response in Plants

It has been reported that, upon being subjected to an abiotic stress, the general cellular process and regulation for activating plant secondary metabolite starts with an extracellular or intracellular signal, which is then recognized by a receptor on the surface of the plasma membrane. In turn, this initiates a signal transduction cascade that leads to activation or de novo synthesis of transcription factors to regulate gene expression involved in the plant secondary metabolism [1].

Different signaling molecules, such as ethylene (ET), jasmonic acid (JA), reactive oxygen species (ROS), salicylic acid, and abscisic acid have been reported to be produced by abiotic stresses as well as to activate plant defense genes, including those from the phenylpropanoid metabolism, triggering the accumulation of phenolic compounds [19,20]. Models recently proposed to explain the physiological and molecular basis of these compounds state that phenolic biosynthesis under several stress conditions including wounding alone, and in combination with UV light (types A, B, and C) stresses or in combination with ET and JA treatments, is an event mainly mediated by ROS as a signal molecule to activate the plant primary and secondary metabolism, while ET and JA modulate ROS levels, although they may also play a mild role in triggering the expression of certain genes related to the synthesis of secondary metabolites [7,20]. Moreover, it has been reported that extracellular adenosine triphosphate is the primary signal that triggers the wound-response in plants [21].

Additionally, ET and methyl jasmonate (MeJA) treatments have shown to upregulate the expression of *CYP79* genes related to glucosinolate biosynthesis in *Arabidopsis*. It was concluded that MeJA was a potent inducer of both *CYP79F* and *CYP79B* genes, placing it as the most potent elicitor of glucosinolate biosynthesis [22].

On the other hand, when stress occurs, Ca^{2+} influx, alkalinization of the apoplast, and protein phosphorylation, among other intracellular events, lead to the synthesis of NADPH oxidase and trigger an oxidative burst in the plant. NADPH oxidase is the key source of the early and sustained accumulation of ROS; it is responsible for the reduction of O_2 to O_2^- and with the presence of SOD enzyme, O_2^- is converted to H_2O_2. It has been shown that ROS molecules, both O_2^- and H_2O_2, are involved in plant stress-induced defense responses [1]. Expression of genes related to plant defense pathway has been demonstrated as consequences of pathogen attack, chilling injury, wounding, and excess light [23,24].

Nitric oxide (NO) is another signaling molecule involved in adaptive plant responses to some specific abiotic stress conditions, particularly low mineral nutrient supply, drought, salinity, and high UVB radiation. NO is also an important component of the mechanisms responsible of coordinating and regulating Ca^{2+} and ROS signaling in plant defense [25]. Reaction with metal centers, thiols, oxygen molecule, and free radicals constitutes the way through which NO modulates plant responses that trigger the biosynthesis and accumulation of phytochemicals such as glucosinolates and phenolic compounds [25].

3. UV Radiation as an Abiotic Stressor

Plants need to evolve different mechanisms for UV-protection and repair. These mechanisms include deposition of UV-absorbing phenolic compounds in the outer epidermal layers and the production of antioxidant systems, action of reparative enzymes such as DNA photolyases, and expression of genes involved in both UV-protection and repair [17].

UVR regulates different aspects of metabolism, modulates biochemical composition and thus, promotes the synthesis and accumulation of secondary metabolites, such as phenolic compounds and glucosinolates [26–28]. Phenolics and carotenoids provide UV-absorbing sunscreen that limits penetration of UVB into leaf tissues. Although glucosinolates are not directly involved in UV protection, UV-mediated effects on glucosinolates are conceivable as they are involved in the common plant defense response, regulated by the signaling pathways involved in the perception of UVB [29,30]. In the following sections, the reported effects of UVB and UVA radiation on phytochemical accumulation are reviewed separately.

3.1. Mechanisms and Effects of UVB Radiation on Phytochemical Biosynthesis

Plant responses to UVR are likely to involve specific UV photoreceptors and signal transduction processes, which lead to the regulation of gene transcription [31,32]. Between the two solar UVR ranges (UVA and UVB) that reach the Earth, UVB radiation is the most harmful. Hence, numerous efforts have been focused on assessing the effects of UVB on plants. In this context, there is currently an extensive body of data concerning UVB-mediated cellular damage, as well as regulatory responses mediated by the UVB photoreceptor UV resistant locus8 (UVR8) [33].

Two main types of signaling pathways have been proposed regarding how plants perceive UVB radiation and how they regulate secondary plant metabolism. One pathway is not specific to UVB and implies that UVB-induced oxidative stress responses (rather than photomorphogenic responses to UVR) may result from damage to molecules and/or the accumulation of signaling molecules such as ROS and wound or defense-related molecules including jasmonic acid, salicylic acid, nitric oxide, and ethylene. In turn, this leads to over-expression of stress-related genes, normally induced by wound and defense signaling pathways (e.g., *PR-1*, *PR-2*, *PR-5* and the defense gene *PDF1.2*) [31,34,35]. In contrast, the signaling pathways that mediate responses to UVB as a signal appear to be UVB-specific and to result in UV-protection or morphological changes [24]. In this type of signaling, the cytosolic UVR8 photoreceptor seems to play a major role. In the presence of UVB, UVR8 monomerizes and interacts with the multifunctional E3 ubiquitin ligase constitutively photomorphogenic 1 (COP1) and translocate into the nucleus where they prevent the degradation of the photomorphogenic transcription factor elongated hypocotyl 5 (HY5). Successively, HY5 and its homolog (HYH) control expression of a range of key elements involved in UV acclimation response and UV protection, such as genes encoding enzymes of the phenylpropanoid pathway, including phenylalanine ammonia lyase (PAL), chalcone synthase (CHS) and flavonol synthase (FLS) [23,32,34,36–41]. Thus, the UVR8 photoreceptor is required for the induction of genes with important functions in UV protection. In addition, there is also the possibility that the aforementioned mechanisms are not solitary and that UVR regulates gene expression by combination of these mechanisms [9].

Moreover, it has been reported that UVB-induced signaling molecules, such as NO, exert a protective role against oxidative stress, alleviating UVB-induced photodamage [19]. For instance, UVB radiation increases both ROS and NO in plants. Then, NO reduces ROS levels and upregulates the expression of several genes involved in phenolic biosynthesis (i.e., the maize transcription factor ZmP and MYB12, its *Arabidopsis* functional homolog and their target genes *CHS* and *CHI–chalcone isomerase*). Thus, biosynthesis and accumulation of some flavonoids and anthocyanins are induced to absorb UVB and scavenge ROS [42]. UVB radiation has also been reported to stimulate ET production in plants. NO and ROS have also been implicated in UVB-induced ethylene production in maize seedlings [43].

The effect of UVB radiation on phenolic accumulation has been studied in several fruits and vegetables and, although not all phenolic compounds are similarly induced, flavonoids and flavonoid glycosides are generally more responsive to UVB than phenolic acids [41]. The accumulation of specific flavonoid glycosides appears to be an intrinsic part of the UVB response, with expression of several UDP- glucosyltransferases being directly controlled by UVB [37].

Table 1 summarizes the main finding of studies performed in several plant models suggesting the use of UVB as an abiotic stress to elicit the biosynthesis and accumulation of phytochemicals such as phenolic compounds, glucosinolates, ascorbic acid and betalains [9,27,29,44–63]. For instance, it has been shown that supplementation with UVB radiation in carrots, apples, grapes and flowering plants induces the accumulation of phenolic compounds due to an UVB-induced upregulation of key genes encoding enzymes of the phenylpropanoid pathway [9,44,46,47].

Table 1. Effect of UVB radiation on the biosynthesis of bioactive phytochemicals and on stress responses in different horticultural crops.

Plant Species	UVB Treatment Parameters and Storage Conditions	Phytochemical Evaluated	Main Findings	Reference
Red grapes (*Vitis vinifera*)	30–510 W, ID: 20–60 cm, 5 s–30 min post-harvest exposure. Storage at 20 °C.	Resveratrol and other PCs	Maximum resveratrol content per standard serving (200 g) was 3 mg (11-fold higher than untreated grapes) and was achieved 3 d after irradiation of grapes placed 40 cm below UVB lamps (510 W), for 30 s. Content of other PCs remained unaltered.	[44]
Apples (skin) (*Malus domestica*)	0.16–0.2 W m^{-2} + 15–20 µmol m^{-2} s^{-1}(visible light), ID: 50 cm, 10 or 20 °C, for 72 h.	PCs	Apples inner facing (when on the tree) accumulated more anthocyanins and Q gly than outer facing ones. Fruit maturity and lower temperature (10 °C vs. 20 °C) prevented UVB-induced phenolic accumulation. Chlorogenic acid increased (~75–500%) using UVB at both temperatures in four of five cultivars evaluated.	[45,46]
Shredded Carrot (*Daucus carota*)	1.5 kJ m^{-2}, 162 s, ID: 17.5 cm. Storage at 15 °C for 72 h.	PCs	PAL activity of UV-B and non-irradiated samples was increased by approximately 760% after 72 h. An accumulation of up to 30% in PC content after 72 h was observed versus control samples,	[47]
Wounded Carrot (*Daucus carota*)	10.44 W m^{-2}, for 0, 1, and 6 h, ID: 50 cm. Post-storage at 15 °C for 4 d.	PCs	After 6 h of UVB radiation, total phenolic content of carrot pies increased 3-fold, AOX capacity increased 7-folds, and PAL activity increased by 90-fold. Chlorogenic acid and a derivative, ferulic acid and isocoumarin were significantly accumulated after 6 h.	[9]
Rapeseed leaves (*Brassica napus* subsp. *napus*)	13 kJ m^{-2} d^{-1}, 3 h/d (at noon) for 16 d, pre-harvest.	Flavonoids	Total K and Q gly increased by ~150% and 70% in cvs. Paroll and Stallion, respectively. UV-B induced a specific increase in Q gly relative to K gly with a 36- and 23-fold increase in cvs Paroll and Stallion. Q and K 3-sophoroside-7-gly and 3-(2-E-sinapoylsophoroside)-7-gly appeared after UVB exposure.	[48]
Kale (*Brassica oleracea* var. *sabellica* L.)	0.22–0.88 kJ m^{-2} d^{-1} + 72 µmol m^{-2} s^{-1} for 1 d, ID: 30 cm, 15 °C, 24 h period of acclimatization before harvest.	Flavonol gly and HA derivatives	Q gly decreased under UVB. For K gly in the investigated UV-B range, monoacylated K tetragly decreased (46–63%), monoacylated K trigly increased depending on the acylation pattern (up to 96%), and monoacylated K digly increased strongly (197–441%) at the highest dose. The HA gly, diSg and S-Fg were enhanced by 49% and 88% in a dose-dependent manner.	[27]
Kale (*Brassica oleracea* var. *sabellica* L.)	1–5 daily doses of 0.25 kJ m^{-2} d^{-1}, for 1 h with 23 h acclimatization intervals, during 5 d (total dose: 1.25 kJ m^{-2} d^{-1}), ID: 60 cm, 5 or 15 °C.	PCs	All Q gly increased with increased UVB radiation, while K gly responded dependent on the HA residue. PCs containing a catechol structure favored in the response to UVB. 11 (of 20) PCs (e.g., monoacylated Q gly) were influenced by the interaction UV-B-temperature. Enhanced mRNA expression of *flavonol 3′-hydroxylase* showed an interaction of UV-B and temperature (highest at 0.75 kJ m^{-2}, 15 °C).	[49]
Silver birch seedlings (*Betula pendula*)	7.3–8.5 kJ m^{-2} d^{-1}, ID: 40 cm, 9 h/d for 10 d.	PCs	UVB induced production of K gly, chlorogenic acids, HA derivatives and Q gly, such as Q-3-galactoside and Q-3-rhamnoside. Leaf area was reduced by UV-B radiation.	[50]
Pak choi (*Brassica napus* subsp. *chinensis*)	0.35 W m^{-2}, 16 h/d for 7 d, pre-harvest, 9 or 22 °C.	PCs	UVB induced a 4-fold higher content in total flavonoids at 22 °C, but not at 9 °C. K-gly acylated with caffeic or coumaric acid (and not ferulic, hydroxyferulic, or sinapic acid) responded to UVB exposure. HA derivatives increased by 2-fold at lower temperatures (9 °C) and did not change at 22 °C.	[51]
Nasturtium (*Nasturtium officinale*)	0.075 and 0.15 W h m^{-2}, for 1 or 2 h, post-harvest, acclimatization period of 2 and 22 h, 20 °C.	Total PCs and GLSs	Plant response to UVB exposure is organ-, plant tissue age-, and phytochemical-specific. At lowest dose and adaptation time, GP increased by 6-, 3- and 2-fold in seeds, leaves and inflorescences, respectively. At highest dose and adaptation time, GP increased by 3- and 6-fold in inflorescences and leaves, but decreased in seeds. At both doses, total PC concentration in leaves and seeds decreased after 2 h, but increased to reach control levels after 22 h.	[52]
Brocoli sprouts (*Brassica oleracea* var. *italica*)	1 kJ m^{-2} d^{-1}, 5 h/d for 5 d, pre-harvest, ID: 40 cm.	GLSs	UV-B induced higher (<2-fold) aliphatic (but not indole) GLS levels when compared with untreated sprouts. This had a negative effect on the growth of aphid *Myzus persicae* and attack by caterpillar *Pieris brassicae*. L.	[29]
Broccoli sprouts (*Brassica oleracea* var. *italica*)	0.6 kJ m^{-2} d^{-1}, 4 h, 24 h period of acclimatization before harvest, ID: 40 cm.	Flavonoids, GLSs, carotenoids, Chls	Increases in K, Q, GRA and 4-MGBS were observed, each of them by up to roughly 2-fold. B-carotene and Chls levels remained unaffected. Increased expression of genes associated with salicylate (4- to 5-fold) and jasmonic acid (3- to 4- fold) signaling defense pathways were observed.	[29]

Table 1. *Cont.*

Plant Species	UVB Treatment Parameters and Storage Conditions	Phytochemical Evaluated	Main Findings	Reference
Broccoli (*Brassica oleracea* var. *italica*)	20 kJ m^{-2} d^{-1} + 19 µmol m^{-2} s^{-1}, 12 h/d for 3 d, post-harvest, during storage at 10 or 18 °C.	GLSs	UVB treatments applied during storage did not influence total and individual GLS levels in broccoli flower buds. Only exposure for 3 d to visible light (25 µmol m^{-2} s^{-1}) increased aliphatic GLSs stored at 10 °C, or indolic GLSs (4-HGBS and 4-MGBS) at 18 °C.	[53]
Broccoli (*Brassica oleracea* var. *italica*)	2.2, 8.8 or 16.4 kJ m^{-2} d^{-1} for 76 d (from planting until maturity stage), ID: 15 cm, post-harvest storage for 60 d at 0 °C.	PCs, GLSs, carotenoids, Chls	Total carotenoid and Chl contents decreased (up to 40 and 70%, respectively) with the increased dose of UV and storage time. The highest UVB dose increased ascorbic acid (~115%) and total PC content (~74%) prior storage. Sinigrin (~2-fold) and GP (~3-fold) contents increased by increased UVB dose and storage time.	[54]
Lettuce (*Lactuca sativa*)	4.2 W, 4 h/d for 6 d; or gradually increased irradiation from 1–7 h/d over 6 d, ID: 20 cm, 20 °C.	Total PCs	Repeated or gradual UVB exposure yielded ~3.6–3.2 times more total PCs, respectively, than the controls did 2 d after UVB exposure. These treatments boosted the antioxidant capacity by 80 and 45%, respectively, 2 d after UVB treatments. Both treatments inhibited lettuce growth.	[55]
Prickly pear (*Opuntia ficus-indica*)	6.4 W m^{-2} for 0, 15, 90, and 180 min and stored for 24 h at 16 °C. ID: 45 cm. Whole and wounded fruit samples were subjected to UVB radiation.	Betalains, total PCs, ascorbic acid	UVB radiation for 15 min was the best treatment to induce the accumulation of bioactive compounds. UVB radiation (15 min) of the wounded tissue induced an immediate accumulation of ascorbic acid (54–58%) and betalains (33–40%) in the peel and pulp of the fruit. After 24 h, the pulp of irradiated whole fruits showed the highest accumulation of betalains (49.8%) and phenolics (125.8%) as compared with the control, whereas the stored wounded tissue treated with UVB presented accumulation of ascorbic acid in the peel (84.6%) and pulp (67.2%).	[56]
Prickly pear (*Opuntia ficus-indica*)	6.4 W m^{-2} for 15 min and stored for 24 h at 16 °C. ID: 45 cm. Whole and wounded fruit samples were subjected to UVB radiation.	Betalains (indicaxantin and betanin)	UVB radiation applied in the whole tissue induced an immediate accumulation of indicaxanthin after treatment, obtaining increases of 106.5% and 325.8% in the peel and pulp, respectively. After storage, the tissue treated with UVB radiation and wounding before storage showed a synergistic effect on the accumulation of betanin in the peel (315.0%) and indicaxanthin in the pulp (447.0%).	[57]
Broccoli sprouts (*Brassica oleracea* var. *italica*)	Seven-day-old broccoli sprouts were exposed to UVB (9.47 W m^{-2}) alone and combined with methyl jasmonate	GSLs	UVB increased the content of aliphatic and indole glucosinolates, such as glucoraphanin (78%) and 4-methoxy-glucobrassicin (177%).	[58]
Broccoli sprouts (*Brassica oleracea* var. *italica*)	Seven-day-old broccoli sprouts were exposed to UVB (3.34 W m^{-2})	GSLs and PCs	After 24 h of UVB treatment, sprouts showed increases in 4-methoxy-glucobrassicin, glucobrassicin, and glucoraphanin by 170%, 78%, and 73%, respectively. Moreover, increases in gallotannic acid (~48%), 5-sinapoyl-quinic acid (~121%), and sinapoyl malate (~12%) were UVB-induced.	[59]
Peaches (*Prunus persica* L., cv. Fairtime)	Fruits were exposed to 10- or 60-min UVB treatment (1.39 and 8.33 kJ m^{-2}, respectively), and sampled at different time points from the exposure	PCs	After 24 h of 60-min UVB exposures, flavanols, flavones, flavonols, and dihydroflavonols increased by 123%, 70%, 55%, and 50% compared to the control group. Specifically, after 24 h of UVB treatment, the 60-min UVB exposure increased spinacetins, isorhamnetins, and kaempferols by 61%, 448%, and 95%, respectively.	[60]
Peaches (*Prunus persica* L., cv. Batsch)	The fruit were placed in a chamber at 20 °C followed by UVB (58 µw/cm^2) irradiation for 2 days	PCs	Cyanidin 3-glucoside reached (0.31 mg/100 g FW) after UVB treatment, value that was 3-fold higher than the fruits stored under dark conditions.	[61]
Apples (skin) (*Malus domestica*)	UVB lamp provided 1.69 W m^{-2} at fruit height. The treatment lasted 36 h (219 kJ m^{-2}).	PCs	Hydroxycinnamic acids (feruloyl glucoside, cryptochlorogenic and chlorogenic acids) showed an increase of 38% following 36 h of treatment and maintained higher values in the treated samples during storage as well as anthocyanins. At the end of the storage time (21 d) flavonols were 64% higher in the UVB-treated apples than the control, indicating that UVB treatment decreased flavonoid loss during storage.	[62]
Table grapes (*Vitis vinifera* × *Vitis labrusca* cv. Summer Black)	Grapes were exposed to 3.6 kJ m^{-2} UVB irradiation	PCs	Samples showed increses in the content of gallic acid (19.05%), protocatechuic acid (64.6%), naringenin (67.7%), quercetin 3-glucoside (13.86%), catechin (54.15%), epicatechin (23.89%), *trans*-resveratrol (23.53%), and *trans*-piceid (31.56%) after UVB treatment as compared with the control.	[63]

Abbreviations: ID: irradiation distance; PC, phenolic compound; GLS, glucosinolate, Chl, chlorophyll, K, kaempferol; Q, quercetin; gly, glycosides; HA, hydroxycinnamic acid; GP, glucotropaeolin; GRA, glucoraphanin; 4-HGBS, 4-hydroxy-glucobrassicin; 4-MGBS, 4-methoxy-glucobrassicin; diSg, disinapoylgentiobiose; S-Fg, sinapoyl-feruloylgentiobiose.

Quercetin and its glycosides, as well as kaempferol and most kaempferol glycosides, were shown to be enhanced after additional UVB radiation in broccoli, canola (*Brassica napus*), and table grapes [48,60,63]. In kale (*Brassica oleracea* var. *sabellica*), monoacylated kaempferol tetraglucosides decreased following exposure to a single, low dose of UVB, whereas the monoacylated kaempferol diglucosides increased strongly under the same dose. Additionally, the hydroxycinnamic acid glycosides disinapoyl-gentiobiose and sinapoyl-feruloyl-gentiobiose were enhanced in a dose-dependent manner under UVB [27].

Tegelberg et al. [50] observed an increase in caffeoylquinic acid in silver birch exposed to slightly above-ambient UVB radiation, while Lancaster et al. [45] showed similar results in apples (*Malus domestica*).

The response of flavonoid glycosides is dependent on the type of phenolic acid that is acylated to the flavonol glycoside (mainly hydroxycinnamic acids). In pak choi (*Brassica campestris* ssp. *chinensis*), total flavonoid levels increased with exposure to additional UVB but kaempferol glycosides acylated with ferulic, hydroxyferulic, or sinapic acid did not respond to UVB light [51]. Likewise, in peaches, cyanidin 3-glucoside reached (0.31 mg/100 g FW) after UVB treatment, a value that was 3-fold higher than in the fruits stored under dark conditions [61]. In kale, the structural characteristics of the hydroxycinnamic acids themselves have an impact on the response to UVB [49]. While the levels of caffeic acid and hydroxyferulic acid monoacylated kaempferol triglycosides (containing a catechol structure) were increased with exposure to higher UVB radiation, the ferulic and sinapic acid monoacylated kaempferol triglycosides (no catechol structure) were not affected. In canola, the nonacylated kaempferol-3-*O*-sophoroside-7-*O*-D-glucoside increased with the additional UVB, while the sinapic acid monoacylated kaempferol glycoside did not respond [48]. Moreover, in peaches after 24 h of 60-min UVB exposures, flavanols, flavones, flavonols, and dihydroflavonols increased by 123%, 70%, 55% and 50% compared to the control group. Specifically, after 24 h of UVB treatment, the 60-min UVB exposure increased spinacetins, isorhamnetins, and kaempferols by 61%, 448% and 95%, respectively [60].

Hydroxycinnamic acids are known scavengers to ROS induced by UVB radiation. In kale, the hydroxycinnamic acid derivatives (caffeoylquinic acid disinapoyl-gentiobiose and sinapoyl-feruloyl-gentiobiose) were hardly affected by subsequent doses of UVB radiation [49]. However, a single moderate UVB dose led to a slight decrease in caffeoylquinic acid but an increase in disinapoyl-gentiobiose and sinapoyl-feruloyl-gentiobiose [27]. Likewise, in apples, hydroxycinnamic acids (feruloyl glucoside, cryptochlorogenic and chlorogenic acids) showed an increase of 38% following 36 h of treatment and maintained higher values in the treated samples during storage as well as anthocyanins. At the end of the storage time (21 d) flavonols were 64% higher in the UVB-treated apples than in the control, indicating that UVB treatment decreased flavonoid loss during storage. [62].

Regarding glucosinolates, reports on supplemental UVB effects on accumulation on individual glucosinolate in *Brassicaceae* plants are rare. However, it has been proposed that at higher doses, UVB induces JA defense and wound signaling, while lower UVB levels induces SA pathway signaling response and expression of genes encoding for pathogenesis-related proteins (e.g., PR-1, PR-2, PR-4, PR-5, PDF1.2), triggering alterations in the plant defense metabolism [28]. Application of low doses of UVB for 5 days increased aliphatic glucosinolate levels in broccoli sprouts, leading to a decreased plant susceptibility to insect attacks [29]. Likewise, low doses of UVB have been shown to increase levels of aliphatic glucosinolates (GRA and 4-MGBS) in *Arabidopsis thaliana* and broccoli sprouts and an aromatic glucosinolate (glucotropaeolin) in nasturtium (*Tropaeolum majus*) [29,52].

Less is known about the effect of UVB on carotenoids and chlorophylls, however, studies in soybean (*Glycine max*) plants and bean (*Phaseolus vulgaris*) leaves suggest that UVB radiation damages the chloroplast's photosystem II (PSII), enhances lipid peroxidation, ion leakage and H_2O_2 content. Nevertheless, the plant may be able to counteract these effects by producing NO, which has been shown to prevent ion leakage, lipid oxidation and chlorophyll loss; and to induce transcript levels of antioxidant enzymes such as superoxide dismutase (SOD), catalase (CAT), ascorbate peroxidase (APX) and hemeoxygenase (HO) [19].

Although the application of UVB stress as an approach to enhance the phytochemical content has been reported in *Brassica* plants, including a report in broccoli sprouts [29,53,54,59], research mainly focuses on the mature vegetables, on either glucosinolate or phenolic enhancement, on postharvest treatments [48,52], and on the accumulation of defensive metabolites to provide plant resistance against insects and pathogens, acting as natural pesticides [29,64], rather than nutraceutical-related applications.

3.2. Mechanisms and Effects of UVA Radiation on Phytochemical Biosynthesis

The UVA component of sunlight has traditionally been considered to be damaging for photosynthesis, with the PSII complex being its main target. Since solar radiation contains much more UVA than UVB, it has been recently suggested that UVA radiation could be the most detrimental component of sunlight for photosynthetic reactions, despite the lower quantum efficiency of UVA-mediated photoinhibitory damage compared to that caused by UVB exposure [33].

Despite the evident relevance of UVA radiation on plant morphology, physiology, biochemistry, and photosynthesis, there is a lack of scientific studies to elucidate the signaling mechanisms governing such responses. Available reports (Table 2) in species such as peppermint (*Mentha piperita*) and lettuce (*Lactuca sativa*) have revealed that UVA radiation, similarly to UVB, can trigger the accumulation of leaf total phenolics [9,58–70]. Changes in individual phenolic compounds, rather than total phenolic content, have been described in *Betula pendula* [69] and *Arbutus unedo* [70], where quercetin (and its derivatives) and gallic acid derivatives were induced, respectively.

Table 2. Effect of UVA (alone or in combination with UVB) radiation on the biosynthesis of bioactive phytochemicals and on stress responses in different horticultural crops.

Plant Species	UVA Treatment Parameters and Storage Conditions	Phytochemical Evaluated	Main Findings	Reference
Peppermint (*Mentha × piperita*)	3 h of 126 µmol m^{-2} s^{-1}(WL), 9 h of 46 µmol m^{-2} s^{-1}(UVA) + WL, 3 h WL, 9 h dark. Or 15 h WL, 9 h UVA. For 18 d.	Total PCs, Chls, essential oils (e.g., menthol)	Treatment with UVA + WL during the day increased leaf area (by 64%), total PC (2-fold) and total essential oil (by 24%) content, while UVA during the night decreased these parameters (by 14, 23 and 38%), when compared to controls. UVA did not affect total Chl content. An interference of UVA with phytochrome is suggested.	[65]
Lettuce (*Lactuca sativa*)	3.7 W, continuous radiation for 7 d, ID: 20 cm, 20 °C.	Total PCs, total ACNs	UVA treatment induced shoot growth (1–2 fold) at days 5–7. UVA caused accumulation of PCs (30% at day 3) and ACNs (4-fold at day 3) until 4 d of treatment, consistent with an increase in *PAL* gene expression (2.4-fold) and PAL activity.	[66]
Rosa hybrida, Fuchsia hybrida	15.9 µmol m^{-2} s^{-1}(of UVA) + 227 µmol m^{-2} s^{-1}(of WL), 12 h/d, for 6–7 weeks, 20 °C.	Flavonoids, carotenoids, Chls	Supplemental UVA did not affect plant morphology of either species. In rosa and fuchsia leaves, it induced increases in levels of Chls a (28 and 7%) and b (25 and 10%), the carotenoids antheraxanthin (65 and 23%), lutein (25 and 3%) and β-carotene (18 and 5%), K (93 and 313%) and Q (77 and 119%) aglycones, and up to 2-fold increases in individual Q derivatives in both species and K derivatives in rose leaves. Some K derivatives in fuchsia increased >2-fold or were newly induced.	[67]
Tomato seedlings (*Solanum lycopersicum*)	7 W m^{-2}, for 24 h total during cultivation, with harvest at different time points (0, 1, 3, 6, 12 and 24 h), 24 °C.	ACNs	UVA induced ACN production by up to 4-fold in tomato seedlings (at 24 h) and by ~2-fold in the fruit epidermis (at 6 h). ACN increased gradually in the hypocotyls, with maximum levels (3-fold) at 12 h. In the cotyledons, ACN content increased (4-fold) at 1 h after UVA exposure, was reduced afterward, and increased again beginning at 3 h. UVA increased PAL expression in a time-dependent manner.	[68]
Wounded Carrot (*Daucus carota*)	12.73 W m^{-2}, for 0, 1, and 6 h at room temperature, ID: 50 cm. Post- storage at 15 °C for 4 d.	PCs	After 6 h of UVA radiation, total phenolic content of carrot pies increased 1-fold, AOX capacity increased 2-folds, and PAL activity increased by 34-fold. Maximum accumulation of the individual PCs chlorogenic acid (3-fold), ferulic acid (~1-fold), 3,5-dicafeoylquinic acid (<1-fold) were observed after 6 h of treatment.	[9]
Broccoli (*Brassica oleracea* var. *italica*)	Field experiment. 14 and 985 kJ m^{-2} d^{-1} of ambient UVB and UVA. From germination until plant age of 27 or 41 d.	PCs and GLSs	Increased concentrations of total flavonoids + HAs (54.4% compared to control plants) after UV exposure was observed. 4-HGBS was the only GLS increased by UV exposure.	[64]

Table 2. *Cont.*

Plant Species	UVA Treatment Parameters and Storage Conditions	Phytochemical Evaluated	Main Findings	Reference
Silver birch seedlings (*Betula pendula*)	Field experiment (30 d). Treatments: Ambient (no filter); UVA 100%, UVB 100%; UVA 100%, UVB 50%; UVA 50%, UVB 50%; UVA 100%, UVB 0%; UVA 50%, UVB 0%; and UVA 0%, UVB 0%. UV doses: 4.48, 3.74, 2.20, 1.98, 0.38, 0.26 and 0 kJ m^{-2} d^{-1}, respectively, FD: 15 cm.	PCs	Epidermal flavonoids decreased when UVB was excluded, and transcripts of *PAL* and *HYH* were expressed at lower levels. UVA linearly accumulated Q-3-galactoside and Q-3-arabinopyranoside and had a quadratic effect on *HYH* expression. There were strong positive correlations between *PAL* expression and accumulation of 4 flavonols under the UV treatments. Chlorogenic acids were not affected by UV treatments.	[69]
Arbutus unedo	Field experiment (1 year). Treatments: 97% UVB reduction (UVA: 0.33–1.29 MJ m^{-2} d^{-1}), 95% UVA + UVB reduction (UV0) or near-ambient UV levels (UVB/A, UVB: 4.2–34.4 kJ m^{-2} d^{-1}).	PCs	Leaves exposed to near-ambient UV radiation had less total flavanol content (1.32-fold) than those developed with almost no UV exposure.UVA radiation increased (1.4-fold) the leaf content of theogallin, a gallic acid derivative. Quercitrin, the major Q derivative, increased by 1.32- and 1.26-fold with UVB/A and UVA exposure, respectively.	[70]
Broccoli sprouts (*Brassica oleracea* var. *italica*)	Seven-day-old broccoli sprouts were exposed to UVA (9.47 W m^{-2}) alone and combined with methyl jasmonate. ID: 45 cm.	GSLs, PCs, carotenoids, and chlorophyll	UVA + methyl jasmonate increased the total glucosinolate content by 154%. MJ induced the biosynthesis of indole glucosinolates, especially neoglucobrassicin (538%), showing a synergistic effect with UVA stress. UVA increased the content of phenolics such as kaempferol glucoside (25.4%) and gallic acid (57%). UVA increased lutein (~23%), chlorophyll b (31%), neoxanthin (34%), and chlorophyll a (67%).	[58]
Broccoli sprouts (*Brassica oleracea* var. *italica*)	Seven-day-old broccoli sprouts were exposed for 120 min UVA (3.15 W m^2). Harvest 2 h post-treatment. ID: 45 cm.	GSLs and PCs	UVA radiation and harvest 2 h afterwards induced the accumulation of 4-O-caffeoylquinic acid (42%), 1-sinapoyl-2,2-diferulolyl-gentiobiose (61%), gallic acid hexoside I (14%), and gallic acid derivative (48%).	[59]
Peaches (*Prunus persica* L., cv. Batsch)	The fruit were placed in a chamber at 20 °C followed by UVA (1000 μw/cm^2) irradiation for 2 days	PCs	Cyanidin 3-glucoside reached (0.61 mg/100 g FW) after UVB treatment, value that was 4-fold higher than the fruits stored under dark conditions.	[61]

Abbreviations: WL, white (visible) light; ID: irradiation distance; FD: filter distance; PC, phenolic compound; Chl, chlorophyll; PAL, phenylalanine ammonia lyase, ACN, anthocyanin; K, kaempferol; Q, quercetin.

On the other hand, scientific data is needed regarding the effects of UVA radiation on the accumulation of total and individual glucosinolates, carotenoids, and other phytochemicals. One study, evaluating the effect of supplemental UVA radiation in leaves of *Rosa hybrida* and *Fuchsia hybrida* on levels of various antioxidants, indicates that UVA induced small increments in levels of chlorophylls *a* and *b*, and the carotenoids antheraxanthin, lutein and β-carotene, and high increments in the flavonols quercetin, kaempferol and their derivatives [67]. The authors conclude that the major protection towards UVA radiation in *R. hybrida* and *F. hybrida* leaves originates from absorption of radiation, and not from ROS scavenging [67], although the mechanisms involved are not fully understood.

It has been stated that the UVA-mediated changes in phenolic composition are likely to be controlled at multiple levels of gene regulation. At the transcription level, UVA radiation induces transcript accumulation of genes involved in the phenylpropanoid pathway including *PAL*, *CHS*, and *dihydroflavonol 4-reductase* (*DFR*) [33]. Post-transcriptionally, the activity of PAL has been increased by UVA in lettuce [66] and tomato (*Solanum lycopersicum*) [67]. Unlike UVB radiation, there is limited information on the specific genetic components associated with UVA signaling in plants. Regarding the latter, early studies in leaves of *A. thaliana* suggest the likelihood that they involve the action of a photoreception system that contains three known classes of photoreceptors: (i) phytochromes (PHY) for far-red and red lights, (ii) crytochromes (CRY) for UVA/blue light, and (iii) phototropins (PHOT) for UVA light [33,71]. For instance, the UVA-induced transcription and expression of key flavonoid biosynthesis genes (e.g., *CHS*) in leaves of *A. thaliana* could be initiated via UVA absorption through CRY1, given that functional CRY1 is required for the expression of *CHS* [72]. In addition, the authors stated that the UVA photo-transduction pathway may interact synergistically with UVB-induced pathways to produce transient signals and may function additively to stimulate *CHS* promoter function [72]. Thus, once UVA light has been perceived, these UVA specific photoreceptors may interact with COP1 and HY5 in a similar manner to UVR8 to further regulate the plant's secondary metabolism [71]. Furthermore, studies with an *A. thaliana* UVB photoreceptor mutant generated the unexpected finding that UVB-specific photoreceptor UVR8 has an impact on UVA-mediated changes in plant metabolites, as UVR8 is likely to interact with UVA/blue light signaling pathways to moderate UVB-driven transcripts [73].

4. Conclusions and Further Research

The reviewed literature shows that UVR is an effective tool to biofortify horticultural crops with nutraceuticals. Further research in this topic should be focused on increasing our understanding of the molecular and physiological mechanisms governing the UVA stress response, since most reports in the literature have assessed the effect of UVB radiation. UVR could be an easy technology to treat fruits and vegetables before eating them even at home, thus UV chambers for residential use could be an interesting piece of equipment to improve the nutraceutical content of crops in the kitchen.

Author Contributions: Conceptualization, D.A.J.-V., M.M.-R. and J.B.; investigation, D.A.J.-V., M.M.-R. and J.B.; literature search, D.A.J.-V., M.M.-R. and J.B.; writing—original draft preparation, D.A.J.-V., M.M.-R. and J.B.; writing—review and editing, D.A.J.-V., M.M.-R. and J.B.; supervision, D.A.J.-V., M.M.-R. and J.B. The manuscript was written through contributions of all authors. All authors have read and agreed to the published version of the manuscript.

Funding: This research received no external funding.

Institutional Review Board Statement: Not applicable.

Informed Consent Statement: Not applicable.

Data Availability Statement: Not applicable.

Acknowledgments: Authors acknowledge the scholarship from the Consejo Nacional de Ciencia y Tecnología (CONACyT) to Melissa Moreira-Rodríguez, and the support of Tecnológico de Monterrey, Bioprocess Research Group.

Conflicts of Interest: The authors declare no conflict of interest.

References

1. Cisneros-Zevallos, L.; Jacobo-Velázquez, D.A. Controlled abiotic stresses revisited: From homeostasis through hormesis to extreme stresses and the impact on nutraceuticals and quality during pre- and postharvest applications in horticultural crops. *J. Agric. Food Chem.* **2020**, *68*, 11877–11879. [CrossRef] [PubMed]
2. Pérez-Balibrea, S.; Moreno, D.A.; García-Viguera, C. Improving the phytochemical composition of broccoli sprouts by elicitation. *Food Chem.* **2011**, *129*, 35–44. [CrossRef]
3. Santana-Gálvez, J.; Cisneros-Zevallos, L.; Jacobo-Velázquez, D.A. A practical guide for designing effective nutraceutical combinations in the form of foods, beverages, and dietary supplements against chronic degenerative diseases. *Trends Food Sci. Technol.* **2019**, *88*, 179–193. [CrossRef]
4. Gastélum-Estrada, A.; Serna-Saldívar, S.O.; Jacobo-Velázquez, D.A. Fighting the COVID-19 pandemic through biofortification: Innovative approaches to improve the immunomodulating capacity of foods. *ACS Food Sci. Technol.* **2021**, *1*, 480–486. [CrossRef]
5. Jacobo-Velázquez, D.A. Definition of biofortification revisited. *ACS Food Sci. Technol.* 2022; *in press*.
6. Torres-Contreras, A.M.; Nair, V.; Cisneros-Zevallos, L.; Jacobo-Velázquez, D.A. Stability of bioactive compounds in broccoli as affected by cutting styles and storage time. *Molecules* **2017**, *22*, 636. [CrossRef]
7. Torres-Contreras, A.M.; Nair, V.; Senés-Guerrero, C.; Pacheco, A.; González-Agüero, M.; Ramos-Parra, P.A.; Cisneros-Zevallos, L.; Jacobo-Velázquez, D.A. Chemical genetics applied to elucidate the physiological role of stress-signaling molecules on the wound-induced accumulation of glucosinolates in broccoli. *Plants* **2021**, *10*, 2660. [CrossRef]
8. Villarreal-García, D.; Nair, V.; Cisneros-Zevallos, L.; Jacobo-Velázquez, D.A. Plants as biofactories: Postharvest stress-induced accumulation of phenolic compounds and glucosinolates in broccoli subjected to wounding stress and exogenous phytohormones. *Front. Plant Sci.* **2016**, *7*, 45. [CrossRef]
9. Surjadinata, B.B.; Jacobo-Velázquez, D.A.; Cisneros-Zevallos, L. UVA, UVB and UVC light enhances the biosynthesis of phenolic antioxidants in fresh-cut carrot through a synergistic effect with wounding. *Molecules* **2017**, *22*, 668. [CrossRef]

10. Torres-Contreras, A.M.; Nair, V.; Cisneros-Zevallos, L.; Jacobo-Velázquez, D.A. Plants as biofactories: Stress-induced production of chlorogenic acid isomers in potato tubers as affected by wounding intensity and storage time. *Ind. Crop. Prod.* **2014**, *62*, 61–66. [CrossRef]
11. Torres-Contreras, A.M.; Nair, V.; Cisneros-Zevallos, L.; Jacobo-Velázquez, D.A. Effect of exogenous amylolytic enzymes on the accumulation of chlorogenic acid isomers in wounded potato tubers. *J. Agric. Food. Chem.* **2014**, *62*, 7671–7675. [CrossRef] [PubMed]
12. He, C.; Jacobo-Velázquez, D.A.; Cisneros-Zevallos, L.; Davis, F.T. Hypobaria and hypoxia affects phytochemical production, gas exchange, and growth of lettuce. *Photosynthetica* **2013**, *51*, 465–473. [CrossRef]
13. Hollósy, F. Effects of ultraviolet radiation on plant cells. *Micron* **2002**, *33*, 179–197. [CrossRef]
14. Loconsole, D.; Santamaria, P. UV lighting in horticulture: A sustainable tool for improving production quality and food safety. *Horticulturae* **2021**, *7*, 9. [CrossRef]
15. Darré, M.; Vicente, A.R.; Cisneros-Zevallos, L.; Artés-Hernández, F. Postharvest ultraviolet radiation in fruit and vegetables: Applications and factors modulating its efficacy on bioactive compounds and microbial growth. *Foods* **2022**, *11*, 653. [CrossRef]
16. Jansen, M.A.K.; Gaba, V.; Greenberg, B.M. Higher plants and UV-B radiation: Balancing damage, repair and acclimation. *Trends Plant. Sci.* **1998**, *3*, 131–135. [CrossRef]
17. Jenkins, G.I.; Brown, B.A. UV-B Perception and Signal Transduction. In *Light and Plant Development*; Blackwell Publishing Ltd.: Oxford, UK, 2007; pp. 155–182.
18. Kunz, B.A.; Cahill, D.M.; Mohr, P.G.; Osmond, M.J.; Vonarx, E.J. Plant responses to UV radiation and links to pathogen resistance. *Int. Rev. Cytol.* **2006**, *255*, 1–40. [CrossRef]
19. Jacobo-Velázquez, D.A.; Martínez-Hernández, G.B.; Rodríguez, S.; Cao, C.-M.; Cisneros-Zevallos, L. Plants as biofactories: Physiological role of reactive oxygen species on the accumulation of phenolic antioxidants in carrot tissue under wounding and hyperoxia stress. *J. Agric. Food Chem.* **2011**, *59*, 6583–6593. [CrossRef]
20. Jacobo-Velázquez, D.A.; González-Agüero, M.; Cisneros-Zevallos, L. Cross-talk between signaling pathways: The link between plant secondary metabolite production and wounding stress response. *Sci. Rep.* **2015**, *5*, 8608. [CrossRef]
21. Gastélum-Estrada, A.; Hurtado-Romero, A.; Santacruz, A.; Cisneros-Zevallos, L.; Jacobo-Velázquez, D.A. Sanitizing after fresh-cutting carrots reduces the wound-induced accumulation of phenolic antioxidants compared to sanitizing before fresh-cutting. *J. Sci. Food Agric.* **2020**, *100*, 4995–4998. [CrossRef]
22. Mikkelsen, M.D.; Larsen-Petersen, B.; Glawischnig, E.; Bøgh-Jensen, A.; Andreasson, E.; Halkier, B.A. Modulation of CYP79 genes and glucosinolate profiles in *Arabidopsis* by defense signaling pathways. *Plant Physiol.* **2003**, *131*, 298–308. [CrossRef] [PubMed]
23. Cisneros-Zevallos, L.; Jacobo-Velázquez, D.A.; Pech, J.C.; Koiwa, H. Signaling Molecules Involved in the Postharvest Stress Response of Plants. In *Handbook of Plant and Crop Physiology*, 3rd ed.; CRC Press: Boca Raton, FL, USA, 2014; pp. 259–276.
24. Vranová, E.; Inzé, D.; Van Breusegem, F. Signal transduction during oxidative stress. *J. Exp. Bot.* **2002**, *53*, 1227–1236. [CrossRef] [PubMed]
25. Simontacchi, M.; Galatro, A.; Ramos-Artuso, F.; Santa-María, G.E. Plant survival in a changing environment: The role of nitric oxide in plant responses to abiotic stress. *Front. Plant Sci.* **2015**, *6*, 977. [CrossRef] [PubMed]
26. Jansen, M.A.K.; Hectors, K.; O'Brien, N.M.; Guisez, Y.; Potters, G. Plant stress and human health: Do human consumers benefit from UV-B acclimated crops? *Plant Sci.* **2008**, *175*, 449–458. [CrossRef]
27. Neugart, S.; Zietz, M.; Schreiner, M.; Rohn, S.; Kroh, L.W.; Krumbein, A. Structurally different flavonol glycosides and hydroxycinnamic acid derivatives respond differently to moderate UV-B radiation exposure. *Physiol. Plant.* **2012**, *145*, 582–593. [CrossRef]
28. Schreiner, M.; Martínez-Abaigar, J.; Glaab, J.; Jansen, M. UV-B Induced secondary plant metabolites. *Opt. Photonik* **2014**, *9*, 34–37. [CrossRef]
29. Mewis, I.; Schreiner, M.; Nguyen, C.N.; Krumbein, A.; Ulrichs, C.; Lohse, M.; Zrenner, R. UV-B irradiation changes specifically the secondary metabolite profile in broccoli sprouts: Induced signaling overlaps with defense response to biotic stressors. *Plant Cell Physiol.* **2012**, *53*, 1546–1560. [CrossRef]
30. Schreiner, M.; Mewis, I.; Huyskens-Keil, S.; Jansen, M.A.K.; Zrenner, R.; Winkler, J.B.; O'Brien, N.; Krumbein, A. UV-B-induced secondary plant metabolites-potential benefits for plant and human health. *CRC Crit. Rev. Plant. Sci.* **2012**, *31*, 229–240. [CrossRef]
31. Jenkins, G.I. Signal transduction in responses to UV-B radiation. *Annu. Rev. Plant Biol.* **2009**, *60*, 407–431. [CrossRef]
32. Tilbrook, K.; Arongaus, A.B.; Binkert, M.; Heijde, M.; Yin, R.; Ulm, R. The UVR8 UV-B Photoreceptor: Perception, Signaling and Response. *Arab. Book* **2013**, *11*, e0164. [CrossRef]
33. Verdaguer, D.; Jansen, M.A.K.; Llorens, L.; Morales, L.O.; Neugart, S. UV-A radiation effects on higher plants: Exploring the known unknown. *Plant Sci.* **2017**, *255*, 72–81. [CrossRef] [PubMed]
34. Brosché, M.; Strid, A. Molecular events following perception of ultraviolet-B radiation by plants. *Physiol. Plant.* **2003**, *117*, 1–10. [CrossRef]
35. Mackerness, S.; Surplus, S.L.; Blake, P.; John, C.F.; Buchanan-Wollaston, V.; Jordan, B.R.; Thomas, B. Ultraviolet-B-induced stress and changes in gene expression in Arabidopsis thaliana: Role of signalling pathways controlled by jasmonic acid, ethylene and reactive oxygen species. *Plant Cell Environ.* **1999**, *22*, 1413–1423. [CrossRef]
36. Brown, B.A.; Cloix, C.; Jiang, G.H.; Kaiserli, E.; Herzyk, P.; Kliebenstein, D.J.; Jenkins, G.I. A UV-B-specific signaling component orchestrates plant UV protection. *Proc. Natl. Acad. Sci. USA* **2005**, *102*, 18225–18230. [CrossRef]

37. Brown, B.A.; Jenkins, G.I. UV-B signaling pathways with different fluence-rate response profiles are distinguished in mature *Arabidopsis* leaf tissue by requirement for UVR8, HY5, and HYH. *Plant Physiol.* **2008**, *146*, 576–588. [CrossRef]
38. Christie, J.M.; Arvai, A.S.; Baxter, K.J.; Heilmann, M.; Pratt, A.J.; O'Hara, A.; Jenkins, G.I. Plant UVR8 photoreceptor senses UV-B by tryptophan-mediated disruption of cross-dimer salt bridges. *Science* **2012**, *335*, 1492–1496. [CrossRef]
39. Favory, J.J.; Stec, A.; Gruber, H.; Rizzini, L.; Oravecz, A.; Funk, M.; Seidlitz, H.K. Interaction of COP1 and UVR8 regulates UV-B-induced photomorphogenesis and stress acclimation in *Arabidopsis*. *EMBO J.* **2009**, *28*, 591–601. [CrossRef]
40. Kaiserli, E.; Jenkins, G.I. UV-B promotes rapid nuclear translocation of the Arabidopsis UV-B–specific signaling component UVR8 and activates its function in the nucleus. *Plant Cell* **2007**, *19*, 2662–2673. [CrossRef]
41. Schreiner, M.; Mewis, I.; Neugart, S.; Zrenner, R.; Glaab, J.; Wiesner, M.; Jansen, M.A.K. UV-B Elicitation of Secondary Plant Metabolites. In *III-Nitride Ultraviolet Emitters*; Dreyer, C., Mildner, F., Eds.; Springer Series in Materials Science; Springer International Publishing: Cham, Switzerland, 2016; Volume 227, pp. 387–414.
42. Tossi, V.; Amenta, M.; Lamattina, L.; Cassia, R. Nitric oxide enhances plant ultraviolet-B protection up-regulating gene expression of the phenylpropanoid biosynthetic pathway. *Plant. Cell Environ.* **2011**, *34*, 909–921. [CrossRef]
43. Wang, Y.; Xu, W.J.; Yan, X.F.; Wang, Y. Glucosinolate content and related gene expression in response to enhanced UV-B radiation in *Arabidopsis*. *Afr. J. Biotechnol.* **2011**, *10*, 6481–6491.
44. Cantos, E.; Espín, J.C.; Tomás-Barberán, F.A. Postharvest induction modeling method using UV irradiation pulses for obtaining resveratrol-enriched table grapes: A new "functional" fruit? *J. Agric. Food Chem.* **2001**, *49*, 5052–5058. [CrossRef] [PubMed]
45. Lancaster, J.E.; Reay, P.F.; Norris, J.; Butler, R.C. Induction of flavonoids and phenolic acids in apple by UV-B and temperature. *J. Hort. Sci. Biotech.* **2000**, *75*, 142–148. [CrossRef]
46. Reay, P.; Lancaster, J. Accumulation of anthocyanins and quercetin glycosides in "Gala" and "Royal Gala" apple fruit skin with UV-B–visible irradiation: Modifying effects of fruit maturity, fruit side, and temperature. *Sci. Hortic.* **2001**, *90*, 57–68. [CrossRef]
47. Formica-Oliveira, A.C.; Martínez-Hernández, G.B.; Díaz-López, V.; Artés, F.; Artés-Hernández, F. Effects of UV-B and UV-C combination on phenolic compounds biosynthesis in fresh-cut carrots. *Postharvest Biol. Technol.* **2017**, *127*, 99–104. [CrossRef]
48. Olsson, L.C.; Veit, M.; Weissenböck, G.; Bornman, J.F. Differential flavonoid response to enhanced UV-B radiation in *Brassica napus*. *Phytochemistry* **1998**, *49*, 1021–1028. [CrossRef]
49. Neugart, S.; Fiol, M.; Schreiner, M.; Rohn, S.; Zrenner, R.; Kroh, L.W.; Krumbein, A. Interaction of moderate UV-B exposure and temperature on the formation of structurally different flavonol glycosides and hydroxycinnamic acid derivatives in kale (*Brassica oleracea* var. *sabellica*). *J. Agric. Food Chem.* **2014**, *62*, 4054–4062. [CrossRef]
50. Tegelberg, R.; Julkunen-Tiitto, R.; Aphalo, P.J. Red: Far-red light ratio and UV-B radiation: Their effects on leaf phenolics and growth of silver birch seedlings. *Plant Cell Environ.* **2004**, *27*, 1005–1013. [CrossRef]
51. Harbaum-Piayda, B.; Walter, B.; Bengtsson, G.B.; Hubbermann, E.M.; Bilger, W.; Schwarz, K. Influence of pre-harvest UV-B irradiation and normal or controlled atmosphere storage on flavonoid and hydroxycinnamic acid contents of pak choi (*Brassica campestris* L. ssp *chinensis* var. *communis*). *Postharvest Biol. Technol.* **2010**, *56*, 202–208. [CrossRef]
52. Schreiner, M.; Krumbein, A.; Mewis, I.; Ulrichs, C.; Huyskens-Keil, S. Short-term and moderate UV-B radiation effects on secondary plant metabolism in different organs of nasturtium (*Tropaeolum majus* L.). *Innov. Food Sci. Emerg. Technol.* **2009**, *10*, 93–96. [CrossRef]
53. Rybarczyk-Plonska, A.; Hagen, S.F.; Borge, G.I.A.; Bengtsson, G.B.; Hansen, M.K.; Wold, A.B. Glucosinolates in broccoli (*Brassica oleracea* L. var. *italica*) as affected by postharvest temperature and radiation treatments. *Postharvest Biol. Technol.* **2016**, *116*, 16–25. [CrossRef]
54. Topcu, Y.; Dogan, A.; Kasimoglu, Z.; Sahin-Nadeem, H.; Polat, E.; Erkan, M. The Effects of UV radiation during the vegetative period on antioxidant compounds and postharvest quality of broccoli (*Brassica oleracea* L.). *Plant Physiol. Biochem.* **2015**, *93*, 56–65. [CrossRef] [PubMed]
55. Lee, M.J.; Son, J.E.; Oh, M.M. Growth and phenolic compounds of *Lactuca sativa* L. grown in a closed-type plant production system with UV-A −B, or −C lamp. *J. Sci. Food Agric.* **2013**, *94*, 197–204. [CrossRef]
56. Ortega-Hernández, E.; Welti-Chanes, J.; Jacobo-Velázquez, D.A. Effects of UVB light, wounding stress, and storage time on the accumulation of betalains, phenolic compounds, and ascorbic acid in red prickly pear (*Opuntia ficus-indica* cv. Rojo Vigor). *Food Bioprocess Technol.* **2018**, *11*, 2265–2274. [CrossRef]
57. Ortega-Hernández, E.; Nair, V.; Serrano-Sandoval, S.N.; Welti-Chanes, J.; Cisneros-Zevallos, L.; Jacobo-Velázquez, D.A. Wounding and UVB light synergistically induce the postharvest biosynthesis of indicaxanthin and betanin in red prickly pears. *Postharvest Biol. Technol.* **2020**, *167*, 111247. [CrossRef]
58. Moreira-Rodríguez, M.; Nair, V.; Benavides, J.; Cisneros-Zevallos, L.; Jacobo-Velázquez, D.A. UVA, UVB light, and methyl jasmonate, alone or combined, redirect the biosynthesis of glucosinolates, phenolics, carotenoids, and chlorophylls in broccoli sprouts. *Int. J. Mol. Sci.* **2017**, *18*, 2330. [CrossRef] [PubMed]
59. Moreira-Rodríguez, M.; Nair, V.; Benavides, J.; Cisneros-Zevallos, L.; Jacobo-Velázquez, D.A. UVA, UVB light doses and harvesting time differentially tailor glucosinolate and phenolic profiles in broccoli sprouts. *Molecules* **2017**, *22*, 1065. [CrossRef]
60. Santin, M.; Castagna, A.; Miras-Moreno, B.; Rocchetti, G.; Lucini, L.; Hauser, M.-T.; Ranieri, A. Beyond the visible and below the peel: How UV-B radiation influences the phenolic profile in the pulp of peach fruit. A biochemical and molecular study. *Front. Plant Sci.* **2020**, *11*, 579063. [CrossRef]

61. Zhao, Y.; Dong, W.; Wang, K.; Zhang, B.; Allan, A.C.; Lin-Wang, K.; Chen, K.; Xu, C. Differential sensitivity of fruit pigmentation to ultraviolet light between two peach cultivars. *Front. Plant Sci.* **2017**, *8*, 1552. [CrossRef]
62. Assumpção, C.F.; Hermes, V.S.; Pagno, C.; Castagna, A.; Mannucci, A.; Sgherri, C.; Pinzino, C.; Ranieri, A.; Flôres, S.H.; Rios, A.d.O. Phenolic enrichment in apple skin following post-harvest fruit UV-B treatment. *Postharvest Biol. Technol.* **2018**, *138*, 37–45. [CrossRef]
63. Sheng, K.; Shui, S.; Yan, L.; Liu, C.; Zheng, L. Effect of postharvest UV-B or UV-C irradiation on phenolic compounds and their transcription of phenolic biosynthetic genes of table grapes. *J. Food Sci. Technol.* **2018**, *55*, 3292–3302. [CrossRef]
64. Kuhlmann, F.; Müller, C. Independent responses to ultraviolet radiation and herbivore attack in broccoli. *J. Exp. Bot.* **2009**, *60*, 3467–3475. [CrossRef] [PubMed]
65. Maffei, M.; Canova, D.; Bertea, C.M.; Scannerini, S. UV-A effects on photomorphogenesis and essential-oil composition in *Mentha piperita*. *J. Photochem. Photobiol. B Biol.* **1999**, *52*, 105–110. [CrossRef]
66. Lee, H.S.; Castle, W.S.; Coates, G.A. High-performance liquid chromatography for the characterization of carotenoids in the new sweet orange (Earlygold) grown in Florida, USA. *J. Chromatogr. A* **2001**, *913*, 371–377. [CrossRef]
67. Helsper, J.P.F.G.; de Vos, C.H.R.; Maas, F.M.; Jonker, H.H.; van den Broeck, H.C.; Jordi, W.; Pot, C.S.; Keizer, L.C.P.; Schapendonk, A.H.C.M. Response of selected antioxidants and pigments in tissues of *Rosa hybrida* and *Fuchsia hybrida* to supplemental UV-A exposure. *Physiol. Plant.* **2003**, *117*, 171–178. [CrossRef]
68. Guo, J.; Wang, M.H. Ultraviolet A-specific induction of anthocyanin biosynthesis and PAL expression in tomato (*Solanum lycopersicum* L.). *Plant Growth Regul.* **2010**, *62*, 1–8. [CrossRef]
69. Morales, L.O.; Tegelberg, R.; Brosché, M.; Keinänen, M.; Lindfors, A.; Aphalo, P.J. Effects of solar UV-A and UV-B radiation on gene expression and phenolic accumulation in *Betula Pendula* leaves. *Tree Physiol.* **2010**, *30*, 923–934. [CrossRef]
70. Nenadis, N.; Llorens, L.; Koufogianni, A.; Díaz, L.; Font, J.; Gonzalez, J.A.; Verdaguer, D. Interactive effects of UV radiation and reduced precipitation on the seasonal leaf phenolic content/composition and the antioxidant activity of naturally growing *Arbutus unedo* plants. *J. Photochem. Photobiol. B* **2015**, *153*, 435–444. [CrossRef]
71. Gyula, P.; Schäfer, E.; Nagy, F. Light perception and signalling in higher plants. *Curr. Opin. Plant Biol.* **2003**, *6*, 446–452. [CrossRef]
72. Fuglevand, G.; Jackson, J.A.; Jenkins, G.I. UV-B, UV-A, and blue light signal transduction pathways interact synergistically to regulate chalcone synthase gene expression in *Arabidopsis*. *Plant Cell* **1996**, *8*, 2347–2357. [CrossRef]
73. Morales, L.O.; Brosché, M.; Vainonen, J.; Jenkins, G.I.; Wargent, J.J.; Sipari, N.; Strid, A.; Lindfors, A.V.; Tegelberg, R.; Aphalo, P.J. Multiple roles for UV RESISTANCE LOCUS8 in regulating gene expression and metabolite accumulation in Arabidopsis under solar ultraviolet radiation. *Plant Physiol.* **2013**, *161*, 744–759. [CrossRef]

MDPI
St. Alban-Anlage 66
4052 Basel
Switzerland
Tel. +41 61 683 77 34
Fax +41 61 302 89 18
www.mdpi.com

Horticulturae Editorial Office
E-mail: horticulturae@mdpi.com
www.mdpi.com/journal/horticulturae

www.ingramcontent.com/pod-product-compliance
Lightning Source LLC
Chambersburg PA
CBHW061620210926
51850CB00009B/2388
* 9 7 8 3 0 3 6 5 8 3 7 4 7 *